Progress in Brain Research
Volume 234

Sport and the Brain: The Science of Preparing, Enduring and Winning, Part B

Serial Editor

Vincent Walsh
Institute of Cognitive Neuroscience
University College London
17 Queen Square
London WC1N 3AR UK

Editorial Board

Mark Bear, *Cambridge, USA.*
Medicine & Translational Neuroscience

Hamed Ekhtiari, *Tehran, Iran.*
Addiction

Hajime Hirase, *Wako, Japan.*
Neuronal Microcircuitry

Freda Miller, *Toronto, Canada.*
Developmental Neurobiology

Shane O'Mara, *Dublin, Ireland.*
Systems Neuroscience

Susan Rossell, *Swinburne, Australia.*
Clinical Psychology & Neuropsychiatry

Nathalie Rouach, *Paris, France.*
Neuroglia

Barbara Sahakian, *Cambridge, UK.*
Cognition & Neuroethics

Bettina Studer, *Dusseldorf, Germany.*
Neurorehabilitation

Xiao-Jing Wang, *New York, USA.*
Computational Neuroscience

Progress in Brain Research
Volume 234

Sport and the Brain: The Science of Preparing, Enduring and Winning, Part B

Edited by

Mark R. Wilson
*School of Sport and Health Sciences,
University of Exeter,
Exeter, United Kingdom*

Vincent Walsh
*Institute of Cognitive Neuroscience,
University College London,
London, United Kingdom*

Beth Parkin
*University of Westminster,
London, United Kingdom*

ACADEMIC PRESS
An imprint of Elsevier

GUELPH HUMBER LIBRARY
205 Humber College Blvd
Toronto, ON   M9W 5L7

Academic Press is an imprint of Elsevier
50 Hampshire Street, 5th Floor, Cambridge, MA 02139, United States
525 B Street, Suite 1800, San Diego, CA 92101-4495, United States
The Boulevard, Langford Lane, Kidlington, Oxford OX5 1GB, United Kingdom
125 London Wall, London, EC2Y 5AS, United Kingdom

First edition 2017

Copyright © 2017 Elsevier B.V. All rights reserved.

No part of this publication may be reproduced or transmitted in any form or by any means, electronic or mechanical, including photocopying, recording, or any information storage and retrieval system, without permission in writing from the publisher. Details on how to seek permission, further information about the Publisher's permissions policies and our arrangements with organizations such as the Copyright Clearance Center and the Copyright Licensing Agency, can be found at our website: www.elsevier.com/permissions.

This book and the individual contributions contained in it are protected under copyright by the Publisher (other than as may be noted herein).

Notices
Knowledge and best practice in this field are constantly changing. As new research and experience broaden our understanding, changes in research methods, professional practices, or medical treatment may become necessary.

Practitioners and researchers must always rely on their own experience and knowledge in evaluating and using any information, methods, compounds, or experiments described herein. In using such information or methods they should be mindful of their own safety and the safety of others, including parties for whom they have a professional responsibility.

To the fullest extent of the law, neither the Publisher nor the authors, contributors, or editors, assume any liability for any injury and/or damage to persons or property as a matter of products liability, negligence or otherwise, or from any use or operation of any methods, products, instructions, or ideas contained in the material herein.

ISBN: 978-0-12-811825-2
ISSN: 0079-6123

For information on all Academic Press publications
visit our website at https://www.elsevier.com/books-and-journals

Publisher: Zoe Kruze
Acquisition Editor: Sam Mahfoudh
Editorial Project Manager: Ana Claudia A. Garcia
Production Project Manager: James Selvam
Cover Designer: Mark Rogers

Typeset by SPi Global, India

Contributors

Ana M. Abreu
Universidade Europeia, Laureate International Universities, Lisbon, Portugal

Salvatore M. Aglioti
Sapienza Università di Roma; IRCCS, Fondazione Santa Lucia, Rome, Italy

Osman H. Ahmed
Faculty of Health and Social Sciences, Bournemouth University, Bournemouth; The FA Centre for Disability Football Research, St George's Park, Burton-Upon-Trent, United Kingdom

Duarte Araújo
CIPER, Faculdade de Motricidade Humana, Universidade de Lisboa, Lisboa, Portugal

Phillip G. Bell
Northumbria University, Newcastle upon Tyne, Tyne and Wear; GSK Human Performance Lab, Brentford, United Kingdom

Sarah E. Browne
Northumbria University, Newcastle upon Tyne, Tyne and Wear; GSK Human Performance Lab, Brentford, United Kingdom

Tim Buszard
Institute of Sport, Exercise and Active Living, Victoria University, Melbourne; Game Insight Group, Tennis Australia, Richmond, VIC, Australia

Matteo Candidi
Sapienza Università di Roma; IRCCS, Fondazione Santa Lucia, Rome, Italy

Chiao-Yun Chen
National Chung Cheng University, Chiayi, Taiwan

Chia N. Chiu
Institute of Cognitive Neuroscience, National Central University, Jhongli, Taiwan

Keith Davids
Centre for Sports Engineering Research, Sheffield Hallam University, Sheffield, United Kingdom

Akbar de Medici
Institute of Sport and Exercise Health, University College London, London, United Kingdom

Vicente L. del Campo
Sport Sciences Faculty, University of Extremadura, Caceres, Spain

David Edwards
Cognacity, London, United Kingdom; University of Zululand, KwaDlangezwa, South Africa

Javid J. Farahani
Institute of Cognitive Neuroscience, University College London, London, United Kingdom

Damian Farrow
Institute of Sport, Exercise and Active Living, Victoria University, Melbourne; Skill Acquisition, Australian Institute of Sport; Game Insight Group, Tennis Australia, Richmond, VIC, Australia

David Fletcher
School of Sport, Exercise and Health Sciences, Loughborough University, Loughborough, United Kingdom

Mark J. Flynn
GSK Human Performance Lab, Brentford, United Kingdom

Shona L. Halson
Australian Institute of Sport, Canberra, Australia

David J. Harris
School of Sport and Health Sciences, University of Exeter, Exeter, United Kingdom

Crystal F. Haskell-Ramsay
Northumbria University, Newcastle upon Tyne, Tyne and Wear, United Kingdom

Nicola J. Hodges
University of British Columbia, Vancouver, BC, Canada

Katy Hornby
Institute of Sport and Exercise Health, University College London, London, United Kingdom

Merel C.J. Hoskens
Faculty of Health, Sport and Human Performance, University of Waikato, Hamilton, New Zealand; Vrije Universiteit Amsterdam, Amsterdam, The Netherlands

Glyn Howatson
Northumbria University, Newcastle upon Tyne, Tyne and Wear, United Kingdom; Water Research Group, North West University, Potchefstroom, South Africa

Karen Howells
School of Education, Childhood, Youth and Sport, The Open University, Milton Keynes, United Kingdom

Amir H. Javadi
Institute of Cognitive Neuroscience, University College London, London, United Kingdom

Laura E. Juliff
Netball Australia, Fitzroy, VIC, Australia

Kazuyuki Kanosue
Faculty of Sport Sciences, Waseda University, Tokorozawa, Saitama, Japan

April Karlinsky
University of British Columbia, Vancouver, BC, Canada

Christie Kohut
Faculty of Kinesiology, University of Calgary, Calgary, AB, Canada

Bhavesh Kumar
Institute of Sport and Exercise Health, University College London, London, United Kingdom

Mike Loosemore
Institute of Sport and Exercise Health, University College London, London, United Kingdom

Hegoda Levansri Makalanda
The Royal London Hospital, London, United Kingdom

Rich S.W. Masters
Faculty of Health, Sport and Human Performance, University of Waikato, Hamilton, New Zealand; School of Public Health, The University of Hong Kong, Hong Kong, China

Nobuaki Mizuguchi
Faculty of Sport Sciences, Waseda University, Tokorozawa, Saitama; Faculty of Science and Technology, Keio University, Yokohama, Kanagawa; The Japan Society for the Promotion of Science, Chiyoda-ku, Tokyo, Japan

Neil G. Muggleton
Institute of Cognitive Neuroscience; Brain Research Center, College of Health Science and Technology, National Central University, Jhongli, Taiwan; Institute of Cognitive Neuroscience, University College London; Goldsmiths, University of London, New Cross, London, United Kingdom

Barry V. O'Neill
GSK Human Performance Lab, Brentford; Institute of Cognitive Neuroscience, University College London, London, United Kingdom

Beth L. Parkin
Department of Psychology, University of Westminster; ICN, UCL, London, United Kingdom

Tim Rogers
Cognacity, London, United Kingdom

Brendan Ryley
Faculty of Kinesiology, University of Calgary, Calgary, AB, Canada

Mustafa Sarkar
School of Science and Technology, Nottingham Trent University, Nottingham, United Kingdom

Ludovic Seifert
CETAPS EA3832, Faculty of Sport Sciences, University of Rouen Normandy, Mont-Saint-Aignan, France

Richard Sylvester
Institute of Sport and Exercise Health, University College London; National Hospital of Neurology and Neurosurgery, London, United Kingdom

Pedro Teques
N2i, Polytechnic Institute of Maia, Castelo da Maia; CIPER, Faculdade de Motricidade Humana, Universidade de Lisboa, Lisboa, Portugal

Tina van Duijn
Faculty of Health, Sport and Human Performance, University of Waikato, Hamilton, New Zealand

Ben Vandervies
Faculty of Kinesiology, University of Calgary, Calgary, AB, Canada

Joan N. Vickers
Faculty of Kinesiology, University of Calgary, Calgary, AB, Canada

Samuel J. Vine
School of Sport and Health Sciences, University of Exeter, Exeter, United Kingdom

Vincent Walsh
ICN, UCL, London, United Kingdom

Katie Warriner
Rugby Football Union, Twickenham, United Kingdom

Mark R. Wilson
School of Sport and Health Sciences, University of Exeter, Exeter, United Kingdom

Karen Zentgraf
Goethe University, Frankfurt, Germany

Contents

Contributors v

CHAPTER 1 Quiet eye training improves accuracy in basketball field goal shooting 1
Joan N. Vickers, Ben Vandervies, Christie Kohut, Brendan Ryley
1. Introduction 2
2. Methods 4
   2.1. Participants 4
   2.2. QET and TT Training Protocol 4
3. Results 6
   3.1. Accuracy (FG%) 6
4. Discussion 6
   4.1. Pre- to Postperformance 8
   4.2. Post to Transfer Performance 8
   4.3. Effectiveness of the Elite QE Free Throw in Field Shooting 9
   4.4. Limitations and Recommendations 9
5. Conclusion 10
References 10
Further Reading 12

CHAPTER 2 Sleep, sport, and the brain 13
Shona L. Halson, Laura E. Juliff
1. Introduction 13
2. Background 14
   2.1. Sleep Stages 15
   2.2. Neurophysiology of the Sleep–Wake Cycle 17
3. Sleep and Athletes 18
   3.1. Importance of Sleep in Athletes 19
   3.2. Incidence of Sleep Disruption in Athletes 19
   3.3. Causes of Sleep Disruption in Athletes 20
   3.4. Sleep and Athlete Performance 22
   3.5. Sleep Extension 23
4. Sleep Deprivation and Brain Function 24
   4.1. Cognitive Performance in Athletes 24
   4.2. Cognitive Performance, Memory, and Learning 24

  4.3. Mood 25
  4.4. Molecular Aspects of Sleep Disruption 26
 5. Conclusion 26
 References 27

CHAPTER 3 The resonant system: Linking brain–body–environment in sport performance 33
 Pedro Teques, Duarte Araújo, Ludovic Seifert,
 Vicente L. del Campo, Keith Davids
 1. Introduction 34
 2. How Does an Athlete Obtain Information to Regulate Action? 37
  2.1. Revisiting Misunderstandings About Perception and Internal Representations 37
  2.2. Linking Brain–Body–Environment: The Resonant System 39
 3. Conclusion 47
 References 48
 Further Reading 52

CHAPTER 4 Catching on it early: Bodily and brain anticipatory mechanisms for excellence in sport 53
 Ana M. Abreu, Matteo Candidi, Salvatore M. Aglioti
 1. Visual Anticipation and Intention Reading 54
 2. From an Action Observation Network to Predictive Embodiment 56
 3. "Gut Feeling" in Sports: From Interoception to Intuition 59
 4. Conclusion 62
 Acknowledgments 62
 References 63

CHAPTER 5 Exploring the applicability of the contextual interference effect in sports practice 69
 Damian Farrow, Tim Buszard
 1. Introduction 69
 2. The Influence of Skill Complexity 71
 3. Mechanisms and Practice 73
 4. The Importance of Demonstrating the Transfer of Learning 75
 5. Future Research Directions to Produce Guidelines for Applied Practice 76
 6. Concluding Remarks 80
 Acknowledgments 81
 References 81

CHAPTER 6 Sport, time pressure, and cognitive performance 85
Chia N. Chiu, Chiao-Yun Chen, Neil G. Muggleton
1. Introduction 86
2. Methods 88
   2.1. Study Design 88
   2.2. Ethical Approval and Consent 88
   2.3. Participants 88
   2.4. Test Procedures 89
   2.5. Data Processing and Analysis 90
3. Results 90
   3.1. Participant Profiles 90
   3.2. Accuracy and Response Times 90
   3.3. Drift Diffusion Analysis 93
4. Discussion 96
5. Conclusions 97
Acknowledgments 98
References 98

CHAPTER 7 Effectiveness of above real-time training on decision-making in elite football: A dose–response investigation 101
Javid J. Farahani, Amir H. Javadi, Barry V. O'Neill, Vincent Walsh
1. Introduction 101
2. Methods and Materials 105
   2.1. Participants 105
   2.2. Experimental Design 105
   2.3. Statistical Analysis 107
3. Results 107
4. Discussion 111
Acknowledgments 112
References 112
Further Reading 115

CHAPTER 8 Can athletes benefit from difficulty? A systematic review of growth following adversity in competitive sport 117
Karen Howells, Mustafa Sarkar, David Fletcher
1. Introduction 118
2. Method 121
   2.1. Search Strategy 122
   2.2. Inclusion Criteria 122

    2.3. Sifting of Papers 122
    2.4. Methodological Rigor 123
    2.5. Data Extraction 124
    2.6. Data Analysis 124
 3. Results 124
    3.1. Study Characteristics 130
    3.2. Quality Appraisal 133
    3.3. Adversity-Related Experiences 133
    3.4. Growth-Related Experiences 143
 4. Discussion 150
    4.1. Definitions and Theory 150
    4.2. Adversity-Related Experiences 152
    4.3. Growth-Related Experiences 152
    4.4. Limitations and Future Research 154
 References 154

CHAPTER 9 Effects of acute high-intensity exercise on cognitive performance in trained individuals: A systematic review 161

Sarah E. Browne, Mark J. Flynn, Barry V. O'Neill,
Glyn Howatson, Phillip G. Bell,
Crystal F. Haskell-Ramsay

 1. Introduction 162
 2. Method 164
    2.1. Eligibility Criteria 164
    2.2. Information Sources and Search Strategy 165
    2.3. Study Selection and Data Collection Process 166
    2.4. Quality Assessment 166
    2.5. Analysis 166
 3. Results 167
    3.1. Study Selection 167
    3.2. Quality Assessment 169
    3.3. Descriptive Characteristics of Included Studies 170
    3.4. Effect of Acute High-Intensity Exercise of
         Cognitive Function 170
 4. Discussion 177
    4.1. Considerations for Future Research 181
    4.2. Limitations of the Current Review 182
 5. Conclusion 183
 References 183

CHAPTER 10 Changes in brain activity during action observation and motor imagery: Their relationship with motor learning 189
Nobuaki Mizuguchi, Kazuyuki Kanosue

1. Basic Aspects of Action Observation 190
2. Basic Aspects of Motor Imagery 191
3. Brain Activity Relating to Task Complexity 192
   3.1. Actual Execution 192
   3.2. Action Observation 193
   3.3. Motor Imagery 194
4. Activity Change in Association With Motor Learning 194
   4.1. Actual Execution 194
   4.2. Action Observation 195
   4.3. Motor Imagery 198
5. Conclusion 198
Acknowledgments 199
References 199

CHAPTER 11 Moving concussion care to the next level: The emergence and role of concussion clinics in the UK 205
Osman H. Ahmed, Mike Loosemore, Katy Hornby, Bhavesh Kumar, Richard Sylvester, Hegoda Levansri Makalanda, Tim Rogers, David Edwards, Akbar de Medici

1. Introduction 206
2. Concussion Clinics in the UK 206
3. Multidisciplinary Approach to Concussion Management 208
   3.1. Sports Physician 208
   3.2. Radiologist 209
   3.3. Neurologist 210
   3.4. Physiotherapist 211
   3.5. Psychologist and Psychiatrist 212
4. Conclusions 214
   4.1. Diagnosis 214
   4.2. Imaging 214
   4.3. Multimedia Technologies 214
   4.4. Education/Dissemination 215
   4.5. Concussion Clinics for All 215
References 216

CHAPTER 12 Neurocognitive mechanisms of the flow state 221
David J. Harris, Samuel J. Vine, Mark R. Wilson

1. Introduction 222
   1.1. The Flow State 222
   1.2. Flow in Sport 223
2. Attention 224
   2.1. Automaticity 224
   2.2. Attentional Control 227
   2.3. Attentional Effort 230
   2.4. Self-Awareness 232
   2.5. Attention Summary 234
3. Discussion 235
4. Conclusions 236
   References 237

CHAPTER 13 Discerning measures of conscious brain processes associated with superior early motor performance: Capacity, coactivation, and character 245
Tina van Duijn, Tim Buszard, Merel C.J. Hoskens, Rich S.W. Masters

1. Introduction 246
2. Method 248
   2.1. Participants 248
   2.2. Materials 248
   2.3. Procedure 249
   2.4. Dependent Variables and Data Analysis 250
3. Results 251
   3.1. Correlation Between EEG Coherence and Performance Variables 251
   3.2. Predicting Single Task Accuracy 253
   3.3. Predicting Combined Task Accuracy 253
4. Discussion 256
5. Limitations 258
6. Conclusion 258
   References 258

CHAPTER 14 Action-skilled observation: Issues for the study of sport expertise and the brain 263
April Karlinsky, Karen Zentgraf, Nicola J. Hodges

1. The AON and "Mirror Properties" 264
2. Neurophysiological Methods 268

3. Action Observation and Recruitment of the Simulation
   Circuit: A Review of the Evidence 270
   3.1. Early Action Experiences 270
   3.2. Physical and Imitative Short-Term Practice 271
   3.3. Sport Expertise and Long-Term Visual-Motor
        Experiences 273
4. Summary and Conclusions 281
References 282

CHAPTER 15 Gunslingers, poker players, and chickens 1: Decision making under physical performance pressure in elite athletes 291
Beth L. Parkin, Katie Warriner, Vincent Walsh

1. Introduction 292
2. Method 294
   2.1. Participants 294
   2.2. Protocol 295
   2.3. Physical Pressure Induction 295
   2.4. Decision-Making Tasks 296
   2.5. Analysis 302
3. Results 303
   3.1. Does performance on the BART change under
        pressure? 303
   3.2. Does Performance on the CGT Change Under
        Pressure? 304
   3.3. Does Performance on the Stroop Task Change Under
        Pressure? 307
   3.4. Were Individual Performance Changes Under Pressure
        Consistent Across Tasks? 307
   3.5. Applicability of Group Data to Individuals 309
4. Discussion 309
5. Conclusion 313
References 313

CHAPTER 16 Gunslingers, poker players, and chickens 2: Decision-making under physical performance pressure in subelite athletes 317
Beth L. Parkin, Vincent Walsh

1. Introduction 318
2. Method 320
   2.1. Participants 320
   2.2. Protocol 321

　　　　2.3. Physical Pressure Induction..321
　　　　2.4. Decision-Making Tasks..322
　　　　2.5. Analysis...325
　　3. Results...326
　　　　3.1. Does performance on the BART change under
　　　　　　　pressure?...326
　　　　3.2. Does performance on the cgt change under pressure?......328
　　　　3.3. Does performance on the SSRT Task change under
　　　　　　　pressure?...331
　　　　3.4. Were individual performance changes under pressure
　　　　　　　consistent across tasks?..331
　　　　3.5. Applicability of group data to individuals........................331
　　4. Discussion..332
　　5. Conclusion...335
　　References...335

CHAPTER 17 Gunslingers, poker players, and chickens 3: Decision making under mental performance pressure in junior elite athletes...................339
　　Beth L. Parkin, Vincent Walsh
　　1. Introduction...340
　　2. Method..342
　　　　2.1. Participants..342
　　　　2.2. Protocol..343
　　　　2.3. Mental Pressure Induction..343
　　　　2.4. Decision-Making Tasks..343
　　　　2.5. Analysis...346
　　3. Results...348
　　　　3.1. Does performance on the bart change under pressure?.....348
　　　　3.2. Does performance on the CGT change under pressure?...349
　　　　3.3. Does performance on the Visual Search Task change
　　　　　　　under pressure?..352
　　　　3.4. Performance on the Dual Task...354
　　　　3.5. Were individual performance changes under pressure
　　　　　　　consistent across tasks?..354
　　　　3.6. Applicability of group data to individuals........................354
　　4. Discussion..354
　　5. Conclusion...357
　　References...357

Index..361

CHAPTER

Quiet eye training improves accuracy in basketball field goal shooting

1

Joan N. Vickers[1], Ben Vandervies, Christie Kohut, Brendan Ryley

Faculty of Kinesiology, University of Calgary, Calgary, AB, Canada
[1]*Corresponding author: Tel.: +403-239-5765; Fax: +403-284-3553,*
e-mail address: vickers@ucalgary.ca

Abstract

University students ($N = 240$) were randomly assigned to a quiet eye training (QET) or technical training (TT) group, and their shooting accuracy (%) determined during a pre-, post-, and transfer test in basketball field shooting. Both groups first received lectures on visuomotor processing and the quiet eye (QE), followed by a laboratory in which participants in the QET group were taught how to adopt the QE characteristics of elite free-throw shooters, which stresses optimal gaze control and focus relative to a single target location, while the TT participants were taught elite biomechanics which stresses optimal control of the shooting stance, arms, and hands. Overall, the QET group's accuracy was significantly higher than the TT group, but differences were found due to skill level and defensive pressure. From pre to post, the accuracy of the QET novices increased significantly compared to the TT novices, but declined during transfer. Both the QET and TT intermediates had relatively high accuracy scores during the pre- and posttests, which then declined, as expected, during the transfer test against defensive pressure. However, during transfer the QET group's accuracy remained higher than the TT group and was surprisingly similar to that found in elite competition. It is recommended that novice and intermediate basketball players be taught how to adopt the QE of elite players, rather than learning only the technical/mechanical aspects of shooting. Theoretically, the study is placed within the context of top-down "cognitive control," as proposed by Cavanagh and Frank (2014), and QET studies which show that when learners are taught how to adopt the QE of elite performers, this appears to contribute to a more optimal organization of the neural networks underlying control of the task which, in turn, leads to improved shooting performance.

Keywords

Gaze, Brain, Vision, Attention, Neuromotor control, Expertise

1 INTRODUCTION

Extensive research shows that elite performers in basketball shooting, and other motor tasks, have developed a long-duration quiet eye (QE) (Lebeau et al., 2016; Mann et al., 2007; Rienhoff et al., 2016). The QE is formally defined as a final fixation or tracking gaze that is located on a specific location or object in the task space within 3 degrees of visual angle (or less) for a minimum of 100 ms. The onset of the QE occurs before a critical movement phase, and the quiet eye offset occurs when the gaze moves off the location by 3 degrees of visual angle (or less) for a minimum of 100 ms (Vickers, 1996a,b; Vickers and Williams, 2007; Vine and Wilson, 2011; Wilson et al., 2009). The first QE study carried out in the basketball free throw found that elite athletes with an average accuracy of 78% fixated the hoop earlier before the final shooting action for an average of 972 ms on hits and 806 ms on misses. Near-elite athletes who averaged 56% had a later QE duration, averaging less than 400 ms on hits and misses (Vickers, 1996a,b), findings that have been corroborated in other studies (Fischer et al., 2015; Rienhoff et al., 2013; Vine and Wilson, 2011; Wilson et al., 2009). Theoretically, it is thought the QE provides the time needed to organize the neural networks underlying control of the skill (Cavanagh and Frank, 2014; Vickers, 1996a, 2009, 2012).

Once the QE of experts has been identified, the five QE characteristics (specific external location, early onset, before a critical movement, offset that is task specific, and long duration) are taught to lesser-skilled participants. Quiet eye training (QET) studies reveal significant gains in QE duration and performance over a wide range of motor tasks (for a recent review, see Wilson et al., 2015). QET has proven to be effective in improving both the focus and accuracy of novice and intermediate level nonexpert performers, under both practice and anxiety conditions (Harle and Vickers, 2001; Vine and Wilson, 2011; Wilson et al., 2009). Harle and Vickers (2001) trained basketball players on QE technique during free throws, and compared the results against two control groups that did not receive QET. Results indicated a significant increase in QE duration and free-throw percentage(FTM%) in both the laboratory setting and competition, with FTM% in competition improving from 54% to 76% over two seasons. Vine and Wilson (2011) looked at the impact of QET on the accuracy of novice players performing a basketball free throw in low-pressure and high-pressure situations. Participants were grouped into either a QET or a technical training (TT) control group. The TT group received instruction on the biomechanics of free-throw shooting, but did not receive any QET. The results showed that the QET group improved performance and reduced the negative effects of anxiety performing significantly better than the control group in two retention tests and one high-pressure test designed to elevate levels of cognitive anxiety.

Theoretically, we placed the study within the concept of "cognitive control" as proposed by Cavanagh and Frank (2014, p. 414), who provides extensive EEG evidence showing humans increase frontal lobe activation in situations when there is a need for goal-directed cognitive control, for example, when involved in a complex targeting task involving high levels of error, as occurs in basketball shooting. Optimal visual control is accompanied by heightened midline theta oscillations from the midcingulate cortex and presupplemental motor area. These heightened oscillations radiate throughout the brain and exert top-down control across a broad network reflecting "a common mechanism, *a lingua franca*, for implementing adaptive control in a variety of contexts involving uncertainty about actions and outcomes" (Cavanagh and Frank, 2014, p. 416). Recent EEG studies in shooting and golf lend further support to this view. For example, Doppelmayr et al. (2008) compared the EEG of expert and novice rifle shooters and found that only the experts regulated the time course of theta activation during the 3 s before the shot so that the maximum period of focus was just before the trigger pull. Novices did not exhibit the same activation, but instead kept a relatively constant level for the 3 s prior to shooting. In a study with biathlon shooters, Gallicchio et al. (2016) found that when exercise levels increased to a high, more difficult level, accuracy was higher when there was an increase in frontal midline theta activation, thus providing further insight into a biathlon QE study by Vickers and Williams (2007) who found that shooters who increased their QE duration by an average of 600 ms during conditions of heightened pressure and exercise continued to shoot at a high level, while those who did not shot at a very low level. Similar results have been found in golf putting (Bertollo et al., 2016; Mann et al., 2011). For example, Mann et al. compared the EEG and QE of low- and high-skilled golfers, and found the experts exhibited an increase in the Bereitschaftspotential event-related potential, during a long-duration QE period golfers unlike the nonexperts who had a shorter duration. The authors concluded that "prolonged fixations, particularly during the final fixation that defines the QE, apparently permit the detailed processing of information and cortical organization necessary for effective motor performance" (p. 232). These results suggest that teaching participants to use an optimal QE duration prior to shooting in basketball should increase cognitive control by supplying task-relevant spatial information, which in turn will lead to higher levels of motor performance.

We therefore determined the field goal percentage (FG%) accuracy of 240 university students as they took part in either a QET laboratory designed to teach them how to focus like an elite shooter, or a TT laboratory where they were taught the biomechanics of elite shooting. The class was made up of novices who had never played basketball, and intermediate level players who had played basketball on a team. We reasoned that knowing how to focus using a QE similar to elite players would lead to greater improvement in shooting accuracy than an equal number of minutes spent in receiving TT. Given the higher experience level of the intermediates, we also expected their accuracy to be higher than the novices. Since a QET study has never been carried out in basketball field shooting, a second goal was to determine if teaching participants the QE characteristics of elite free-throw shooters would also provide performance benefits in field shooting. We expected the QET groups to experience greater gains in shooting accuracy than the TT group, pre to post, as well as during a transfer test against defensive pressure. A third goal was to determine if it was possible to introduce the two methods of training (QET, TT) to a large group of university students enrolled in a course with lectures and an experiential laboratory setting, and if these methods can be effectively taught in large group sessions.

2 METHODS
2.1 PARTICIPANTS
Participants were enrolled in a Behavioral Neuroscience course, which included lectures in neuroanatomy, visual motor processing, and visual attention in human movement (Carter, 2014; Corbetta and Shulman, 2002; Corbetta et al., 2008; Vickers, 2007). The five characteristics of the QE in golf putting and the basketball free throw (Vickers, 2007) were explained and demonstrated in lecture. QET training studies in the free throw by Harle and Vickers (2001) and Wilson et al. (2009)were also presented. Prior to the course, participants were assigned to 1 of 10 laboratory groups by the Registrar's office, 24 students per group, for a total of 240 participants. Neither the professor nor the lab instructor were aware of who was assigned to each laboratory, and the students were not aware of what was to be presented in each laboratory. Twenty-seven participants who failed to complete a pre-, post-, and/or transfer test, or did not attend the lab, were removed, leaving a total of 213 participants. Prior to the course, even numbered labs were arbitrarily assigned to the QET group ($n=102$), and the uneven number laboratories were assigned to the TT group ($n=111$).

2.2 QET AND TT TRAINING PROTOCOL
The laboratory was carried out in a large gymnasium with six regulation basketball hoops, and taught by the same instructor, using videos that ensured the same content was provided to each group. Upon arrival in the gymnasium, participants were instructed to take practice shots and warm up for 5 min. They were then shown a video of NBA player Steve Nash, https://www.youtube.com/watch?v=pzQruMtrqQw, performing free throws without a dribble. This video included no instruction on either focus patterns or mechanics. They were paired up based on height, given one regulation size five basketballs per pair, and assigned to a basket, two pairs per basket and practiced taking set shots. The set shot technique is similar to the free throw but is taken from anywhere inside the three-point line, most often against defensive pressure. Participants who preferred to take jump shots were allowed to take the more advanced shot. A pretest of 10 set or jump shots followed from X1 (five shots from each side) without defensive pressure (see Fig. 1).

TT and QET training were then provided as follows. Participants assigned to the TT group watched a video of Michael Jordan, https://www.youtube.com/watch?v=JdTQi4L6khw, explaining and performing free throws without a dribble, using proper technique, and the technical points shown in Table 1 (right). Participants assigned to the QET group viewed a video of the QE of an elite prototype in the free throw, and the five QE characteristics were emphasized as shown in Table 1 (left). In addition, all participants were provided with Table 1 from Vine et al. (2011), which outlined both QET and TT training steps, but were told to omit the three preshot dribbles used in the free throw.

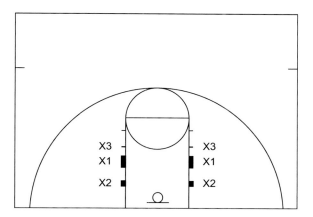

FIG. 1

A regulation basket and the location of shots taken during the pre-, post-, and transfer tests.

Table 1 Quiet Eye Training (Harle and Vickers, 2001; Vickers, 2007; Vine and Wilson, 2011 Adapted) and Technical Training (Michael Jordon and Steve Nash Videos) Instructional Points in the Free Throw

Quiet Eye Training	Technical Training
(1) Take your stance at the line and orient your gaze to the hoop as soon as possible. Fixate the front of hoop even as you carry out your preshot routine	(1) Take a balanced stance at the free-throw line and carry out your preshot routine
(2) Hold the ball in front and fixate a single location on the hoop (front, back, or middle) for about 1 s and say "nothing but net." Visualize the ball going in the basket	(2) Focus on the rim, and if a right-handed shooter, the left hand is the guide and the right hand holds the ball on the fingertips with an L shape elbow pointing to the target
(3) Perform the set shot/jump shot so that the ball and your hands travel up through the midline of your body. The ball will occlude the basket as it enters your visual field	(3) Propel the ball upward by extending from the feet, knees, through to arm and fingertip extension
(4) Shoot using a quick, fluid action	(4) Follow through so your fingertips go toward the hoop
(5) Do not be concerned about holding your gaze on the hoop during the extension phase of the shot	(5) Be confident and positive you will make the shot

After QET or TT instructions, participants took 20 practice shots using the allocated techniques, followed by 10 posttest shots from the same location (X1) used in the pretest. During the transfer test, set or jump shots were taken from two new locations, X2 and X3, under defensive pressure from the outstretched hand of a partner. Small prizes were made available to the winning lab based on the highest transfer FG% average, and to the most improved lab based on FG% improvement from post to transfer. In total, participants completed 10 pretest, 10 posttest, and 10 transfer shots. Total time to complete the laboratory was 50 min. Skill level was determined using the pretest scores resulting in a novice ($n=123$, 0%–30%) and an intermediate ($n=90$, 40% and above) group. The 30% threshold for novices was based on results from previous basketball studies in the free throw (Ryu et al., 2016; Vine and Wilson, 2011). Accuracy scores (%) were analyzed using a group (TT, QET) × skill level (novice, intermediate) ANOVA, with repeated measures across tests (pre, post, transfer). Effect sizes were calculated using Cohen's d. Greenhouse–Geisser epsilon was used to control for violations of sphericity, and the alpha level for significance was set at 0.05 with Bonferroni adjustment to control for Type 1 errors.

3 RESULTS
3.1 ACCURACY (FG%)

Significant main effects were found for group, $F_{(1, 209)}=4.45$, $P=0.04$, $d=0.17$; skill level, $F_{(1, 209)}=174.89$, $P<0.0001$, $d=1.02$; and test, $F_{(2, 418)}=27.54$, $P<0.0001$, $d=0.33$, and the interactions of test × group, $F_{(2, 418)}=40.11$, $P<0.001$, $d=0.16$, and test × group × skill level, $F_{(2, 418)}=3.38$, $P<0.04$, $d=0.19$. Since skill level was based on pretest scores, the critical results were the main effects for group, and the significant interaction of group × skill level × test. Fig. 2A shows that the QET group was significantly more accurate than the TT, averaging 37.2% compared to 31.9% for the TT group. Fig. 2B shows the significant interaction of group × skill level × test. The QET and TT novices had similarly low levels of accuracy in the pretest, but during the posttest the QET group increased their accuracy to 37.2%, which differed from the TT group's 28.1%, $F_{(1, 550.4)}=3.39$, $P<0.002$. During the transfer test, accuracy of both the QET and TT novices declined to 25%; however, both the QET and TT novice's accuracy significantly improved from pre to transfer, $F_{(1, 418)}=20.13$, $P<0.0001$. For the intermediates, the QET and TT groups achieved equally high pre- and posttest scores, range 50.1%–53.2%. During the transfer test, the QET intermediate's accuracy was 40.4%, while the TT group averaged 33.8%, a difference that approached significance, $F_{(1, 550)}=3.65$, $P<0.056$.

4 DISCUSSION

At the outset we had three goals. First, we sought to determine if QET was more effective in improving than TT when teaching novice and intermediate participants the set/jump shot. Second, we wanted to determine if the five QE characteristics

4 Discussion

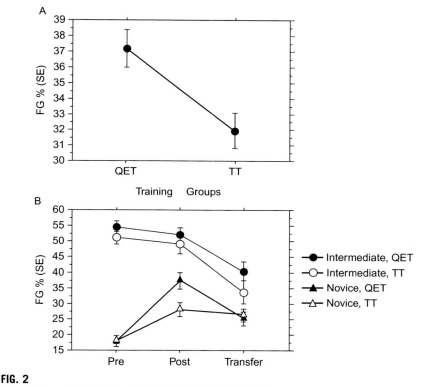

FIG. 2

(A) Field shooting accuracy (%) of the quiet eye trained and technical trained groups. (B) Field shooting accuracy (%) of novices and intermediates in the quiet eye trained and technical trained groups across and pre-, post-, and transfer test.

found in the free throw would also be effective in improving accuracy in the set or jump shot. And third, we wanted to see if teaching a large group of students about the QE and QET within regularly scheduled lectures and a laboratory would be effective in improving accuracy on a pre- and posttest without defense, and a transfer against defensive pressure. Our prediction that QET would lead to greater gains in FG% than TT was upheld; over the three tests the QET group was more accurate than the TT group. Teaching the elite QE characteristics found in the free throw was effective in improving set/jump shot accuracy of a large number of participants within the lecture and laboratory setting. Before discussing the results it is important to be aware of norms for the free throw (FTM%), which is an undefended shot similar to that used in the pre- and posttests, and the FG% against defensive pressure as used in the transfer test. No norms exist for novice or intermediate level basketball players, but during the 2016–17 season, elite male and female athletes on the 50 NCAA teams had FTM% that ranged between 73%–79% for the men, and 74%–80% for the women, while FG% for men ranged between 47%–52%, and for women between 44%–53%. http://stats.ncaa.org/rankings/change_sport_year_div.

4.1 PRE- TO POSTPERFORMANCE

For the novices, the QET and TT groups had similar, very low pretest scores of 17%. In the posttest after QE training, the QET group's accuracy improved to 37% (a gain of 20%) vs 28% for the TT group. Being shown how to focus the QE externally on a single target location similar to elite athletes performing free throws was clearly more beneficial for the novices than being provided additional biomechanical information that typically focuses attention internally on the body. For the intermediate group, both the QET and the TT groups achieved equally high pretest scores (range 52%–54%), equaling the pretest scores (56%) of university athletes in Wilson et al. The relatively high level of accuracy suggests the QE information given in lectures about the QE of elite players may have positively affected both of the higher-skilled groups. Both groups then maintained the same level of accuracy in the posttest, whereas the QET group in Wilson et al. increased their accuracy to 68.0%. The failure of the QET intermediates to increase their accuracy during the posttest may have been due to the small number of practice trials (20) provided, compared to the greater number (360) performed in Wilson et al.

4.2 POST TO TRANSFER PERFORMANCE

During the transfer test, defensive pressure was applied and the accuracy of both the novices and intermediates declined as expected, revealing the difficulty basketball players have in dealing with distractions in the visual field. Novice transfer performance averaged 25.5% for both groups, while the QET and TT intermediates accuracy was 40% and 34%, respectively. These percentages seem low, but when compared to the averages of elite NCAA male and players, which range between 44% and 53%, the 40% achieved by the QET group is noteworthy, as most of the participants did not train regularly or compete, notwithstanding the transfer task they performed was easier than that experienced by elite players in competition. During the transfer test, pressure was applied on both groups in the form of prizes to the highest scoring and most improved laboratories. This may have led to pressure from teammates, an increase in anxiety, and a desire to do well and avoid embarrassment. Pressure-induced anxiety can reduce the influence of the dorsal attentional network, which maintains attention on a critical target goal, while the ventral attentional network is susceptible to shifts in attention and distraction (Corbetta and Shulman, 2002; Corbetta et al., 2008; Moore et al., 2012; Panasiti et al., 2016). The relatively high accuracy level, achieved by the QET intermediates during transfer, suggests they benefited from using a long-duration QE on the front of the hoop which may have exerted greater top-down cognitive control and an ability to ignore the distractions imposed by the defensive player in the visual field.

4.3 EFFECTIVENESS OF THE ELITE QE FREE THROW IN FIELD SHOOTING

We found support for teaching the set and jump shot using the same QE characteristics as found in free-throw shooting, which is performed using a "low style" of shooting, rather than a "high style" that has been found for some athletes in the jump shot (de Oliveira et al., 2007; Oudejans et al., 2002; Vickers, 1996a; Vickers and Williams, 2007). In the low style, the athlete fixates a single location on the hoop for about 1 s (QE), prior to beginning the final extension of shooting arm and wrist (see Table 1 right). Since the ball enters the visual field once the shot begins it occludes the basket for 150–200 ms (de Oliveira et al., 2008; Oudejans et al., 2002; Vickers, 2007). Low-style shooters do not try to refixate the hoop after this, but instead take the shot quickly under open-loop control. The open-loop control system functions when movements are so fast (usually under 200 ms) that there is no time for feedback from the gaze, joints, muscles, emotions, or crowd to affect the movement. When the high style is used, the athlete fixates the hoop during the jump and maintains fixation throughout the final extension of the shooting arm and hand. There is some controversy about whether the "low style" found in the free throw and taught in this study is more accurate than a "high style" in which the athlete refixates the hoop, after the occlusion period. Refixating a location can take between 300 and 400 ms, which is a long period of time in basketball jump shooting. More importantly, when a new fixation is acquired during the jump, it can slow down the final shooting movement, which can place it under closed-loop control, which means the athlete may be susceptible to feedback entering the system from both external distractions (crowd noise, comments) and those arising internally (gaze, muscles, joints, anxiety, etc.).

4.4 LIMITATIONS AND RECOMMENDATIONS

There were a number of limitations in the study, including the limited time dedicated to training and practice, and the small number of shots used for the pre-, post-, and transfer tests. Only half of the participants received each form of training, a problem that could be alleviated if two laboratory sessions are scheduled, in a counterbalanced design, so students are able to experience equal amounts of both types of training. The addition of eye tracking would provide objective QE results, thus bolstering the arguments given above about the importance of QE task-specific information being beneficial to improving accuracy in basketball shooting. It is tempting to suggest that both groups should *not* have been exposed to lectures in visuomotor processing, visual attention, and the QE lecture information prior to the pretest, but the positive gains made through QET training in the current study and others in the past suggest that the information was useful. Learners should be made aware that motor skills are organized 100% in the brain before movements are activated, using task-specific spatial and temporal visual information. Instead of omitting the lecture information on the QE, it is recommended that an equal number of minutes be devoted to TT in lectures, thus providing balance in terms of amount of prior instruction received the two forms of training.

5 CONCLUSION

To our knowledge this is the first QE study to compare the effects of QET and TT on the performance of a large number of novice and intermediate learners in the lecture–laboratory setting. It is also the first to explore the effects of QET and TT, with and without defensive pressure, thus providing insight into the effect distractions in the visual field have on performers. We show that QET led to greater gains in accuracy for novice and intermediate participants than TT. Theoretically, we placed the study within the concept of "cognitive control" as proposed by Cavanagh and Frank (2014) and EEG and QE studies in goal-directed tasks, such as rifle shooting and golf, which show that when visual focus is increased on a critical target location, there is an increase in theta oscillations from the midcingulate cortex and presupplemental motor area, which in turn radiate throughout the brain and exert top-down control on the visuomotor networks controlling movements. Given the large sample size and applied nature of this study we were not able to provide direct QE or EEG evidence, but we hope researchers will design studies where this information can be forthcoming. We recognize that this study is a modest beginning, but one that has led to a number of questions that can be explored in the teaching and research environments. Overall, the study demonstrates that QE training techniques can effectively be taught to large group students in a short time period (i.e., 50 min) in an experiential lab setting, and provide benefits in accuracy for novice and intermediate level basketball players, as well as knowledge about new sources of neural, visual, and perceptual information beneficial to human performance.

REFERENCES

Bertollo, M., di Fronso, S., Filho, E., Conforto, S., Schmid, M., Bortoli, L., Robazza, C., 2016. Proficient brain for optimal performance: the MAP model perspective. PeerJ 4, e2082. http://dx.doi.org/10.7717/peerj.2082.

Carter, R., 2014. The Human Brain Book. Dorling Kindersley/Penguin, London, UK.

Cavanagh, J.F., Frank, M.J., 2014. Frontal theta as a mechanism for cognitive control. Trends Cogn. Sci. 18 (8), 414–421. http://dx.doi.org/10.1016/j.tics.2014.04.012.

Corbetta, M., Shulman, G.L., 2002. Control of goal-directed and stimulus-driven attention in the brain. Nat. Rev. Neurosci. 3 (3), 201–215. http://dx.doi.org/10.1038/nrn755.

Corbetta, M., Patel, G., Shulman, G.L., 2008. The reorienting system of the human brain: from environment to theory of mind. Neuron 58 (3), 306–324. http://dx.doi.org/10.1016/j.neuron.2008.04.017.

de Oliveira, R.F., Huys, R., Oudejans, R.R., van de Langenberg, R., Beek, P.J., 2007. Basketball jump shooting is controlled online by vision. Exp. Psychol. 54 (3), 180–186. http://dx.doi.org/10.1027/1618-3169.54.3.180.

de Oliveira, R.F., Oudejans, R.R.D., Beek, P.J., 2008. Gaze behavior in basketball shooting: further evidence for online visual control. Res. Q. Exerc. Sport 79, 399–404.

Doppelmayr, M., Finkenzeller, T., Sauseng, P., 2008. Frontal midline theta in the pre-shot phase of rifle shooting: differences between experts and novices. Neuropsychologia 46 (5), 1463–1467. http://dx.doi.org/10.1016/j.neuropsychologia.2007.12.026.

References

Fischer, L., Rienhoff, R., Tirp, J., Baker, J., Strauss, B., Schorer, J., 2015. Retention of quiet eye in older skilled basketball players. J. Mot. Behav. 47, 407–414. http://dx.doi.org/10.1080/00222895.2014.1003780. 1–8.

Gallicchio, G., Finkenzeller, T., Sattlecker, G., Lindinger, S., Hoedlmoser, K., 2016. Shooting under cardiovascular load: electroencephalographic activity in preparation for biathlon shooting. Int. J. Psychophysiol. 109, 92–99. http://dx.doi.org/10.1016/j.ijpsycho.2016.09.004.

Harle, S.K., Vickers, J.N., 2001. Training quiet eye improves accuracy in the basketball free throw. Sport Psychol. 15 (3), 289–305.

Lebeau, J.C., Liu, S., Saenz-Moncaleano, C., Sanduvete-Chaves, S., Chacon-Moscoso, S., Becker, B.J., Tenenbaum, G., 2016. Quiet eye and performance in sport: a meta-analysis. J. Sport Exerc. Psychol. 38 (5), 441–457. http://dx.doi.org/10.1123/jsep.2015-0123.

Mann, D.T., Williams, A.M., Ward, P., Janelle, C.M., 2007. Perceptual-cognitive expertise in sport: a meta-analysis. J. Sport Exerc. Psychol. 29 (4), 457–478.

Mann, D., Coombes, S.A., Mousseau, M., Janelle, C., 2011. Quiet eye and the Bereitschaftspotential: visuomotor mechanisms of expert motor performance. Cogn. Process. 12 (3), 223–234. http://dx.doi.org/10.1007/s10339-011-0398-8.

Moore, L.J., Vine, S.J., Cooke, A., Ring, C., Wilson, M., 2012. Quiet eye training expedites motor learning and aids performance under heightened anxiety: the roles of response programming and external attention. Psychophysiology 49 (7), 1005–1015. http://dx.doi.org/10.1111/j.1469-8986.2012.01379.x.

Oudejans, R.R.D., van de Langenberg, R.W., Hutter, R.I., 2002. Aiming at a far target under different viewing conditions: visual control in basketball jump shooting. Hum. Mov. Sci. 21 (4), 457–480. http://dx.doi.org/10.1016/s0167-9457(02)00116-1.

Panasiti, M.S., Pavone, E.F., Aglioti, S.M., 2016. Electrocortical signatures of detecting errors in the actions of others: an EEG study in pianists, non-pianist musicians and musically naive people. Neuroscience 318, 104–113. http://dx.doi.org/10.1016/j.neuroscience.2016.01.023.

Rienhoff, R., Hopwood, M., Fischer, L., Strauss, B., Baker, J., Schorer, J., 2013. Transfer of motor and perceptual skills from basketball to darts. Front. Psychol. 4, 593. http://dx.doi.org/10.3389/fpsyg.2013.00593.

Rienhoff, R., Tirp, J., Strauss, B., Baker, J., Schorer, J., 2016. The 'quiet eye' and motor performance: a systematic review based on Newell's constraints-led model. Sports Med. 46 (4), 589–603. http://dx.doi.org/10.1007/s40279-015-0442-4.

Ryu, D., Mann, D.L., Abernethy, B., Poolton, J.M., 2016. Gaze-contingent training enhances perceptual skill acquisition. J. Vis. 16 (2), 2. http://dx.doi.org/10.1167/16.2.2.

Vickers, J.N., 1996a. Control of visual attention during the basketball free throw. Am. J. Sports Med. 24 (6 Suppl), S93–S97.

Vickers, J.N., 1996b. Visual control when aiming at a far target. J. Exp. Psychol. Hum. Percept. Perform. 22 (2), 342–354.

Vickers, J.N., 2007. Perception, Cognition & Decision Training: The Quiet Eye in Action. Human Kinetics, Champaign, IL.

Vickers, J.N., 2009. Advances in coupling perception and action: the quiet eye as a bidirectional link between gaze, attention, and action. In: Markus Raab, J.G.J., Hauke, R.H. (Eds.), Progress in Brain Research. vol. 174. Elsevier, Amsterdam, pp. 279–288.

Vickers, J.N., 2012. Neuroscience of the quiet eye in golf putting. Int. J. Golf Sci. 1 (1), 2–9.

Vickers, J.N., Williams, A.M., 2007. Performing under pressure: the effects of physiological arousal, cognitive anxiety, and gaze control in biathlon. J. Mot. Behav. 39 (5), 381–394.

Vine, S.J., Wilson, M., 2011. The influence of quiet eye training and pressure on attention and visuo-motor control. Acta Psychol. (Amst) 136 (3), 340–346. http://dx.doi.org/10.1016/j.actpsy.2010.12.008.

Vine, S.J., Moore, L.J., Wilson, M.R., 2011. Quiet eye training facilitates competitive putting performance in elite golfers. Front. Psychol. 2, 8. http://dx.doi.org/10.3389/fpsyg.2011.00008.

Wilson, M., Vine, S.J., Wood, G., 2009. The influence of anxiety on visual attentional control in basketball free throw shooting. J. Sport Exerc. Psychol. 31 (2), 152–168.

Wilson, M., Causer, J., Vickers, J.N., 2015. Aiming for excellence: the quiet eye as a characteristic of expertise. In: Baker, J., Farrow, D. (Eds.), Routledge Handbook of Sport Expertise. Routledge, New York, NY, pp. 22–37.

FURTHER READING

Causer, J., Vickers, J.N., Snelgrove, R., Arsenault, G., Harvey, A., 2014. Performing under pressure: quiet eye training improves surgical knot-tying performance. Surgery 156 (5), 1089–1096. http://dx.doi.org/10.1016/j.surg.2014.05.004.

Gibson, J., 1979. The Ecological Approach to Visual Perception. Houghton-Mifflin, New York.

CHAPTER

Sleep, sport, and the brain

2

Shona L. Halson*,1, Laura E. Juliff†
*Australian Institute of Sport, Canberra, Australia
†Netball Australia, Fitzroy, VIC, Australia
1Corresponding author: Tel.: +61-2-6214-1111, e-mail address: shona.halson@ausport.gov.au

Abstract

The recognition that sleep is one of the foundations of athlete performance is increasing both in the elite athlete arena as well as applied performance research. Sleep, as identified through sleep deprivation and sleep extension investigations, has a role in performance, illness, injury, metabolism, cognition, memory, learning, and mood. Elite athletes have been identified as having poorer quality and quantity of sleep in comparison to the general population. This is likely the result on training times, competition stress/anxiety, muscle soreness, caffeine use, and travel. Sleep, in particular slow wave sleep, provides a restorative function to the body to recover from prior wakefulness and fatigue by repairing processes and restoring energy. In addition, research in the general population is highlighting the importance of sleep on neurophysiology, cognitive function, and mood which may have implications for elite athlete performance. It is thus increased understanding of both the effects of sleep deprivation and potential mechanisms of influence on performance that may allow scientists and practitioners to positively influence sleep in athletes and ultimately maximize performances.

Keywords

Athlete, Performance, Cognitive function, Memory, Sleep deprivation

1 INTRODUCTION

It is becoming increasingly apparent that sleep is one of the primary foundations of athlete performance and well-being. Sleep has been attributed to having an essential role in human health, vital for physical and cognitive performance and well-being (Kölling et al., 2016; Krueger et al., 2008; Simpson et al., 2016). As such, most humans require large amounts of sleep with approximately one-third of human life spent in a state of sleep (Fuller et al., 2006). Despite the biological necessity for sleep, prevalence of acute sleep deprivation is not uncommon within the general population and athletes (Cook et al., 2011; Krueger et al., 2008). Beyond the basic health benefits of sleep (Goel et al., 2009; Savis, 1994), within an athletic population, sleep is described as an important recovery strategy due to its physiological and restorative effects (Halson, 2008), so that adequate sleep has been labeled as a "new frontier in sport performance enhancement" (Leeder et al., 2012; Roky et al., 2012). Regardless, to date a paucity of research exists focused on sleep in athletes with most recommendations developed from sleep data on shift workers and patients with sleep pathologies such as insomnia (Reilly and Edwards, 2007; Walsh, 2009).

2 BACKGROUND

The sleep–wakefulness cycle is described as one of the main discernible biological circadian rhythms; natural fluctuations of physiological and behavioral processes occurring over a 24-h period in the human body (Davenne, 2009; Thun et al., 2015). The circadian rhythm of sleep is evident through alertness associated with daylight and an increasing propensity to sleep occurring during the dark part of a 24-h day (Reilly and Edwards, 2007; Thun et al., 2015). In addition to the circadian drive for sleep a homeostatic mechanism exists in humans where an increasing need for sleep arises after a period of wakefulness (Leger et al., 2008; Thun et al., 2015).

In order to understand the complexities of sleep, there has been a continued increase in sleep-related research. Despite the increase, the function of sleep remains unclear (Zhang et al., 2011). Based on observations of the body and brain during sleep deprivation studies, it is hypothesized that the brain has a chance to "turn off" and repair neuronal connections during sleep (Porkka-Heiskanen et al., 2013; Underwood, 2010). This concept is reinforced when considering the function of sleep at a neurometabolic, somatic, and cognitive level (Frank, 2006). Waking imposes a metabolic and neural cost to the nervous system that on a neurometabolic level is restored in the subsequent sleep period (Halson, 2008). Somatically, sleep is associated with several purposes such as the restoration of tissue and of the immune and endocrine systems (Erlacher et al., 2011; Frank, 2006). While cognitively, sleep plays a role in memory, learning, and synaptic plasticity (Frank, 2006; Halson, 2008). What is known is that sleep is a complex, precisely regulated process consisting of periods of synchronized cortical activity (Porkka-Heiskanen et al., 2013). The events in the brain are accompanied and coordinated with physiological changes in the rest of the body. These physiological changes create a network of programmed physiological activities each night in order for neural, metabolic, and cognitive restoration and repair from prior wakefulness (Porkka-Heiskanen et al., 2013). The specific oscillations and morphological changes in the brain and body allow for the identification of specific sleep stages (Roky et al., 2012).

2.1 SLEEP STAGES

Over a 24-h period, the human body continually cycles through states of wake and sleep (Savis, 1994). The fluctuation of sleep states enables the body to recover and renew from prior wakefulness enabling an individual to awaken feeling fresh and alert (Davenne, 2009; Santos et al., 2007). During sleep, the detection of sleep stages is achieved through identifying changes in electrical activity in the brain (Fig. 1). Specifically, brain electrical frequency also commonly known as brain waves (delta 5–4 Hz; theta 4–8 Hz; alpha 8–12 Hz; beta 13–30 Hz) and amplitude (the strength of the electrical signal) change during each sleep stage (discussed later) (Savis, 1994). Approximately every 90 min during a night of normal sleep, the brain oscillates between two main sleep stages: nonrapid eye movement sleep (NREM) and rapid eye movement sleep (REM) (Fig. 2), with approximately 75% of total sleep time spent in NREM (Bear et al., 2016).

Within NREM sleep, four progressively deeper sleep stages (i.e., Stages 1, 2, 3, and 4) exist (Taylor et al., 1997). Stage 1 "bridges the gap" between waking and sleeping and is characterized by diminished responsiveness to external stimuli

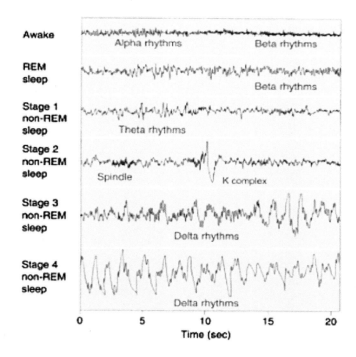

FIG. 1

Electroencephalography (EEG) rhythms that characterize each sleep stage (Bear et al., 2016).

16 CHAPTER 2 Sleep, sport, and the brain

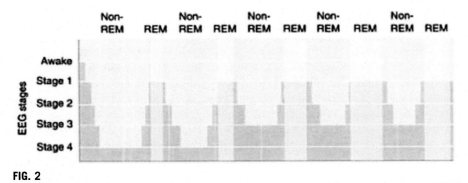

FIG. 2

A night of human sleep: illustrating the cyclical nature of sleep from nonrapid eye movement sleep (NREM) to rapid eye movement (REM) sleep (Bear et al., 2016).

(Savis, 1994). Approximately 50% of total sleep time is spent in Stage 2 and while "deeper" than Stage 1, is not classified as restorative (Savis, 1994). Stages 3 and 4 are commonly grouped together and referred to as either delta, slow wave sleep (SWS), or deep sleep (Davenne, 2009; Martini, 2001; Savis, 1994). For the purposes of this review, the term SWS will be used to represent Stages 3 and 4. Comprising approximately 15%–20% of total sleep time, SWS is electrically characterized by low frequency, high amplitude delta waves as a result of minimal cerebral cortex activity (Savis, 1994; Walsh, 2009). Furthermore, parasympathetic activation results in a decrease (circadian daily lowest) in heart rate, blood pressure, core temperature, respiratory rate, and energy utilization (Davenne, 2009; Martini, 2001; Savis, 1994). On the contrary, growth hormone secretion is the highest during SWS which importantly aids in neural and peripheral cellular restoration (Birzniece et al., 2011; Davenne, 2009). Following progression through each of the four NREM sleep stages, a period of REM sleep occurs.

Approximately 90 min after sleep onset, during the normal sleep cycle, the brain transitions from NREM to REM with greater durations of REM sleep obtained during successive sleep cycles (Savis, 1994). Described as "an active, hallucinating brain in a paralysed body" (Bear et al., 2007), the brain during REM displays similar brain wave patterns (i.e., low-amplitude, high-frequency) to those observed during wakefulness (Davenne, 2009). However, blocking of the corticospinal pathways by the brain stem results in total muscle relaxation during which time myofibril restoration occurs (Davenne, 2009). Concurrently, upregulation of sympathetic activity increases brain temperature, heart rate, cerebral blood flow, and brain protein synthesis (Savis, 1994). This process plays an important role in the formation of memories; the transcription of messenger RNAs involved in protein synthesis and has been suggested to be critical for memory consolidation and persistent forms of brain plasticity in in vitro and in vivo (Gronli et al., 2013; Seibt et al., 2012). Furthermore, during REM it is believed dreaming occurs (Bear et al., 2007). Overall, it is evident throughout a night's sleep there are distinct neural and physiological changes that take place enabling researchers and clinicians to measure or indirectly identify specific sleep stages, which are of scientific and clinical interest.

2.2 NEUROPHYSIOLOGY OF THE SLEEP–WAKE CYCLE

To understand the sleep–wake cycle, it is essential to address the proposed interaction with the ascending arousal system. Behaviorally, sleep is defined as a reversible state involving perceptual disengagement from an environment with the inability to respond to stimuli (Carskadon and Dement, 2011). Distinctively different from a coma, sleep may be instantly terminated through activation of an arousal system in response to a biological signal or physiological stressor (Roky et al., 2012; Schwartz and Roth, 2008). During the 1940s and 1950s, Giuseppe Moruzzi and colleagues sought to examine the influence of brainstem control of waking and arousal, discovering two distinct sleep states: (1) the ascending arousal system in the hypothalamus that promotes wakefulness and (2) sleep promoting neurons in the ventrolateral preoptic nucleus (VLPO) (Schwartz and Roth, 2008). The ascending arousal system comprises a number of monoaminergic cell populations such as noradrenergic, serotoninergic, dopaminergic, histaminergic, and cholinergic neurons, and orexin/hypocretin nuclei located along two branches (Born and Fehm, 2000; Šaponjić, 2009; Schwartz and Roth, 2008). The first branch innervates the thalamus, while the second branch projects into the lateral hypothalamus, basal forebrain, and cerebral cortex (Born and Fehm, 2000; Šaponjić, 2009; Schwartz and Roth, 2008). Discharging in a coordinated and overlapping manner, the monoaminergic systems promote sustained wakefulness (Schwartz and Roth, 2008). Every 24-h however, the arousal system is blocked by inhibitory neurotransmitters γ-aminobutyric acid (GABA) and galanin within the VLPO, downregulating monoaminergic cells leading to sleep (Šaponjić, 2009; Schwartz and Roth, 2008). The mutual inhibition that exists between the arousal system and the sleep producing system creates definitive wake and sleep states (Saper et al., 2005b). Systems such as these are called "flip-flop" switches by electrical engineers as they tend to avoid transitional states, when either side dominates over the other, the switch flips into the alternative state; hence explaining why wake–sleep transitions are often quite abrupt (ability to quickly fall asleep and wake suddenly) (Saper et al., 2005b). A disadvantage to this type of circulatory system is that unwanted perturbations causing physiological arousal (exercise, stress, etc.) may turn off the alternative state abruptly without warning (Saper et al., 2005a).

While primarily active during sleep, the VLPO also receives afferents from each of the monoaminergic systems (Saper et al., 2005b). Noradrenaline and serotonin can inhibit the VLPO (Saper et al., 2005b) ultimately resulting in a downregulation of the sleep promoting GABA system by the very arousal system it blocks during sleep (Saper et al., 2005b; Schwartz and Roth, 2008). Studies in the 1970s and 1980s clarified the nature of this pathway and identified neurotransmitters originating from the ascending arousal system (Saper et al., 2005b). Indeed, occupational studies have observed increased activation of monaminergic cells of the ascending arousal system following physiological stressors that led to suppression of sleep, ultimately disrupting the sleep–wake cycle (Hansen et al., 2012).

It should be noted that the ascending arousal system is proposed to comprise only one component of the sleep–wake cycle and exactly how much influence it has on the cycle remains unknown. Mathematical models have been developed in an attempt to explain the mechanisms and dynamics associated with the sleep–wake cycle (Rowsell et al., 2011). The most current model by Phillips and Robinson proposes links between homeostatic and circadian processes as well as the monoaminergic neurons in the ascending arousal system that promote wake and sleep neurons in the VLPO area of the hypothalamus promoting sleep (Sargent et al., 2014a; Skeldon et al., 2014). This model highlights the complexity of the sleep–wake cycle in relation to the homeostatic need to sleep as a function of prolonged wakefulness, the circadian drive of the daily cycles of sleep and wakefulness, as well as the brain circulatory system (Davenne, 2009; Reilly and Edwards, 2007; Saper et al., 2005a).

The rapid expansion of knowledge in the past decade has increased understanding of the basic circuitry underlying the sleep–wake cycle and the regulation of the two primary endogenous systems mentioned previously; the circadian pacemaker located within the suprachiasmatic nucleus of the hypothalamus and the homeostatic drive for sleep regulation (Saper et al., 2005a). Although on the surface there is an increased understanding of the sleep–wake cycle, at a deeper level the circuitry remains a challenge to comprehend due to the complexity of the systems adaptation to ever changing environments and variables (Saper et al., 2005a). It is clear that homeostatic and circadian drives for sleep can be overcome and desynchronized for brief periods when external events demand a sympathetic response. For an elite level, athlete possible sympathetic influencers such as training and competition may present as external perturbations to the sleep system. As it has been previously suggested by Davenne (2009) athletes are highly sensitive to any disruptive factors that can desynchronize their circadian rhythms when compared to sedentary individuals.

3 SLEEP AND ATHLETES

With most sleep research focused on clinical populations (Bonnet and Arand, 2010; Chen et al., 2011; Freedman and Papsdorf, 1976), little research exists in other populations such as athletes where the restorative properties of sleep are likely to be of importance (Dickinson and Hanrahan, 2009). Indeed, sleep has been recognized as an important component of both the psychological and physiological state of an athlete (Dickinson and Hanrahan, 2009; Erlacher et al., 2011). Data regarding the sleep habits of athletes are ambiguous with the majority of research examining sleep and performance from anecdotal or nonathletic populations (Dickinson and Hanrahan, 2009; Savis, 1994). Furthermore, while a great amount is understood regarding the duration of sleep obtained by adults in the general population, the ideal quality and quantity of sleep is yet to be elucidated for athletes (Leeder et al., 2012).

Athletes may in fact require more sleep than nonathletes to recover from the stresses of training and competition and to consolidate what was learnt from training (Famodu, 2014). In 2005, the first known review on sleep in sport (Postolache et al., 2005) concluded little is known regarding the relationship between sleep and performance in athletes. Since then research in the area has increased with recent athlete sleep monitoring studies revealing far from ideal sleep durations and quality (Lastella et al., 2014a; Samuels, 2009). However, sleep research in athletes remains a challenge as conjecture exists regarding data collected from laboratory-based sleep observations in athletes, as this data are likely artificial and not reflective of the demands made in real performance settings (Leger et al., 2008). Regardless, with team and individual athlete rankings, athlete selections, and sponsorship deals dependent on consistent high-level performances, it is important researchers understand the sleep demands of athletes in order to optimize an athlete's recovery and overall exercise performance (Dickinson and Hanrahan, 2009; Halson, 2008; Mougin et al., 1996).

3.1 IMPORTANCE OF SLEEP IN ATHLETES

Sleep has anecdotally been suggested as the single best recovery strategy available to an athlete (Halson, 2008; Leeder et al., 2012). Despite the significance of this statement, research into athlete recovery has heavily focused on techniques such as hydrotherapy, compression garments, massage, and nutritional interventions with sleep and napping frequently overlooked (Davies et al., 2010; Leeder et al., 2012). While indirectly these recovery techniques may aid sleep by decreasing inflammation (Bleakley and Davison, 2010; Vaile et al., 2008) and pain (Herrera et al., 2010; Vaile et al., 2008) as well as modulating body temperature (Halson, 2008), research aimed specifically at sleep as a recovery strategy is needed. Sleep, in particular SWS, provides a restorative function to the body to recover from prior wakefulness and fatigue by repairing processes and restoring energy (Taylor et al., 1997). This repair process ensures the body is refreshed and prepared for full functioning in the subsequent wake period. In addition, the impact and importance of sleep for an athlete's performance may also be demonstrated through studies where athletes who experienced nonrestorative sleep exhibited increased tiredness, decreased mood, and decreased cognitive functions such as impaired reaction times and decision-making abilities (discussed later) (Lastella et al., 2014a; Samuels, 2009).

3.2 INCIDENCE OF SLEEP DISRUPTION IN ATHLETES

Sleep studies on athlete populations in the literature are limited with most information from studies using subjective or non-PSG objective measures. Recently, an actigraphy monitoring study of 46 Great Britain Olympic squad members by Leeder et al. (2012) found athletes experienced poorer markers of sleep compared to age and sex-matched nonathletic individuals (sleep efficiency: $80.6 \pm 6.4\%$ and $88.7 \pm 3.6\%$, respectively); however, the athletes remained within a healthy sleep range (Leeder et al., 2012). Following this, an Australian study utilizing actigraphy revealed 124 international and national athletes slept on average 6 h and 42 min per night, well below the recommended 8 h (Lastella et al., 2015). Interestingly, this study highlighted differences between team and individual sport athletes, with team sport athletes reported sleeping 30 min longer than individual sport athletes (Lastella et al., 2015). In agreement, a study analyzing 70 elite Australian athletes over 2 weeks during a normal training block observed 88% of the sleep periods fell below the recommended 8 h of sleep per night and 60% were below 7 h, in addition to 76% of the sleep periods below 90% sleep efficiency (Sargent et al., 2014b). Further breakdown of the results indicated significant differences in sleep/wake behavior on training days compared with the rest days. On nights prior to training days, time spent in bed was significantly shorter ($08:18 \pm 01:12$ h), and the amount of sleep obtained was significantly less ($06:30 \pm 01:18$ h) compared with nights prior to rest days ($08:42 \pm 01:36$; $06:48 \pm 01:42$, respectively) (Sargent et al., 2014b). Together the above three studies indicate growing evidence suggesting elite athletes do not obtain sufficient sleep during normal training phases (Sargent et al., 2014b). The variation and extent to which an athlete's sleep may be disturbed is highly individual and should be taken into consideration as some athletes will be more likely to incur sleep disturbances than others. Despite many proposed explanations to account for possible variances in athlete sleep study findings, all of the above monitoring studies provide a useful platform for researchers to further explore sleep and the causes of sleep disruption in specific athlete populations.

3.3 CAUSES OF SLEEP DISRUPTION IN ATHLETES

Numerous internal and external factors such as temperature, late competitions, anxiety, training volume, whole body stiffness, technology, anticipation prior to competition and psychological stressors have anecdotally been suggested to negatively impact sleep in athletes (Blumert et al., 2007; Venter, 2008). For traveling athletes additional factors such as time zone changes (Eagles and Lovell, 2016), unfamiliar surroundings (bedroom), altitude (Sargent et al., 2013), stress of travel, and noise may make sufficient sleep difficult to obtain (Davenne, 2009; Savis, 1994). Furthermore, it is believed that it is not necessary to elicit total sleep deprivation to influence performance, as even fragmented sleep (e.g., where individuals are awoken at regular intervals throughout a sleep period in that it results in disrupted sleep, Cote et al., 2003) has been shown to compromise athletic performances (Blumert et al., 2007; Savis, 1994; Venter, 2008). Current research has sought to monitor athletes' sleep following numerous training phases (Taylor et al., 1997; Teng et al., 2011), different training schedules (Lastella et al., 2015; Sargent et al., 2014a), at altitude (Sargent et al., 2013), during and following competition (Lastella et al., 2014b; Leger et al., 2008; Netzer et al., 2001), following interventions such as cold water immersion (Fullagar et al., 2015; Robey et al., 2014) and during hot and cold environments (Montmayeur et al., 1994) in an attempt to comprehend the scenarios athletes are exposed to and the effects these may have on subsequent sleep and performance.

Athletes may also be of adolescent age (13–19 year range) whereby they may exhibit different circadian rhythms to adults (Carskadon, 2005). During this phase of development, individuals may experience a delay in their body clock, resulting in the desire to go to bed later and wake later.

Owing to athletes subjecting their bodies to large physical demands on a regular basis it is believed they may require more sleep than the average person to repair and recuperate (Halson, 2013). Following a high training block of increased load, Teng et al. (2011) reported 28 male cyclists decreased total sleep time from 7.3 to 6.9 h and sleep efficiency from 86.3% to 84.3%, while sleep activity increased from 15.4 average count/epoch to 18.4 average count/epoch, compared with a baseline week. Similar results were found in seven national swimmers following increased training volumes with increased sleep movements, increased time spent in SWS, and decreased mood compared with a taper period (Taylor et al., 1997). In addition to changes in sleep variables noted during the different training phases, the authors also observed the amount of sleep an athlete obtained may have been dictated by their training schedule (Lastella et al., 2015). The earlier the swimmers were required to start training, the less sleep they obtained the night before (5 h and 24 min) compared with the nights prior to no training days (7 h and 5 min) (Sargent et al., 2014a). In contrast to the sleep disturbances observed during increased training volumes, a study of 16 aerobic-based athletes found sleep architecture changes following a day of no exercise (SWS reduced by 15.5 ± 7, REM sleep increase by 17.9 ± 5.7 min) compared with a training day indicating a taper phase may also have effect on an athlete's sleep architecture (Hague et al., 2003). These studies raise an interesting question as to whether certain training periods/environments illicit reduced sleep variables and should be further explored, as this may be an important consideration for staff when programming training.

To explore sleep habits during competition, Erlacher et al. (2011) questioned 632 German athletes with results indicating 65.8% had been subject to poor sleep in the nights prior to an important competition with 79.7% experiencing problems falling asleep due to nervousness and thoughts about the upcoming competition (Erlacher et al., 2011). Further analysis of this cohort confirmed differences in team and individual sport athletes with greater sleep problems observed in individual athletes when compared with team sport athletes (Erlacher et al., 2011). These differences were suggested to be a direct consequence of the type of sport, with individual athletes likely to experience increased pressure due to the individual nature of their events (Erlacher et al., 2011). These findings of disturbed sleep prior to competition in individual sport athletes is supported with objective data indicating endurance cyclists took 30 min longer to fall asleep the night before competition in addition to experiencing reduced sleep duration on competition night (6.5 ± 0.9 h) compared with a baseline night (7.4 ± 0.6 h) (Lastella et al., 2014b).

While the awareness of the importance of sleep for athletes is increasing, many athletes fail to prioritize sleep and indeed view sleep as a "flexible commodity" that can be exchanged for more enjoyable activities (Halson, 2016). For many athletes, there is a distinct lack of awareness and concomitant dedication to achieving optimal sleep. The high prevalence of smart phone, computer games, and social media use has proliferated this issue.

Insomnia, as defined by The American Academy of Sleep Medicine, is difficulty initiating or maintaining sleep, or nonrestorative sleep accompanied by daytime impairment such as fatigue, or difficulties with memory or concentration (Montmayeur et al., 1994). While an athlete may not be diagnosed with insomnia, it is important for athletes, coaches, and support staff to recognize that insomnia-like symptoms may present during crucial periods within a season and are likely to influence overall performance. These studies have provided a baseline for exploring sleep in athletes and create a platform to raise further research enquiries.

3.4 SLEEP AND ATHLETE PERFORMANCE

As previously highlighted, a lack of research exists examining the influence of sleep on performance. As such, this section will review information from existing athlete data to provide a theoretical construct for the influence that sleep disruption may have on an athletic population. In an attempt to understand the importance of sleep in athletes, studies have utilized sleep deprivation protocols. Total sleep deprivation refers to a period of sustained wakefulness greater than 24-h whereas partial sleep deprivation is where a participants sleep is restricted to less than or equal to 6 h (Lastella et al., 2014a).

Sleep deprivation in an athletic population has been shown to severely affect both physiological (Abedelmalek et al., 2013b; Skein et al., 2013) and psychological (Blumert et al., 2007; Reilly and Piercy, 1994) well-being. Most sleep deprivation studies have highlighted an altered mood state affecting motivation as a primary outcome of lack of sleep (Reilly and Edwards, 2007; Savis, 1994). Sinnerton and Reilly explored swim performance following 2.5 h of sleep compared with normal sleep (around 8 h) in eight swimmers over four consecutive days. The results showed decrements in mood states following reduced sleep with increased fatigue, anger, depression, and tension and no change in back or grip strength, lung function, or swim times (Reilly and Edwards, 2007). Reilly and Piercy (1994) found comparable results following 3 h of sleep per night for three nights in eight male participants. The participants displayed impaired mood states of confusion, vigor, and fatigue which, the authors implied may have led to the decreased power performance variables (bench press and leg press) (Reilly and Piercy, 1994).

Studies enforcing total sleep deprivation in athletes, while not entirely ecologically valid, have demonstrated reduced countermovement jumps (CMJ), intermittent sprint times, and volume in team sport athletes (Skein et al., 2011, 2013; Souissi et al., 2013). A night of total sleep deprivation decreased and delayed the recovery of lower-body power; mean and peak CMJ distances ($p=0.10–0.16$; $d=0.95–1.05$) up to 16 h following a competitive rugby league match in 11 athletes compared with a control night of normal sleep (8 h) (Skein et al., 2013). Furthermore, knee extensor maximal voluntary contraction was significantly lower postgame following total sleep deprivation ($p=0.02$; $d=0.67–0.76$). Similarly, Skein et al. (2011) reported that 30 h of sleep deprivation in 10 male team sport athletes resulted in reduced intermittent sprint performance with mean sprint times slower following sleep deprivation (2.78 ± 0.17 s) compared with a control night (2.74 ± 0.15 s). In addition, less mean and total distances were covered by 8 of the 10 team sport athletes following the sleep deprivation night ($p=0.01$) (Skein et al., 2011).

Along with total sleep deprivation, partial sleep deprivation studies have displayed reduced maximal anaerobic power variables in football players (Abedelmalek et al., 2013a) and judo athletes (Souissi et al., 2013). Partial sleep deprivation of 4 h toward the end of a night's sleep resulted in decreased muscle strength and power with relative decreases ranging between 2.2% and 9.3% for peak power and between 2.8% and 7.3% for mean power during a Wingate cycle test in the afternoon when compared with a night of normal sleep in 12 judo athletes (Souissi et al., 2013). A similar study of 12 male soccer players following 4.5 h sleep found a significant decrease in peak power and mean power ($p<0.001$) during the Wingate cycle test after partial sleep deprivation compared with a reference night of normal sleep (8 h) (Abedelmalek et al., 2013a). These studies highlight the importance sleep has on athletic performance.

In contrast, Mougin et al. (1996) found following a 30 s Wingate cycle test, performance was able to be maintained following a delayed bedtime until 3 am compared to a night of 8 h sleep suggesting maximal performance may be maintained under partial sleep loss conditions (Mougin et al., 1996). This finding suggests athletes may be able to overcome sleep loss to produce maximal all out efforts; however, they may not be able to maintain these levels during sustained, repeated effort exercise (Halson, 2008; Reilly and Piercy, 1994). Following a series of weight-lifting tasks using traditional weight training exercises (biceps curl, bench press, deadlift, and leg press), Reilly and Edwards (2007) found deterioration in all four exercises particularly after the second night of restricted sleep. From this study, it was apparent that the greatest impairments in performance were seen with sustained exercise later in the session implying a cumulative fatigue effect following sleep restriction (Reilly and Edwards, 2007). Regardless, considering the adverse effects of the above sleep deprivation studies on mood, cognitive, and athletic performance, it is evident that total and partial sleep deprivation has the ability to negatively impact athletes (Lastella et al., 2014a).

3.5 SLEEP EXTENSION

On the contrary, the importance of sleep for athlete performance can be demonstrated through the extension of sleep duration. A group of Stanford University swimmers extended their sleep to 10 h per night for 6–7 weeks from their normal sleep durations varying between 6 and 8 h with swim results indicating superior performances (Mah et al., 2008). Specifically, athletes swam a 15-m sprint 0.51 s faster, reacted 0.15 s quicker off the blocks, and improved turn time by 0.10 s (Mah et al., 2008). The authors from this study stated "the sleep extension results begin to elucidate the importance of sleep on athletic performance and, more specifically, how sleep is a significant factor in achieving peak athletic performance." For modern day athletes who have demanding work and training schedules it may not be feasible to simply extend sleep durations. It therefore remains to be answered whether comparable results may be induced by increasing sleep quality for a given sleep duration through exploring effective sleep interventions (Fowler et al., 2014).

4 SLEEP DEPRIVATION AND BRAIN FUNCTION
4.1 COGNITIVE PERFORMANCE IN ATHLETES

Partial sleep deprivation has been found to negatively influence cognitive function and psychomotor performance through reduced focus (Davenne, 2009), determination (Underwood, 2010), processing (Underwood, 2010), logical thinking (Underwood, 2010), and vigilance (Davenne, 2009). For example, 60 dart players demonstrated decreased alertness, increased fatigue, and decreased accuracy (frequently missing the target) following 3–4 h of sleep compared with a normal night of 7–8 h sleep (Edwards and Waterhouse, 2009). These results led the researchers to conclude acute partial sleep deprivation of one night decreases overall psychomotor performance in dart players (Edwards and Waterhouse, 2009).

Both total (Taheri and Arabameri, 2012) and partial sleep deprivation (Jarraya et al., 2013) have been shown to compromise the transmission speed of impulses from the brain to the working muscles, affecting reflex, and reaction times (Underwood, 2010). Four hours of partial sleep deprivation at the end of the night (athletes slept from 22:00 to 03:00) decreased attentional capacities and increased reaction times (593.00 ± 8.27 ms) in 12 handball goal keepers compared with a minimum of 8 h sleep (398.28 ± 4.91 ms) (Jarraya et al., 2013). Mean choice reaction times were also shown to be compromised following a night of total sleep deprivation (281.65 ± 31 ms) compared with a night of habitual sleep (244 ± 39 ms) in college athletes (Taheri and Arabameri, 2012). Together these findings suggest that adequate sleep is important for peak performance as executive function tasks are particularly sensitive to sleep (Taheri and Arabameri, 2012). For many athletes, total sleep deprivation and fragmented sleep is of concern, as athletes are reliant on the ability to make fast, accurate decisions, and execute skills effectively for optimal performances (Cook et al., 2011).

4.2 COGNITIVE PERFORMANCE, MEMORY, AND LEARNING

While the functions of sleep are widely debated and are yet to be fully described, researchers have examined the effects of sleep deprivation over a single night, reduced sleep for several hours per night, or disturbance of sleep continuity. The consequences of sleep deprivation provide insights into the importance of sleep and why it is considered fundamental to athlete health and well-being. Although scientific research investigating the consequences of poor sleep is rapidly increasing, data on the specific effects in athletes is limited. Therefore, much of the information regarding the importance of sleep for athletes is derived from minimal data on performance, but is primarily inferred from data from the general population.

Cognitive performance decrements are a well-known consequence of sleep deprivation and/or restriction in the general population and have been demonstrated to occur at the subcellular through to complex behaviors (van Someren et al., 2015). Research in adults has shown that sleep deprivation has the greatest effect on cognitive function, in comparison to sleep restriction, and these effects are the largest for vigilance tasks (de Bruin et al., 2017). Other indices of cognitive function such as attention, memory, and reasoning have also been shown to be reduced in some sleep deprivation studies with deficits accumulating following persistent sleep restriction (de Bruin et al., 2017).

Much research has focused on the effects of sleep deprivation on brain plasticity and memory, demonstrating that sleep has a powerful effect on both initial learning, and as well as subsequent long-term memory consolidation (Abel et al., 2013). Stickgold (2013) describes that with the exception of rapid (2–4 h) consolidation, many forms of memory processing occurs preferentially, if not exclusively, during sleep. Sleep can result in additional memory stabilization and enhancement, selective item consolidation, item integration, and multiitem generalizations (Stickgold, 2013).

A recent systematic review of the effects of sleep manipulation on cognitive function in adolescents found that (1) sleep directly after learning improves memory consolidation, (2) sleep extension in adolescents with chronic sleep reduction contributes to improvements in working memory, and (3) sleep deprivation shows clear decrements in vigilance tasks (De Bruin et al., 2017). These findings may be particularly relevant for adolescent athletes who are often required to wake early for training prior to attending school and who are also likely to be in a phase of athlete skill development.

While the majority of cutting edge scientific research in cognitive function and memory is conducted in nonathletes, it is possible to infer the influence of poor sleep on athlete learning and development. This may be particularly important for athletes learning new skills and memorizing "plays" or strategies in team sports.

4.3 MOOD

Sleep loss has been shown to influence emotion regulation and research has identified a role for sleep in various mood disorders, vulnerability to psychopathology, and overall poor psychological functioning (Ong et al., 2016). In a review on social interactions, emotion, and sleep, Beattie et al. (2015) report that sleep deprivation diminishes emotional expressivity and recognition as well as increases emotional reactivity. These changes in the experience and expression of emotions are typically seen as increased lability and irritability (Coles et al., 2015). Data also exist regarding increased risk taking (Lei et al., 2016) and impulsivity (Demos et al., 2016) following sleep deprivation.

Sleep deprivation has also been linked to increased anxiety, with insomniacs and patients with delayed sleep phase disorder demonstrating higher anxious symptoms

(Coles et al., 2015). A bidirectional relationship appears to exist, with disruptions in sleep being a result of anxiety (as seen in athletes precompetition), but also anxiety can precede or occur at the same times as sleep disturbances (Coles et al., 2015).

Whether lower emotion regulation, increased impulsivity, and increased anxiety influences an elite athletes performance is unknown. However, it may be postulated that these concerns may be of importance both in and out of the athletic arena.

4.4 MOLECULAR ASPECTS OF SLEEP DISRUPTION

Many of the cognitive impairments that occur with poor sleep appear to be associated with memory and learning processes that require the hippocampus (Kreutzmann et al., 2015). Therefore, it has been suggested that this brain region is most sensitive to sleep loss. Recent research has suggested that sleep deprivation may impair hippocampal neuronal plasticity and memory processing by altering cyclic adenosine monophosphate (c-AMP)-protein kinase A activity (Kreutzmann et al., 2015). This may have subsequent effects on transcription, signaling, and receptor expression of glutamate. Ultimately impaired hippocampal plasticity and function may result and potentially contribute to cognitive disorders (Kreutzmann et al., 2015).

Clock genes have been identified which generate circadian rhythms and 43% of all protein-coding genes demonstrate a circadian rhythm (Zhang et al., 2014). Furthermore, the percentage of genes that are transcribed during sleep has been shown to decrease from 6.4% to 1% when sleep is delayed by 4 h every day (Archer et al., 2014). This negative influence of poor sleep on transcription, translation, and genes has implications for inflammatory and immune responses, which may be a primary mediator for the role of sleep in health and well-being (van Someren et al., 2015).

Finally, recent research in mice has shown a positive effect of sleep on increasing interstitial space resulting in facilitated convection of interstitial and cerebral spinal fluid (Xie et al., 2013). This may result in the enhanced removal of neurotoxic waste products such as beta-amyloid (van Someren et al., 2015). This is one example of some of the recent research that aids and develops our understanding of the underlying mechanisms regarding the consequences of sleep deprivation.

5 CONCLUSION

While there is still much to learn about both the reasons why sleep is considered vital to athletes and the subsequent effects when athletes are deprived of sleep, evidence to date supports the fact that sleep is important for optimal athletic performance. Paradoxically, current data suggest that many elite athletes do not experience optimal sleep quality or quantity. From a cognitive perspective, this may have implications for reaction time, memory, learning, and mood. As greater understanding of the role of sleep in multifaceted aspects of human performance is realized, athletes, scientists, and coaches will have additional tools to not only assess sleep behavior but also provide appropriate interventions to enhance athlete performance.

REFERENCES

Abedelmalek, S., Chtourou, H., Aloui, A., Aouichaoui, C., Souissi, N., TABKA, Z., 2013a. Effect of time of day and partial sleep deprivation on plasma concentrations of IL-6 during a short-term maximal performance. Eur. J. Appl. Physiol. 113, 241–248.

Abedelmalek, S., Souissi, N., Chtourou, H., Denguezli, M., Aouichaoui, C., Ajina, M., Aloui, A., Dogui, M., Haddouk, S., Tabka, Z., 2013b. Effects of partial sleep deprivation on proinflammatory cytokines, growth hormone, and steroid hormone concentrations during repeated brief sprint interval exercise. Chronobiol. Int. 30, 502–509.

Abel, T., Havekes, R., Saletin, J.M., Walker, M.P., 2013. Sleep, plasticity and memory from molecules to whole-brain networks. Curr. Biol. 23, R774–R788.

Archer, S.N., Laing, E.E., Moller-Levet, C.S., van der Veen, D.R., Bucca, G., Lazar, A.S., Santhi, N., Slak, A., Kabiljo, R., von Schantz, M., Smith, C.P., Dijk, D.J., 2014. Mistimed sleep disrupts circadian regulation of the human transcriptome. Proc. Natl. Acad. Sci. U.S.A. 111, E682–E691.

Bear, M.F., Connors, B.W., Paradiso, M.A., 2007. Neuroscience: Exploring the Brain. Lippincott Williams & Wilkins, United States.

Bear, M.F., Connors, B.W., Paradiso, M.A., 2016. Neuroscience: Exploring the Brain. Lippincott Williams & Wilkins, United States.

Beattie, L., Kyle, S.D., Espie, C.A., Biello, S.M., 2015. Social interactions, emotion and sleep: a systematic review and research agenda. Sleep Med. Rev. 24, 83–100.

Birzniece, V., Nelson, A.E., Ho, K.K., 2011. Growth hormone and physical performance. Trends Endocrinol. Metab. 22, 171–178.

Bleakley, C.M., Davison, G.W., 2010. What is the biochemical and physiological rationale for using cold-water immersion in sports recovery? A systematic review. Br. J. Sports Med. 44, 179–187.

Blumert, P.A., Crum, A.J., Ernsting, M., Volek, J.S., Hollander, D.B., Haff, E.E., Haff, G.G., 2007. The acute effects of twenty-four hours of sleep loss on the performance of national caliber male collegiate weightlifters. J. Strength Cond. Res. 21, 1146.

Bonnet, M.H., Arand, D.L., 2010. Hyperarousal and insomnia: state of the science. Sleep Med. Rev. 14, 9–15.

Born, J., Fehm, H., 2000. The neuroendocrine recovery function of sleep. Noise Health 2, 25–38.

Carskadon, M.A., 2005. Sleep and circadian rhythms in children and adolescents: relevance for athletic performance of young people. Clin. Sports Med. 24, 319–328.

Carskadon, M.A., Dement, W.C., 2011. Norman human sleep: an overview. In: Principles and Practices of Sleep Medicine, fifth ed. Elsevier Saunders, Canada.

Chen, H.-C., Lin, C.-M., Lee, M.-B., Chou, P., 2011. The relationship between pre-sleep arousal and spontaneous arousals from sleep in subjects referred for diagnostic polysomnograms. J. Chin. Med. Assoc. 74, 81–86.

Coles, M.E., Schubert, J.R., Nota, J.A., 2015. Sleep, circadian rhythms, and anxious traits. Curr. Psychiatry Rep. 17, 73.

Cook, C.J., Crewther, B.T., Kilduff, L.P., Drawer, S., Gaviglio, C.M., 2011. Skill execution and sleep deprivation: effects of acute caffeine or creatine supplementation—a randomized placebo-controlled trial. J. Int. Soc. Sports Nutr. 8, 2.

Cote, K.A., Milner, C.E., Osip, S.L., Ray, L.B., Baxter, K.D., 2003. Waking quantitative electroencephalogram and auditory event-related potentials following experimentally induced sleep fragmentation. Sleep 26, 687–694.

Davenne, D., 2009. Sleep of athletes-problems and possible solutions. Biol. Rhythm. Res. 40, 45–52.

Davies, D., Graham, K., Chow, C.M., 2010. The effect of prior endurance training on nap sleep patterns. Int. J. Sports Physiol. Perform. 5, 87.

de Bruin, E.J., van Run, C., Staaks, J., Meijer, A.M., 2017. Effects of sleep manipulation on cognitive functioning of adolescents: a systematic review. Sleep Med. Rev. 32, 45–57.

Demos, K.E., Hart, C.N., Sweet, L.H., Mailloux, K.A., Trautvetter, J., Williams, S.E., Wing, R.R., Mccaffery, J.M., 2016. Partial sleep deprivation impacts impulsive action but not impulsive decision-making. Physiol. Behav. 164, 214–219.

Dickinson, R.K., Hanrahan, S.J., 2009. An investigation of subjective sleep and fatigue measures for use with elite athletes. J. Clin. Sport Psychol. 3, 224–266.

Eagles, A.N., Lovell, D.I., 2016. Changes in sleep quantity and efficiency in professional rugby union players during home-based training and match play. J. Sports Med. Phys. Fitness. 56 (5), 565–571.

Edwards, B.J., Waterhouse, J., 2009. Effects of one night of partial sleep deprivation upon diurnal rhythms of accuracy and consistency in throwing darts. Chronobiol. Int. 26, 756–768.

Erlacher, D., Ehrlenspiel, F., Adegbesan, O.A., El-Din, H.G., 2011. Sleep habits in German athletes before important competitions or games. J. Sports Sci. 29, 859–866.

Famodu, O., 2014. Effectiveness of Sleep Extension on Athletic Performance and Nutrition of Female Track Athletes. West Virginia University, USA.

Fowler, P., Duffield, R., Vaile, J., 2014. Effects of domestic air travel on technical and tactical performance and recovery in soccer. Int. J. Sports Physiol. Perform. 9, 378–386.

Frank, M.G., 2006. The mystery of sleep function: current perspectives and future directions. Rev. Neurosci. 17, 375–392.

Freedman, R., Papsdorf, J.D., 1976. Biofeedback and progressive relaxation treatment of sleep-onset insomnia: a controlled, all-night investigation. Biofeedback Self Regul. 1, 253–271.

Fullagar, H.H., Duffield, R., Skorski, S., Coutts, A.J., Julian, R., Meyer, T., 2015. Sleep and recovery in team sport: current sleep-related issues facing professional team-sport athletes. Int. J. Sports Physiol. Perform. 10, 950–957.

Fuller, P.M., Gooley, J.J., Saper, C.B., 2006. Neurobiology of the sleep-wake cycle: sleep architecture, circadian regulation, and regulatory feedback. J. Biol. Rhythms 21, 482–493.

Goel, N., Rao, H., Durmer, J.S., Dinges, D.F., 2009. Neurocognitive consequences of sleep deprivation. Semin. Neurol., 29 (4), 320–339.

Gronli, J., Soule, J., Bramham, C.R., 2013. Sleep and protein synthesis-dependent synaptic plasticity: impacts of sleep loss and stress. Front. Behav. Neurosci. 7, 224.

Hague, J.F., Gilbert, S.S., Burgess, H.J., Ferguson, S.A., Dawson, D., 2003. A sedentary day: effects on subsequent sleep and body temperatures in trained athletes. Physiol. Behav. 78, 261–267.

Halson, S.L., 2008. Nutrition, sleep and recovery. Eur. J. Sport Sci. 8, 119–126.

Halson, S.L., 2013. Sleep and the elite athlete. Sports Sci. 26, 1–4.

Halson, S.L., 2016. Stealing sleep: is sport or society to blame? Br. J. Sports Med. 50, 381.

Hansen, Å.M., Thomsen, J.F., Kaergaard, A., Kolstad, H.A., Kaerlev, L., Mors, O., Rugulies, R., Bonde, J.P., Andersen, J.H., Mikkelsen, S., 2012. Salivary cortisol and sleep problems among civil servants. Psychoneuroendocrinology 37, 1086–1095.

Herrera, E., Sandoval, M.C., Camargo, D.M., Salvini, T.F., 2010. Motor and sensory nerve conduction are affected differently by ice pack, ice massage, and cold water immersion. Phys. Ther. 90, 581.

Jarraya, M., Jarraya, S., Chtourou, H., Souissi, N., Chamari, K., 2013. The effect of partial sleep deprivation on the reaction time and the attentional capacities of the handball goalkeeper. Biol. Rhythm. Res. 44, 503–510.

Kölling, S., Ferrauti, A., Pfeiffer, M., Meyer, T., Kellmann, M., 2016. Sleep in sports: a short summary of alterations in sleep/wake patterns and the effects of sleep loss and jet-lag. Dtsch. Z. Sportmed. 67, 35–38.

Kreutzmann, J.C., Havekes, R., Abel, T., Meerlo, P., 2015. Sleep deprivation and hippocampal vulnerability: changes in neuronal plasticity, neurogenesis and cognitive function. Neuroscience 309, 173–190.

Krueger, J.M., Rector, D.M., Roy, S., van Dongen, H.P.A., Belenky, G., Panksepp, J., 2008. Sleep as a fundamental property of neuronal assemblies. Nat. Rev. Neurosci. 9, 910–919.

Lastella, M., Lovell, G.P., Sargent, C., 2014a. Athletes' precompetitive sleep behaviour and its relationship with subsequent precompetitive mood and performance. Eur. J. Sport Sci., (Suppl. 1), S123–S130.

Lastella, M., Roach, G.D., Halson, S.L., Martin, D.T., West, N.P., Sargent, C., 2014b. Sleep/wake behaviour of endurance cyclists before and during competition. J. Sports Sci., 1–7.

Lastella, M., Roach, G.D., Halson, S.L., Sargent, C., 2015. Sleep/wake behaviours of elite athletes from individual and team sports. Eur. J. Sport Sci. 15, 94–100.

Leeder, J., Glaister, M., Pizzoferro, K., Dawson, J., Pedlar, C., 2012. Sleep duration and quality in elite athletes measured using wristwatch actigraphy. J. Sports Sci. 30, 541–545.

Leger, D., Elbaz, M., Raffray, T., Metlaine, A., Bayon, V., Duforez, F., 2008. Sleep management and the performance of eight sailors in the Tour de France a la voile yacht race. J. Sports Sci. 26, 21–28.

Lei, Y., Wang, L., Chen, P., Li, Y., Han, W., Ge, M., Yang, L., Chen, S., Hu, W., Wu, X., Yang, Z., 2016. Neural correlates of increased risk-taking propensity in sleep-deprived people along with a changing risk level. Brain Imaging Behav. Epub ahead of print Dec 14.

Mah, C., Mah, K., Dement, W., 2008. Extended sleep and the effects on mood and athletic performance in collegiate swimmers. Sleep, A128. Amer Acad Sleep Medicine.

Martini, F., 2001. Fundamentals of Anatomy and Physiology. Prentice Hall, USA.

Montmayeur, A., Buguet, A., Sollin, H., Lacour, J.-R., 1994. Exercise and sleep in four African sportsmen living in the Sahel. Int. J. Sports Med. 15, 42–45.

Mougin, F., Bourdin, H., Simon-Rigaud, M., Didier, J., Toubin, G., Kantelip, J., 1996. Effects of a selective sleep deprivation on subsequent anaerobic performance. Int. J. Sports Med. 17, 115–119.

Netzer, N.C., Kristo, D., Steinle, H., Lehmann, M., Strohl, K.P., 2001. REM sleep and catecholamine excretion: a study in elite athletes. Eur. J. Appl. Physiol. 84, 521–526.

Ong, A.D., Kim, S., Young, S., Steptoe, A., 2016. Positive affect and sleep: a systematic review. Sleep Med. Rev. pii: S1087-0792(16)30068-5.

Porkka-Heiskanen, T., Zitting, K.M., Wigren, H.K., 2013. Sleep, its regulation and possible mechanisms of sleep disturbances. Acta Physiol. (Oxf.) 208 (4), 311–328.

Postolache, T.T., Hung, T.-M., Rosenthal, R.N., Soriano, J.J., Montes, F., Stiller, J.W., 2005. Sports chronobiology consultation: from the lab to the arena. Clin. Sports Med. 24, 415–456.

Reilly, T., Edwards, B., 2007. Altered sleep-wake cycles and physical performance in athletes. Physiol. Behav. 90, 274–284.

Reilly, T., Piercy, M., 1994. The effect of partial sleep deprivation on weight-lifting performance. Ergonomics 37, 107–115.

Robey, E., Dawson, B., Halson, S., Gregson, W., Goodman, C., Eastwood, P., 2014. Sleep quantity and quality in elite youth soccer players: a pilot study. Eur. J. Sport Sci. 14 (5), 410–417.

Roky, R., Herrera, C.P., Ahmed, Q., 2012. Sleep in athletes and the effects of Ramadan. J. Sports Sci. 30, 75–84.

Rowsell, G.J., Coutts, A.J., Reaburn, P., Hill-Haas, S., 2011. Effect of post-match cold-water immersion on subsequent match running performance in junior soccer players during tournament play. J. Sports Sci. 29, 1–6.

Samuels, C., 2009. Sleep, recovery, and performance: the new frontier in high-performance athletics. Phys. Med. Rehabil. Clin. N. Am. 20, 149–159.

Santos, R., Tufik, S., De Mello, M., 2007. Exercise, sleep and cytokines: is there a relation? Sleep Med. Rev. 11, 231–239.

Saper, C.B., Cano, G., Scammell, T.E., 2005a. Homeostatic, circadian, and emotional regulation of sleep. J. Comp. Neurol. 493, 92–98.

Saper, C.B., Scammell, T.E., Lu, J., 2005b. Hypothalamic regulation of sleep and circadian rhythms. Nature 437, 1257–1263.

Šaponjić, J., 2009. Neurochemical mechanisms of sleep regulation. Glas Srp. Akad. Nauka Med., 50, 97–109.

Sargent, C., Schmidt, W.F., Aughey, R.J., Bourdon, P.C., Soria, R., Claros, J.C.J., Garvican-Lewis, L.A., Buchheit, M., Simpson, B.M., Hammond, K., 2013. The impact of altitude on the sleep of young elite soccer players (ISA3600). Br. J. Sports Med. 47, i86–i92.

Sargent, C., Halson, S., Roach, G.D., 2014a. Sleep or swim? Early-morning training severely restricts the amount of sleep obtained by elite swimmers. Eur. J. Sport Sci. 14 (Suppl. 1), S310–S315.

Sargent, C., Lastella, M., Halson, S.L., Roach, G.D., 2014b. The impact of training schedules on the sleep and fatigue of elite athletes. Chronobiol. Int. 31, 1160–1168.

Savis, J., 1994. Sleep and athletic performance: overview and implications for sport psychology. Sports Psychol. 8, 111–125.

Schwartz, J.R.L., Roth, T., 2008. Neurophysiology of sleep and wakefulness: basic science and clinical implications. Curr. Neuropharmacol. 6, 367–378.

Seibt, J., Dumoulin, M.C., Aton, S.J., Coleman, T., Watson, A., Naidoo, N., Frank, M.G., 2012. Protein synthesis during sleep consolidates cortical plasticity in vivo. Curr. Biol. 22, 676–682.

Simpson, N., Gibbs, E., Matheson, G., 2016. Optimizing sleep to maximize performance: implications and recommendations for elite athletes. Scand. J. Med. Sci. Sports 27 (3), 266–274.

Skein, M., Duffield, R., Edge, J., Short, M.J., Mündel, T., 2011. Intermittent-sprint performance and muscle glycogen after 30 h of sleep deprivation. Med. Sci. Sports Exerc. 43, 1301–1311.

Skein, M., Duffield, R., Minett, G.M., Snape, A., Murphy, A., 2013. The effect of overnight sleep deprivation after competitive rugby league matches on postmatch physiological and perceptual recovery. Int. J. Sports Physiol. Perform. 8, 556–564.

Skeldon, A.C., Dijk, D.J., Derks, G., 2014. Mathematical models for sleep-wake dynamics: comparison of the two-process model and a mutual inhibition neuronal model. PLoS One. 9 (8), e103877.

Souissi, N., Chtourou, H., Aloui, A., Hammouda, O., Dogui, M., Chaouachi, A., Chamari, K., 2013. Effects of time-of-day and partial sleep deprivation on short-term maximal performances of judo competitors. J. Strength Cond. Res. 27, 2473–2480.

Stickgold, R., 2013. Parsing the role of sleep in memory processing. Curr. Opin. Neurobiol. 23, 847–853.

Taheri, M., Arabameri, E., 2012. The effect of sleep deprivation on choice reaction time and anaerobic power of college student athletes. Asian J. Sports Med. 3, 15.

Taylor, S.R., Rogers, G.G., Driver, H.S., 1997. Effects of training volume on sleep, psychological, and selected physiological profiles of elite female swimmers. Med. Sci. Sports Exerc. 29, 688.

Teng, E., Lastella, M., Roach, G., Sargent, C., 2011. In: The effect of training load on sleep quality and sleep perception in elite male cyclists. Little Clock, Big Clock: Molecular to Physiological Clocks. Australasian Chronobiology Society, Melbourne, Australia, 05-10.

Thun, E., Bjorvatn, B., Flo, E., Harris, A., Pallesen, S., 2015. Sleep, circadian rhythms, and athletic performance. Sleep Med. Rev. 23, 1–9.

Underwood, J., 2010. Sleep now clearly a predictor of performance. Coaches Plan 17, 31–34.

Vaile, J., Halson, S., Gill, N., Dawson, B., 2008. Effect of hydrotherapy on the signs and symptoms of delayed onset muscle soreness. Eur. J. Appl. Physiol. 102, 447–455.

van Someren, E.J., Cirelli, C., Dijk, D.J., van Cauter, E., Schwartz, S., Chee, M.W., 2015. Disrupted sleep: from molecules to cognition. J. Neurosci. 35, 13889–13895.

Venter, R., 2008. Sleep for performance and recovery in athletes. Contin. Med. Educ. 26, 331.

Walsh, J.K., 2009. Enhancement of slow wave sleep: implications for insomnia. J. Clin. Sleep Med. 5, S27.

Xie, L., Kang, H., Xu, Q., Chen, M.J., Liao, Y., Thiyagarajan, M., O'Donnell, J., Christensen, D.J., Nicholson, C., Iliff, J.J., Takano, T., Deane, R., Nedergaard, M., 2013. Sleep drives metabolite clearance from the adult brain. Science 342, 373–377.

Zhang, J., Ronald, C.W., Kong, A.P.S., Yee-So, W., Li, A.M., Lam, S.P., Li, S.X., Yu, M.W.M., Ho, C.S., Chan, M.H.M., Zhang, B., Wing, Y.K., 2011. Relationship of sleep quantity and quality with 24-hour urinary catecholamines and salivary awakening cortisol in healthy middle aged adults. Sleep 34, 225–233.

Zhang, R., Lahens, N.F., Ballance, H.I., Hughes, M.E., Hogenesch, J.B., 2014. A circadian gene expression atlas in mammals: implications for biology and medicine. Proc. Natl. Acad. Sci. U.S.A. 111, 16219–16224.

CHAPTER 3

The resonant system: Linking brain–body–environment in sport performance

Pedro Teques[*,†,1], Duarte Araújo[†], Ludovic Seifert[‡], Vicente L. del Campo[§], Keith Davids[¶]

[*]*N2i, Polytechnic Institute of Maia, Castelo da Maia, Portugal*
[†]*CIPER, Faculdade de Motricidade Humana, Universidade de Lisboa, Lisboa, Portugal*
[‡]*CETAPS EA3832, Faculty of Sport Sciences, University of Rouen Normandy, Mont-Saint-Aignan, France*
[§]*Sport Sciences Faculty, University of Extremadura, Caceres, Spain*
[¶]*Centre for Sports Engineering Research, Sheffield Hallam University, Sheffield, United Kingdom*
[1]*Corresponding author: Tel.: +351-229-866-026, e-mail address: pteques@ipmaia.pt*

Abstract

The ecological dynamics approach offers new insights to understand how athlete nervous systems are embedded within the body–environment system in sport. Cognitive neuroscience focuses on the neural bases of athlete behaviors in terms of perceptual, cognitive, and motor functions defined within specific brain structures. Here, we discuss some limitations of this traditional perspective, addressing how athletes functionally adapt perception and action to the dynamics of complex performance environments by continuously perceiving information to regulate goal-directed actions. We examine how recent neurophysiological evidence of functioning in diverse cortical and subcortical regions appears more compatible with an ecological dynamics perspective, than traditional views in cognitive neuroscience. We propose how athlete behaviors in sports may be related to the tuning of resonant mechanisms indicating that perception is a dynamic process involving the whole body of the athlete. We emphasize the important role of metastable dynamics in the brain–body–environment system facilitating continuous interactions with a landscape of affordances (opportunities for action) in a performance environment. We discuss implications of these ideas for performance preparation and practice design in sport.

This work was partly supported by the Fundação para a Ciência e Tecnologia, under Grant UID/DTP/UI447/2013 to CIPER—Centro Interdisciplinar para o Estudo da Performance Humana (unit 447).

Keywords

Attunement, Dynamical systems theory, Embodied cognition, Neural resonance, Sports

1 INTRODUCTION

Nervous system function enables a multicellular organism to maintain adaptive contact with its environment, through its everyday behaviors, such locomotion, object interception, and its interactions with other organisms. Such organisms can achieve these fundamental behavioral goals by using information to regulate actions and acting to create information (Gibson, 1979; Reed, 1989). In similar vein, to successfully perform in sport, athletes need to become attuned to informational variables that specify goal achievement, e.g., how to adapt gait to place a leading foot on a take-off board in the long jump or how to regulate arm and hand movements to catch a ball (e.g., Davids et al., 2002). To achieve these performance goals, an athlete's nervous system regulates propulsive forces and maintains equilibrium under gravity, utilizes reactive forces, and negotiates different support surfaces, while interacting with key features, objects, and others, in a dynamic environment (Chow et al., 2015; Davids et al., 2008).

Behavioral neuroscience research, addressing the relationship between brain and behavior, has often adopted a monodisciplinary perspective on sport performance. For instance, cognitive sport psychologists traditionally study how the brain processes information to produce internal representations of external phenomena, or how knowledge is acquired, how memories are stored and retrieved, seeking to describe how these cognitive processes are correlated with "activity" in different brain regions and neuronal networks (e.g., Tenenbaum et al., 2009). Movement scientists study how a voluntary plan of action is transformed into patterns of muscular contraction that move a performer's limbs or eyes to produce movements (e.g., Brown et al., 2006). This division of labor among behavioral neuroscientists has also been reflected in the sport sciences, with sport psychologists, sports biomechanists, and physiologists typically investigating processes and systems in isolation. One implication of this traditional focus on mental representations in the brain has been sport pedagogy's obsession with a template or "common optimal pattern" for performance behaviors (Brisson and Alain, 1996), toward which all learners should aspire, whether they are eye movements (cf. Davids and Araújo, 2016) or sport actions (cf. Davids et al., 2015).

Additionally, problems have arisen in the neuroscientific research concerned with localizing functions to specific brain structures. These research aims contrast with strong evidence suggesting that the functioning of a brain region serves an embodied agent acting in a dynamic and complex environment, with many opportunities for action (Cisek and Pastor-Bernier, 2014). Psychological studies aligned with recent embodied theories, from a behavioral or neuroimaging viewpoint, introduce a key point on sport expertise suggesting that sensorimotor experiences are grounded in actions. Beilock (2008) argued that, for a better understanding of how athletes perceive visual information to interact with an environment, it is mandatory to incorporate previous motor experiences involving the individual. Theories of embodied cognition assume that the mind includes connections, not only with the brain, but also with the body and environment (Shapiro, 2011), breaking down the distinction between perception and action espoused in ecological theories (e.g., Gibson, 1979).

Gray (2014) has provided empirical evidence that the perception of athletes is embodied because it is mediated by the physical properties of objects and their ability to interact with them (e.g., perception is influenced by action-related variables, such as skill level, difficulty, or goals of the task). Nevertheless, Gray (2014) revealed the boundaries of embodied perception at a practical level to enhance skill performance. He showed that expert athletes could use speed or size information of objects to modify their visual behaviors or movement initiation during interactions.

Theories of embodied cognition have also been influential in research on action observation and predictions in sport. For example, Renden et al. (2014) found that previous motor experiences helped referees and players in detecting deceptive movements in football compared to novices and fans who merely watched games. Similarly, Pizzera and Raab (2012) showed how judgments in match officiating were dependent on previous motor and visual experiences in that sport.

This body of data suggests that descriptive tools deployed by behavioral neuroscientists must do justice to the continuous interactions of brain–body–environment from which behavior emerges (Keil et al., 2000; Kiverstein and Miller, 2015). They must account for ways in which an athlete is able to expertly coordinate his/her behavior with a dynamically changing environment (Araújo et al., 2006). For example, Naito and Hirose (2014) found a reduced load on the medial wall of motor-cortical foot regions in the brain of the expert soccer player Neymar da Silva Santos Júnior, when compared to other professional and amateur soccer players. They argued that Neymar may efficiently control foot movements, probably by largely conserving motor-cortical neural resources. This search for brain connectivity patterns that underlie use of perception and action in performance has been the focus of neuroscience research into behavioral contexts like sport (Kiverstein and Miller, 2015).

Considering the dynamic complexity of human movement in sport (Balague et al., 2013), this emphasis on accumulation of discrete facts about human functioning often encourages a neuroscientific *atomism*. This development of neuroscientific knowledge and understanding has been primarily focused on description of specific brain modules or networks which is believed to be useful in explaining the holistic and dynamic nature of the human experience in sport. In fact, the organization of functional performance behaviors is underpinned by the continuous dynamical relations between intentions, emotions anticipations, perception, and action in each individual (Davids et al., 1994, 2013). The need for theories of brain and behavior in psychology and neuroscience, and associated models of integrated, dynamical interactions between these performance-regulating processes, has long been recognized in sport science (Davids et al., 1994, 2001).

This integrated view contrasts with research on sport performance from a cognitivist neuroscience perspective, which has tended to focus on how an athlete decides between discrete situations during observations, using particular deliberations before acting (e.g., "if-then" rules to deliberate between options, or deliberating situational probabilities) (McPherson and Kernodle, 2003; Williams and Ward, 2007). Rather, during performance, actions of individual performers generate perceptual information about their intentions, which, in turn, constrains the emergence of further actions (Davids et al., 2015). Exemplifying this theoretical rationale, Seifert et al. (2014) reported how skilled ice-climbers used a large range of actions, including interlimb coordination patterns that crossed the midline of the body on a vertical icefall, leading them to seek, explore, and exploit different sources of visual and kinesthetic information, and supporting different body positions on the ice to maintain equilibrium and achieve a rapid, safe traversal. In contrast, symmetrical use of limbs with ice tools and crampons led beginners to adopt a secure supportive body position (X-shaped) on the ice surface, constrained by fear of falling.

In short, individuals have a range of skills and abilities, honed through continuous interactions with an environment, and live under the constraints of both (Warren, 2006). In order to develop our understanding of athletes' continuous, exploratory interactions, and their role in learning and performance in sport, there is a need to consider the *ecological neurodynamics of* human behavior.

Walsh (2014) has noted the potential of this embodied and embedded approach to grasp some "pesky things" (e.g., affordances) for neuroscience in explaining the role of different subsystems in regulating performance in sport. This is a significant theoretical framework for analyzing behavioral adaptations to brain–body–environment constraints in sport (Seifert et al., 2014, 2016b). An embodied and embedded approach to the neuroscience of cognitions, perception, and action could flourish if analyses of athletes' performance behaviors are predicated on principles of neurobiology, ecological psychology, and dynamical systems theory, rather than representational cognitive neuroscience premises (e.g., Davids, 2009; Seifert et al., 2014, 2016a; Tognoli and Kelso, 2014). In particular, Kelso (2008, 2012) emphasized that multistable coordination dynamics exist at many levels and provide a platform for understanding creative brain–behavior couplings. Multistability refers to "functional equivalence" (Kelso, 2012) and is also captured in the neurobiological system property of degeneracy (Edelman and Gally, 2001; Tononi et al., 1999). This is defined as "the ability of elements that are structurally different to perform the same function or yield the same output" (Edelman and Gally, 2001, p. 13763). The role of degeneracy has been extensively exemplified in performance of bimanual coordination (Kelso, 1995) and postural regulation tasks (Bardy et al., 2002). Tognoli and Kelso (2014) also highlighted transitions and metastability in brain and behavior, revealing a subtle blend of integration and segregation in neural components. Segregation reflects tendencies for brain regions and behaviors to express their individual autonomy and specialized functions, which coexist with tendencies to couple and coordinate globally when performing multiple functions (integration).

These findings support the idea that, in an ecological neurodynamics approach, organisms and their performance environments form complex, adaptive, and dynamical systems (Kiverstein and Miller, 2015; Seifert et al., 2016a). These key ideas have important implications for considering skill acquisition, expertise, and talent development in sport since ecological neurodynamics is concerned with understanding the relationship between the athlete (not just its nervous system) and key properties of a performance environment. Applying these ideas in studies of body–environment relations provides an enhanced focus on athletes' actions, perception, and cognitions revealing the adaptive contributions of psychological, physical, and emotional capacities and their interactions with a performance setting in sport (Araújo and Davids, 2009). In the remainder of this paper, we review insights on brain and behavior from an ecological neurodynamics perspective. Our discussion is organized into three parts intended to overview knowledge from behavioral neuroscience about the *how* of perception, cognition, and action in sport, seeking to explain *how* athletes detect and use information available in a performance environment. We revisit research on connectivity patterns in the brain, challenging the view that the brain uses sensory information to build and update internal representations of the world. We discuss *how* a resonance hypothesis can overcome limitations of traditional cognitive neuroscience to link behaviors to neural functioning during sport performance. We explore the hypothesis that perception implies the resonance of the environment as well as a resonance of the brain to information from the environment, arguing that perception is a dynamic process involving the whole body of the athlete. In the second part of the paper, we discuss *how* skilled individuals functionally adapt perception and action to interacting constraints during performance. In the final part of the paper, we conclude that athletes' cognitive processes depend on the whole organism in its practical and skilled engagement with an environment full of action possibilities.

2 HOW DOES AN ATHLETE OBTAIN INFORMATION TO REGULATE ACTION?

In this section, we discuss how the mechanism of resonance facilitates organisms with a nervous system in detecting and using relevant affordances from a landscape available in performance environments.

2.1 REVISITING MISUNDERSTANDINGS ABOUT PERCEPTION AND INTERNAL REPRESENTATIONS

Psychological and computational theories often propose that perceiving is a matter of constructing a mental representation from sensory inputs (e.g., Frank et al., 2014; Marr, 1982; Schack and Mechsner, 2006). In cognitive neuroscience, it is assumed that such internal representations form schemas (scripts, plans, and the like), integrating emotional structures, and cognitive processes and subsystems (e.g., knowledge architecture, long-term working memory), motor structures (motor programs for coordination, for example), and activating cortical areas (Tenenbaum et al., 2009). The effectiveness of these processes is based on the system's capacity to encode and access information relevant to the task being performed (Tenenbaum & Land, 2009). Over the years, however, attempts to interpret neural data from this perspective have encountered several challenges. For example, evidence has highlighted the presence of continuous and dynamic loops, either within or between hierarchical levels, involving both feedforward and feedback interactions within and between the so-called memorial and perceptual systems (e.g., Damasio, 1989; Henson and Gagnepain, 2010; Meyer and Damasio, 2009; Peterson et al., 2012). In fact, the acts of perceiving, being attentive or memorizing, are not neatly attached to any single level of representation. What is crucial for this traditional approach is not what is perceived, but the nature of what is represented, and the fact that the content of all representations results from a dynamical set of feedforward–feedback interactions (Nadel and Peterson, 2013).

However, the perceptual system is not composed of representations of earlier events (Gibson, 1966a,b, 1979). In ecological neurodynamics, the process by which an athlete becomes aware of what occurs around him/her is direct perception. This claim follows from research seeking to show that properties that are perceived can be fully specified by information available to the perceiver from a moving point of observation (Lee, 1976). Runeson's (1977) argument on "smart" perceptual mechanisms shows how one might conceptualize perceiving as the direct registration of higher-order information without appealing to constructive, mediating mental processes. Runeson et al. (2000) also showed that, while perceivers may not initially utilize specifying information, with experience they learn to do so, which in turn results in higher levels of task performance (see Araújo et al., 2009a; Renshaw et al., 2016, for applications in sport).

The process by which an athlete is made aware of things outside his/her direct perception is a form of (external) representation, what Gibson (1966a) called "mediated perception" (p. 234). Based on the ideas of William James, Gibson (1966a) evidenced this difference, by distinguishing between *knowledge of acquaintance* and *knowledge about*. He proposed that *knowledge about* involves perception, which is indirect or mediated by language, symbols, pictures, judgments, and verbal instructions, all of which can facilitate analogical reasoning and verbal communication of what an information source means (see Araújo et al., 2009b; Zourbanos et al., 2015, for examples in sport). This type of knowledge may be used when coaches verbally instruct an athlete on how to deal with an adversary or about a strategy to adopt in a team game like soccer. This type of knowledge and perception are clearly overused in traditional sport pedagogy, which emphasizes the typical methodological cycle of coach instruction–coach feedback-correction and so on (as highlighted later in Fig. 3). *Knowledge of acquaintance*, in contrast, describes how an organism can perceive the surrounding layout of its performance environment in the scale of its body and action capabilities (Araújo et al., 2009b). This type of knowledge is used to regulate behaviors during continuous interactions with objects, events, other people, surfaces and features of the terrain in the environment.

It can be argued that the key functioning principle of the central nervous system (CNS) is not that of information processing. Rather, it develops networks of neurons, which, through continuous integration and interactions with other parts of the body and environment, may achieve performance goals. From this point of view, it is clear that neurons do not create maps of the environment, inner models, or representations, which would somehow correspond to perceptions and actions. If the nervous system must somehow reproduce or represent the organization of the environment, this would mean that the representations in the nervous system would always lag behind the occurrence of events in the environment (Jarvilehto, 1998). Rather neurons and other parts of the nervous system continuously form patterns and interconnections to subserve performance during environmental interactions (Edelman, 1992; Kelso, 1995). These pattern forming tendencies are ubiquitous in neurobiological systems (Kelso, 1995, 2012). They result in relatively stable neuronal interconnections emerging when task goals are successfully achieved, being subtly refined and adapted through exposure to new experiences and continuous learning and development (Edelman, 1992; Seifert et al., 2016b).

To summarize, the assumption of internal representations underpinning perception does not appear to be well supported by the evidence of a dynamical set of feedforward–feedback interactions, either within or between hierarchical levels of the CNS and brain. Neuroanatomical organizations are temporary tendencies, only relatively stable and self-organizing to capture the intertwined relations between perception, actions, and cognitions (Kelso, 1995). This kind of continuous and dynamic looping, involving both feedforward and feedback interactions in the brain, provides a useful starting point for describing a functional mechanism underlying what Gibson (1966a,b) called the *resonance* of a perceptual system to ecological information. Rather than postulating that the brain constructs information from sensory system input (McCulloch and Pitts, 1943), or that the organism responds to environmental stimulation (Miller et al., 1960), Gibson (1966a) proposed that the CNS resonates to information. For him, resonance is a central process in explaining *how* perception of the environment emerges through continuous interactions of an organism.

2.2 LINKING BRAIN–BODY–ENVIRONMENT: THE RESONANT SYSTEM

Gibson claimed that resonance had a role in the use of information from environmental patterns of energy (Gibson, 1966a). However, there is considerable uncertainty about what the term *resonance* means. This lack of clarity is reflected in the subject matter of ecological psychology (e.g., Shaw, 2003), in attempts to understand perception and human behavior (e.g., Sartori and Betti, 2015), and is reflected in the lack of research on resonance in sport performance contexts. Consequently, ecological psychology researchers have proposed *novel frameworks*, predicated on an embodied brain–body–environment, without ever making concrete references to the concept of resonance (Bruineberg and Rietveld, 2014; Raja, in press). Some have argued that resonance remains an intriguing and stimulating metaphor, requiring clarification (Gordon, 2004). In this section, we propose to clarify the concept of resonance as used in J.J. Gibson's ecological approach to visual perception (Gibson, 1966a,b, 1979), since Gibson's key ideas have been somewhat susceptible to misunderstandings and apparent inconsistencies (Costall and Morris, 2015). The Gibsonian concept of resonance is analyzed here with reference to its implications for understanding the role of perception and action in sport performance.

Gibson (1966a) proposed that the nervous system, including the brain, resonates to information. Phenomenologically, he explained resonance using a metaphor of detecting resonant frequencies by a radio station, i.e., given that many frequencies reach a receiver from the antenna, proper tuning of the receiver causes an electric current in it to resonate in response to some incoming signals, and not others. In the case of visual perception, for example, the radio waves in this metaphor stand for light structured by surrounding optic arrays in the environment. Peripheral sensory organs, somewhat like the antenna of a radio mast, let the signals pass through, and the radio must be *tuned in* to the information. Tuning is accomplished by arranging the frequencies to be the same or nearly the same in the signals (Michaels and Carello, 1981). For example, if a soccer player in midfield seeks a teammate in space near the opposition goal, the ball carrier needs to be tuned in to available information in this performance environment for an affordance inviting a long pass. According to Gibsonian ideas, when the teammate is perceived in space near the opposition goal, the midfield player's perceptual system resonates to that information for an affordance.

More than 40 years after the original proposal of this idea, with the ongoing development of neurophysiological technology (e.g., functional magnetic resonance imaging, electroencephalography, positron emission tomography), a *resonance mechanism* is much more than a metaphor (e.g., Fitzgibbon et al., 2014; Leonetti et al., 2015). Traditional approaches to cognitive neuroscience seek to identify neurophysiological activation in the CNS correlated with behaviors interpreted as evidence for an individual reproducing an internally represented action. In contrast, resonance can explain how the nervous system can resonate to the invariants of a global energy array providing information about a performance environment (Gibson, 1966b). In sport, athletes can maintain functional contact with their performance environment through the *resonance mechanism*, which is tuned (e.g., synchronized) to the patterns of structured surrounding energy (information) obtained (used) through active perception of the affordances of the performance environment. This description captures how the brain–body–environment system are highly integrated and interconnected, making it problematic to study performance behaviors in sport away from typical contexts (e.g., in laboratory studies).

Importantly, resonance is not an accomplishment of the brain but involves all the body (sub)systems involved in perceiving and acting in the environment. Gibson (1966a, p. 268) proposed that:

the five modes of attention, listening, smelling, tasting, touching, and looking, are specialized in one respect and unspecialized in another. They are specialized for vibration, odor, chemical contact, mechanical contact, and ambient light, respectively, but they are redundant for the information in these energies whenever it overlaps. Their ways of orienting, adjusting, and exploring are partly constrained by anatomy, but partly free.

This description neatly captures the relationship between structure and function in the nervous system, suggesting how important it is to study performance behaviors in context when actions, perception, and cognition are intertwined in helping athletes maintain contact with their environment. This key idea was illustrated by the research of Seifert et al. (2014) who showed how expert icefall climbers tended to simultaneously and continuously use kinesthetic, haptic, acoustic, and visual information to regulate their actions during performance, while beginners tended to be dominated by overreliance on one source of information (Fig. 1).

Beginners in climbing typically rely on information from the depth of an ice tool blade entering the surface of a frozen waterfall, while experts, in contrast, perceive haptic information from vibrations of the ice tool, acoustic feedback from the sound

FIG. 1

Example of ice tool angles exemplifying that the beginner (*right panel*) mainly located his two ice tools horizontally (*gray*), while the expert climber (*left panel*) alternated horizontal, oblique (*white*) and vertical (*black*) angle locations of his ice tools, perhaps reflecting the property of degeneracy in the integrated perception–action system.

of the blade, and vision of the anchorage (and from the color of the ice to decide on an affordance for a secure and functional anchorage during their traversal upwards) (Seifert et al., 2014). Those various sources of information (e.g., size, depth, and shape of the holes in icefalls) specify a range of functional actions for expert climbers, whereas beginners tend to perceive global, structural icefall characteristics (e.g., existing holes in icefall), which may not specify actions. Those findings suggested that expert climbers better exploited the property of degeneracy in their perception–action systems, opening up more opportunities to perceive affordances and vice versa. Degeneracy, advocated in studies of neurobiology and cognitive system anatomy (e.g., Edelman and Gally, 2001; Price and Friston, 2002; Whitacre, 2010), can also be observed in perception–action systems (Seifert et al., 2016a), allowing performers to explore different perceptual–motor solutions and facilitating discovery of functional patterns of coordination. With regards to skill acquisition, degeneracy signifies that, with practice, learners can structurally vary their perception and action system organization (exploiting inherent system tendencies for multistability and metastability) without compromising function. This tendency with increasing expertise in sport supports the adaptive flexibility of movement coordination patterns in satisfying task constraints (Seifert et al., 2013, 2016a). These data and ideas suggest that affordance perception and neurobiological system degeneracy may evolve together during the acquisition of skill and expertise in sport.

It is likely that what Gibson had in mind in the explanation of resonance was not the notions of conduits and incoming messages in an electrical–mechanical system, since he also challenged the use of the term "sensory" to explain nerve function (Gibson, 1966a). In fact, Gibson (1979) suggested that "the nervous system operates in circular loops and that information is never conveyed but extracted by the picking up of invariants over time" (p. 235). Instead of reflexes (e.g., Schmidt and Lee, 2011) or stimulus–response compatibility effects (e.g., Richez et al., 2016), these loops, through which animals engage in information transactions with their environments, resonate concurrently in several structures of the nervous system, as well as resonate in the neuromuscular system, and even more broadly through the environment–organism coupling (Shaw, 2002). For example, Davids et al. (2004) reinterpreted data from a group of international soccer players demonstrating that wearing textured insoles in soccer boots enhanced tactile information from the sole of the foot and increased movement discrimination capacity in ankle inversion sensitivity tests to levels similar to those in barefoot conditions. The reinterpretation was based on notions of stochastic resonance and functional variability induced in the sensorimotor system by textured insoles, acting as a form of "essential noise" to enhance the accuracy of foot positioning. It was convincingly argued that stochastic resonance can actually enhance perception of information to support motor performance. Movement system variability, conceived as noise-induced resonance can facilitate the detection of information to regulate functional performance behaviors.

These theoretical insights from Gibson came to be evidenced 40 years later by studies in humans using neuroimaging technology (e.g., Fitzgibbon et al., 2014; Leonetti et al., 2015; Wykowska and Schubö, 2012). This body of research identified a neural mechanism that allows a direct matching between perception and action, called interchangeably "motor simulation," "motor resonance," or a "mirror mechanism" (Barchiesi and Cattaneo, 2015). A mirror mechanism is composed by a particular class of visuomotor neurons, originally discovered in area F5 of the premotor cortex in monkeys, which discharge when an individual observes a hand grasping an object. The same population of neurons that controls the action execution of grasping movements becomes active in the observer's motor areas. An important functional aspect of mirror neurons is that the relation between their visual and motor properties is not unified, but progressively diverges into parallel subsystems, each specialized toward the demands of different sensorimotor functions (Rizzolatti and Luppino, 2001). From this perspective, results do not clearly identify specific localizations in the brain (Rizzolatti et al., 2001). Instead, it seems that action-relevant information is both generated by, and reciprocally used to regulate, movement (Milner and Goodale, 1995). Milner and Goodale suggested the existence of separate visual pathways for perception and action: a ventral stream, where cells are sensitive to stimulus features, and a dorsal stream, where cells are sensitive to spatial relationships. The predominant role of the dorsal stream is to mediate visually guided actions. They proposed that the dorsal stream is sensitive to spatial information, not to build a representation of the environment for knowledge acquisition, but because spatial information is critical for specifying the parameters of potential and ongoing actions (see Van der Kamp et al., 2008, for a discussion in sport). In this respect, it seems apparent that the dorsal stream may provide a nervous system platform for an athlete to interact with what Gibson (1966a) termed "knowledge of" the environment during performance in competition and practice. It appears to be notoriously difficult to separate structure and function in the nervous system. This is especially obvious in assigning a specific perceptual or motor function to regions of the brain, where neurons appear to be related to different perceptual functions (Andersen and Buneo, 2003; Culham and Kanwisher, 2001; Wykowska and Schubö, 2012).

Ironically, whenever neurophysiologists have sought to identify specific behavioral functions, such as perception and action, they gradually recognize that such functions are not implemented by particular cortical regions (Pessoa, 2014). An alternative possibility may be not to attempt to explain perception and action from a perspective that focuses exclusively on the overarching control of the human mind. Although neuroimaging technology has developed important tools for studying aspects of perception and action in sport, an ecological explanation of how athletes perceive and act in sport contexts is also critical for enhancing our understanding of performance behaviors. An ecological perspective of brain functioning may be insightful, based on the fact that the brain is part of a body (embodied) that is embedded in an environment forming a comprehensive, integrated system. Of course, what we propose is certainly not to replace neuroscientific knowledge in general, but rather to provide a theoretical framework that is ecologically and biologically plausible for understanding how the structure and function of the nervous system contributes to perception, cognition, and action in sport performance (Fig. 2).

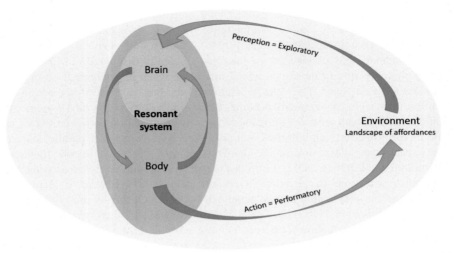

FIG. 2

Linking brain–body–environment. The classification about the ways the individual can obtain information of an exploratory (information is obtained by screening the environment) or performatory (information is obtained by steering and control of performance) nature is based on Gibson (1966a).

The environment of an actor matters significantly because it consists of the affordances of objects, places, other people, and events for an individual (Reed, 1996). Throughout evolution, organisms, and their nervous systems have had to be preoccupied by almost constant interaction with a complex and ever changing environment, which continuously offers a variety of opportunities and invitations for action (Withagen et al., 2012). The relationship between the physical properties of a performance environment and an individual's action capabilities provides a landscape of affordances in sport, such as the trajectory of a ball to catch, hit, or avoid, or a gap to dribble through (Davids et al., 2015). This affordance landscape provides the basis for designing practice tasks in sport, facilitating sport practitioners in simulating key aspects of a competitive performance environment in a field of affordances (Bruineberg & Rietveld, 2014) which are most pertinent and relevant for any individual or group of athletes (Davids et al., 2017) (Fig. 3).

The ability to perceive an affordance is both extero- and proprio-specific, i.e., about the environment as well as about the performer (e.g., a collision in a tackle during rugby union is both environmental and a personal event, and this applies to all encounters between athletes and their surroundings). Even self-perception is not entirely a personal matter. Gibson (1966a) identified that perception is a question of a reciprocal relationship between the individual and the environment, suggesting that "there are many concurrent loops available for the proprioceptive control of action. They seem to be at different levels; some of the loops remain inside the body,

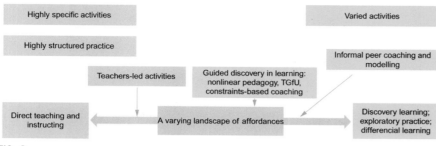

FIG. 3

Different methodologies in the microscale of practice and physical activity (hours, days, weeks, months) for exploring different fields of the affordance landscape in skill acquisition. Traditional practice design has a "default" mode situated at the highly structured end of the affordance landscape, focused on narrow fields by direct teaching/coaching/verbal instructions. Coaches need to focus on *learning design*, moving between varied fields of the affordance landscape based on the resonance capacities (action readiness) of each athlete (Davids et al., 2017).

some pass outside" (p. 36). It has become clear, predicated on Gibsonian ideas, that even such activities as perceiving one's own postural activity or sensing the influence of gravity on movement are based on detection of information from the surrounding patterns of structured energy (Bourrelly et al., 2016; Varlet et al., 2015, see Bardy and Laurent, 1998, for an example in sport).

In fact, resonating implies a reciprocal relationship between a perceiver and a perceiver's environment. "The brain is a self-tuning resonator" Gibson proposed (Gibson, 1966b, p. 146) and achieving resonance implies that the perceiver is tuned to the invariant properties of detected ambient energy. Such structured information invariantly specifies features of a particular substances, surfaces, objects, or events in relation to a particular perceiving–acting agent. Thus, Gibsonian resonance involves the development of perceptual attunement. Throughout growth and development, the organism makes attuned differentiations of affordances in a performance environment relative to the organism's action capabilities (Davids et al., 2015). With learning, experience, and expertise development, each individual becomes increasingly sensitized to the affordances that are available and relevant to goal achievement, so each individual adapts for the detection of better information (Araújo et al., 2017; Fajen et al., 2009) (Fig. 4).

This idea was exemplified in a study by Seifert et al. (2014) which highlighted the better perceptual attunement of expert ice-climbers to visual, acoustic, and haptic information. These specifying information sources allowed the climbers to perceive use-ability (affordances) of holes in an icefall, a dynamic structure, constantly changing through ambient temperature regulation, ice thickness and density, and usage during traversals. The dynamical self-tuning of the human perceptual system supports the idea that movement generates information, and reciprocally, information

FIG. 4

Picture of ice climbing environment illustrating a mix of sections with ice, snow, water, and rock where the climber picks up visual, acoustic, and haptic information through his ice tools, in order to find the best route.

is used to regulate movement, confirmed by Gibson (1979) in arguing that "we must perceive in order to move, but we must also move in order to perceive" (p. 223). Gibson's (1950) original ideas about visual perception suggest that, when an athlete moves relative to a specific context, a global pattern of optical flow is generated at the moving point of observation and corresponds to the direction of athlete movement. Reciprocally, this information can be used to regulate forces applied by the athlete in controlling subsequent movements, which, in turn, generate a new flow field, and so on in a circular causal cycle. Thus, action emerges from the interaction between the neural dynamics of the CNS, the biomechanics of a body in motion and the physical dynamics of the environment (Kugler and Turvey, 1987; Warren, 1990). From this viewpoint, the role of the affordances in co-relation with an athlete's resonant system, including neuroanatomical and physiological constraints (i.e., CNS, muscles, tendons, skeleton), is paramount in performance regulation in sports. A systems orientation in the study of brain and behavior in sport is essential: the structure and function of an athlete's resonating system is best understood in the context where affordances are relevant for goal achievement.

Perceived affordances in a particular performance context tune with resonant dynamics that shape each athlete's action readiness for interacting with the environment (Rietveld et al., 2013). For example, competitive football performance affords different movement patterns in comparison with the team sport of futsal, but there are some fundamental ways in which resonance capacities can provide a basis for exploiting the underlying complementarity between these sports (Travassos, Davids, & Araújo, in press). Task-specific coordination in a movement system emerges from the perception of an affordance, the selection of which is based on exploratory actions. Only if an athlete's resonant system is tuned into specific constraints of a performance context, will he/she be enabled to consistently achieve task goals. In this way, learning design in sport pedagogy can be based on fundamental principles of behavioral neuroscience from an ecological neurodynamics perspective.

3 CONCLUSION

In this chapter, we discussed how the notion of neural resonance is compatible with an ecological dynamics perspective, advocating that available information surrounds an active athlete in a competitive performance environment, where invariants of structured energy arrays resonates in his/her resonant system. An athlete's brain resonates to whatever is invariant under transformation and becomes increasingly attuned to it with recurrence over time, through experience, learning, and practice. Thus, an active, well-trained athlete needs to learn to perceive regulatory information from a specific performance context. It is not imposed, although "the environment provides an inexhaustible reservoir of information" (Gibson, 1966a, p. 269). Perceptual activities like listening (to the sound of a teammate's call for the ball), touching (a racquet handle in tennis), feeling (an opponent's garments and limbs in a judo bout), and looking (at the vaulting box during a run up in gymnastics) are continuous in regulating sport performance. Continuous engagement of the CNS is needed for perception and action in sport performance to emerge. A performance environment in competition and practice needs to be perceived for action, and through actions. Attunement during learning facilitates the detection of the informational structures arising from a specific performance environment. The CNS of an individual athlete can be trained and developed to resonate, through learning designs which simulate specific performance contexts which require him/her to actively use functional perceptual systems for perceiving and realizing affordances in sport performance (Davids et al., 2012).

REFERENCES

Andersen, R.A., Buneo, C.A., 2003. Sensorimotor integration in posterior parietal cortex. Adv. Neurol. 93, 159–177.

Araújo, D., Davids, K., 2009. Ecological approaches to cognition and action in sport and exercise: Ask not only what you do, but where you do it. Int. J. Sport Psychol. 40, 5–37.

Araújo, D., Davids, K., Hristovski, R., 2006. The ecological dynamics of decision making in sport. Psychol. Sport Exerc. 7, 653–676.

Araújo, D., Cordovil, R., Ribeiro, J., Davids, K., Fernandes, O., 2009a. How does knowledge constrain sport performance? An ecological perspective. In: Araújo, D., Ripoll, H., Raab, M. (Eds.), Perspectives on Cognition and Action in Sport. Nova Science Publishers, Hauppauge, NY, pp. 100–120.

Araújo, D., Davids, K., Chow, J., Passos, P., 2009b. The development of decision making skill in sport: an ecological dynamics perspective. In: Araújo, D., Ripoll, H., Raab, M. (Eds.), Perspectives on Cognition and Action in Sport. Nova Science Publishers, New York, pp. 157–170.

Araújo, D., Dicks, M., Davids, K., 2017. Selecting among affordances: a basis for channeling expertise in sport. In: Cappuccio, M.I. (Ed.), Handbook of Embodied Cognition and Sport Psychology. MIT Press, Boston, in press.

Balague, N., Torrents, C., Hristovski, R., Davids, K., Araújo, D., 2013. Overview of complex systems in sport. J. System. Sci. Compl. 26, 4–13.

Barchiesi, G., Cattaneo, L., 2015. Motor resonance meets motor performance. Neuropsychologia 69, 93–104.

Bardy, B.G., Laurent, M., 1998. How is body orientation controlled during somersaulting? J. Exp. Psychol. Human Percep. Perf. 24, 963–977.

Bardy, B.G., Oullier, O., Bootsma, R.J., Stoffregen, T.A., 2002. Dynamics of human postural transitions. J. Exp. Psychol. 28, 499–514.

Beilock, S.L., 2008. Beyond the playing field: Sport psychology meets embodied cognition. Int. Rev. Sport Exerc. Psychol. 1, 19–30.

Bourrelly, A., McIntyre, J., Morio, C., Despretz, P., Luyat, M., 2016. Perception of affordance during short-term exposure to weightlessness in parabolic flight. PLoS One 11, e0153598.

Brisson, T.A., Alain, C., 1996. Should common optimal movement patterns be identified as the criterion to be achieved? J. Motor Behav. 28, 211–223.

Brown, S., Martinez, M.J., Parsons, L.M., 2006. The neural basis of human dance. Cereb. Cortex 16, 1157–1167.

Bruineberg, J., Rietveld, E., 2014. Self-organization, free energy minimization, and optimal grip on a field of affordances. Front. Hum. Neurosci. 8, 599.

Chow, J.Y., Davids, K., Button, C., Renshaw, I., 2015. Nonlinear Pedagogy in Skill Acquisition: An Introduction. Routledge, Taylor & Francis, New York.

Cisek, P., Pastor-Bernier, A., 2014. On the challenges and mechanisms of embodied decisions. Philos. Trans. R. Soc. Lond. B. Biol. Sci. 369, 20130479.

Costall, A., Morris, P., 2015. The "Textbook Gibson": the assimilation of dissidence. Hist. Psychol. 18 (1), 1–14.

Culham, J.C., Kanwisher, N.G., 2001. Neuroimaging of cognitive functions in human parietal cortex. Curr. Opin. Neurobiol. 11, 157–163.

Damasio, A.R., 1989. Time-locked multiregional retroactivation: a systems-level proposal for the neural substrates of recall and recognition. Cognition 33, 25–62.

Davids, K., 2009. The organization of action in complex neurobiological systems. In: Araújo, D., Ripoll, H., Raab, M. (Eds.), Perspectives on Cognition and Action in Sport. Nova Science Publishers, New York, pp. 3–13.

Davids, K., Araújo, D., 2016. What could an ecological dynamics rationale offer Quiet Eye research? Comment on Vickers. Curr. Issues Sport Sci. 1, 104.

Davids, K., Handford, C., Williams, M., 1994. The natural physical alternative to cognitive theories of motor behaviour: an invitation for interdisciplinary research in sports science? J. Sports Sci. 12, 495–528.

Davids, K., Williams, A.M., Button, C., Court, M., 2001. An integrative modeling approach to the study of intentional movement behavior. In: Singer, R.N., Hausenblas, H., Jannelle, C. (Eds.), Handbook of Sport Psychology, second ed. John Wiley & Sons, New York, pp. 144–173.

Davids, K., Savelsbergh, G.J.P., Bennett, S.J., Van der Kamp, J., 2002. Interceptive Actions in Sport: Information and Movement. Routledge, Taylor & Francis, London.

Davids, K., Shuttleworth, R., Button, C., Renshaw, I., Glazier, P., 2004. "Essential noise"—enhancing variability of informational constraints benefits movement control: a comment on Waddington and Adams (2003). Br. J. Sports Med. 38, 601–605.

Davids, K.W., Button, C., Bennett, S.J., 2008. Dynamics of Skill Acquisition: A Constraints-Led Approach. Human Kinetics, Champaign, IL.

Davids, K., Araújo, D., Hristovski, R., Passos, P., Chow, J.Y., 2012. Ecological dynamics and motor learning design in sport. In: Hodges, N., Williams, M. (Eds.), Skill Acquisition in Sport: Research, Theory and Practice, second ed. Routledge, Abingdon, UK, pp. 112–130.

Davids, K., Araújo, D., Vilar, L., Renshaw, I., Pinder, R.A., 2013. An ecological dynamics approach to skill acquisition: implications for development of talent in sport. Tal. Dev. Excel. 5, 21–34.

Davids, K., Araújo, D., Seifert, L., Orth, D., 2015. Expert performance in sport: an ecological dynamics perspective. In: Baker, J., Farrow, D. (Eds.), Routledge Handbook of Sport Expertise. Routledge, London, pp. 130–144.

Davids, K., Güllich, A., Araújo, D., Shuttleworth, R., 2017. Understanding environmental and task constraints on athlete development: analysis of micro-structure of practice and macro-structure of development histories. In: Baker, J., Cobley, S., Schorer, J., Wattie, N. (Eds.), Routledge Handbook of Talent Identification and Development in Sport. Routledge, London.

Edelman, G.M., 1992. Bright Air, Brilliant Fire: On the Matter of the Mind. Basic Books, New York.

Edelman, G.M., Gally, J.A., 2001. Degeneracy and complexity in biological systems. Proc. Natl. Acad. Sci. 98, 13763–13768.

Fajen, B.R., Riley, M.R., Turvey, M.T., 2009. Information, affordances, and the control of action in sport. Int. J. Sports Psychol. 40, 79–107.

Fitzgibbon, B.M., Fitzgerald, P.B., Enticott, P.G., 2014. An examination of the influence of visuomotor associations on interpersonal motor resonance. Neuropsychologia 56, 439–446.

Frank, C., Land, W.M., Popp, C., Schack, T., 2014. Mental representation and mental practice: experimental investigation on the functional links between motor memory and motor imagery. PLoS One 9, e95175.

Gibson, J.J., 1950. The Perception of the Visual World. Houghton Mifflin, Boston.

Gibson, J.J., 1966a. The Senses Considered as Perceptual Systems. Houghton Mifflin, Boston.

Gibson, J.J., 1966b. The problem of temporal order in stimulation and perception. J. Psychol. 62, 141–149.

Gibson, J.J., 1979. The Ecological Approach to Visual Perception. Houghton Mifflin, Boston.

Gordon, I.E., 2004. Theories of Visual Perception. Psychology Press, Taylor & Francis Group, Hove and New York.

Gray, R., 2014. Embodied perception in sport. Int. Rev. Sport Exerc. Psychol. 7, 72–86.

Henson, R.N., Gagnepain, P., 2010. Predictive, interactive multiple memory systems. Hippocampus 20, 1315–1326.

Jarvilehto, T., 1998. The theory of the organism–environment system II: significance of nervous activity in the organism–environment system. Integr. Phys. Behav. Sci. 33, 331–338.

Keil, D., Holmes, P.S., Bennett, S.J., Davids, K., Smith, N.C., 2000. Theory and practice in sport psychology and motor behaviour needs to be constrained by integrative modelling of brain and behaviour. J. Sports Sci. 18, 433–443.

Kelso, J.A.S., 1995. Dynamic Patterns: The Self-Organization of Brain and Behavior. MIT, Cambridge, MA.

Kelso, J.A.S., 2008. An essay on understanding the mind. Ecol. Psychol. 20, 180–208.

Kelso, J.A.S., 2012. Multistability and metastability: understanding dynamic coordination in the brain. Philos. Trans. R. Soc. Lond. B Biol. Sci. 376, 906–918.

Kiverstein, J., Miller, M., 2015. The embodied brain: towards a radical embodied cognitive neuroscience. Front. Hum. Neurosci. 9, 237.

Kugler, P.N., Turvey, M.T., 1987. Information, Natural Law, and the Self-Assembly of Rhythmic Movement. Erlbaum, Hillsdale, NJ.

Lee, D.N., 1976. A theory of visual control of braking based on information about time-to-collision. Perception 5, 437–459.

Leonetti, A., Puglisi, G., Siugzdaite, R., Ferrari, C., Cerri, G., Borroni, P., 2015. What you see is what you get: motor resonance in peripheral vision. Exp. Brain Res. 233, 3013–3022.

Marr, D.C., 1982. Vision. W.H. Freeman, San Francisco.

McCulloch, W.S., Pitts, W.H., 1943. A logical calculus of the ideas immanent in nervous activity. Bull. Math. Biophys. 5, 115–133.

McPherson, S.L., Kernodle, M.W., 2003. Tactics, the neglected attribute of expertise: problem representations and performance skills in tennis. In: Starkes, J.L., Ericsson, K.A. (Eds.), Expert Performance in Sports: Advances in Research on Sport Expertise. Human Kinetics, Champaign, IL, pp. 137–167.

Meyer, K., Damasio, A., 2009. Convergence and divergence in a neural architecture for recognition and memory. Trends Neurosci. 32, 376–382.

Michaels, C.F., Carello, C., 1981. Direct Perception. Prentice-Hall, Englewood Cliffs, NJ.

Miller, G.A., Galanter, E., Pribram, K.H., 1960. Plans and the Structure of Behavior. Holt, Rinehart and Winston, New York.

Milner, A.D., Goodale, M.A., 1995. The Visual Brain in Action. Oxford University Press, Oxford, UK.

Nadel, L., Peterson, M.A., 2013. The hippocampus: part of an interactive posterior representational system spanning perceptual and memorial systems. J. Exp. Psychol. Gen. 142, 1242–1254.

Naito, E., Hirose, S., 2014. Efficient foot motor control by Neymar's brain. Front. Hum. Neurosci. 8, 594.

Pessoa, L., 2014. Understanding brain networks and brain organization. Phys. Life Rev. 11, 400–435.
Peterson, M.A., Cacciamani, L., Barense, M.D., Scalf, P.E., 2012. The peripheral cortex modulates V2 activity in response to the agreement between part familiarity and configuration familiarity. Hippocampus 22, 1965–1977.
Pizzera, A., Raab, M., 2012. Perceptual judgments of sports officials are influenced by their motor and visual experience. J. Appl. Sport Psychol. 24, 59–72.
Price, C.J., Friston, K.J., 2002. Degeneracy and cognitive anatomy. Trends Cogn. Sci. 6, 416–421.
Raja, V., A theory of resonance: towards an ecological cognitive architecture, Mind. Mach, in press. Available from: https://link.springer.com/article/10.1007%2Fs11023-017-9431-8 [Accessed 10 May 2017].
Reed, E.S., 1989. Neural regulation of adaptive behavior. Eco. Psychol. 1, 97–117.
Reed, E.S., 1996. James J. Gibson and the Psychology of Perception. Yale University Press, New Haven and London.
Renden, P.G., Kerstens, R., Oudejans, R.R., Cañal-Bruland, R., 2014. Foul or dive? Motor contributions to judging ambiguous foul situations in football. Eur. J. Sport Sci. 14, 221–227.
Renshaw, I., Araújo, D., Button, C., Chow, J.Y., Davids, K., Moy, B., 2016. Why the constraints-led approach is not teaching games for understanding: a clarification. Phys. Educ. Sport Pedag. 21, 459–480.
Richez, A., Olivier, G., Coello, Y., 2016. Stimulus-response compatibility effect in the near-far dimension: a developmental study. Front. Psychol. 7, 1169.
Rietveld, E., de Haan, S., Denys, D., 2013. Social affordances in context: what is it that we are bodily responsive to? Behav. Brain Sci. 36, 436.
Rizzolatti, G., Luppino, G., 2001. The cortical motor system. Neuron 31, 889–901.
Rizzolatti, G., Fogassi, L., Gallese, V., 2001. Neurophysiological mechanisms underlying the understanding and imitation of action. Nat. Rev. Neurosci. 2, 661–670.
Runeson, S., 1977. On the possibility of "smart" perceptual mechanisms. Scand. J. Psychol. 18, 172–179.
Runeson, S., Juslin, P., Olsson, H., 2000. Visual perception of dynamic properties: cue heuristics versus direct-perceptual competence. Psychol. Rev. 107, 525–555.
Sartori, L., Betti, S., 2015. Complementary actions. Front. Psychol. 6, 557.
Schack, T., Mechsner, F., 2006. Representation of motor skills in human long-term memory. Neurosci. Lett. 391, 77–81.
Schmidt, R.A., Lee, T.D., 2011. Motor control and learning: a behavioral emphasis, fifth ed. Human Kinetics, Champaign, IL.
Seifert, L., Button, C., Davids, K., 2013. Key properties of expert movement systems in sport: an ecological dynamics perspective. Sports Med. 43, 167–168.
Seifert, L., Wattebled, L., Herault, R., Poizat, G., Adé, D., Gal-Petitfaux, N., Davids, K., 2014. Neurobiological degeneracy and affordances detection support functional intra-individual variability of inter-limb coordination in complex discrete task. PLoS One 9, e89865.
Seifert, L., Wattebled, L., Orth, D., L'Hermette, M., Boulanger, J., Davids, K., 2016a. Skill transfer specificity shapes perception and action under varying environmental constraints. Hum. Mov. Sci. 48, 132–141.
Seifert, L., Komar, J., Araujo, D., Davids, K., 2016b. Neurobiological degeneracy: a key property for adaptations of perception and action to constraints. Neurosci. Biobehav. Rev. 69, 159–165.
Shapiro, L., 2011. Embodied Cognition. Routledge, Oxon, UK.

Shaw, R., 2002. Theoretical hubris and the willingness to be radical: Na open letter to James J. Gibson. Ecol. Psychol. 14, 235–247.

Shaw, R., 2003. The agent–environment interface: Simon's indirect or Gibson's direct coupling? Ecol. Psychol. 15, 37–106.

Tenenbaum, G., Land, W.M., 2009. Mental representations as an underlying mechanism for human performance. In: Raab, M., Johnson, J., Heekeren, H. (Eds.), Progress in Brain Research: Mind and Motion—The Bidirectional Link Between Thought and Action 174, Elsevier, Amsterdam, pp. 251–266.

Tenenbaum, G., Hatfield, B., Eklund, R., Land, W., Camielo, L., Razon, S., Schack, T., 2009. Conceptual framework for studying emotions–cognitions–performance linkage under conditions, which vary in perceived pressure. In: Raab, M., Johnson, J., Heekeren, H. (Eds.), Progress in Brain Research: Mind and Motion—The Bidirectional Link Between Thought and Action, vol. 174. Elsevier, Amsterdam, pp. 159–178.

Tognoli, E., Kelso, J.A.S., 2014. The metastable brain. Neuron 81, 35–48.

Tononi, G., Sporns, O., Edelman, G.M., 1999. Measures of degeneracy and redundancy in biological networks. Proc. Natl. Acad. Sci. 96, 3257–3262.

Travassos, B., Davids, K., Araújo, D., (in press). Is futsal a donor sport for football? Exploiting complementarity for early diversification in talent development. Sci. Med. Football.

Van der Kamp, J., Rivas, F., Van Doorn, H., Savelsbergh, G., 2008. Ventral and dorsal contributions in visual anticipation in fast ball sports. Int. J. Sport Psychol. 39, 100–130.

Varlet, M., Bardy, B.G., Chen, F.C., Alcantara, C., Stoffregen, T.A., 2015. Coupling of postural activity with motion of a ship at sea. Exp. Brain Res. 233, 1607–1616.

Walsh, V., 2014. Is sport the brain's biggest challenge? Curr. Biol. 24, R859–R860.

Warren, W.H., 1990. The perception–action coupling. In: Bloch, H., Bertenthal, B.I. (Eds.), Sensory-Motor Organizations and Development in Infancy and Early Childhood: Proceedings of the NATO Advanced Research Workshop on Sensory-Motor Organizations and Development in Infancy and Early Childhood. Springer, Dordrecht, Netherlands.

Warren, W.H., 2006. The dynamics of perception and action. Psych Rev. 113, 358–389.

Whitacre, J.M., 2010. Degeneracy: a link between evolvability, robustness and complexity in biological systems. Theor. Biol. Med. Model. 7, 6.

Williams, A.M., Ward, P., 2007. Anticipation skill in sport: exploring new horizons. In: Tenenabum, G., Eklund, R. (Eds.), Handbook of Sport Psychology. Wiley, New York, pp. 203–223.

Withagen, R., de Poel, H.J., Araújo, D., Pepping, G.J., 2012. Affordances can invite behavior: reconsidering the relationship between affordances and agency. New Ideas Psychol. 30, 250–258.

Wykowska, A., Schubö, A., 2012. Action intentions modulate allocation of visual attention: electrophysiological evidence. Front. Psychol. 3, 379.

Zourbanos, N., Tzioumakis, Y., Araújo, D., Kalaroglou, S., Hatzigeorgiadis, A., Papaioannou, A., Theodorakis, Y., 2015. The intricacies of verbalizations, gestures, and game outcome using sequential analysis. Psychol. Sport Exerc. 18, 32–41.

FURTHER READING

Araújo, D., Diniz, A., Passos, P., Davids, K., 2014. Decision making in social neurobiological systems modeled as transitions in dynamic pattern formation. Adapt. Behav. 22, 21–30.

CHAPTER

Catching on it early: Bodily and brain anticipatory mechanisms for excellence in sport

Ana M. Abreu*, Matteo Candidi[†,‡], Salvatore M. Aglioti[†,‡,1]

Universidade Europeia, Laureate International Universities, Lisbon, Portugal
[†]*Sapienza Università di Roma, Rome, Italy*
[‡]*IRCCS, Fondazione Santa Lucia, Rome, Italy*
[1]*Corresponding author: Tel.: +39-649917601; Fax: +39-649917635,*
e-mail address: salvatoremaria.aglioti@uniroma1.it

Abstract
Programming and executing a subsequent move is inherently linked to the ability to anticipate the actions of others when interacting. Such fundamental social ability is particularly important in sport. Here, we discuss the possible mechanisms behind the highly sophisticated anticipation skills that characterize experts. We contend that prediction in sports might rely on a finely tuned perceptual system that endows experts with a fast, partially unconscious, pickup of relevant cues. Furthermore, we discuss the role of the multimodal, perceptuomotor, multiple-duty cells (mirror neurons) that play an important function in action anticipation by means of an inner motor simulation process. Finally, we suggest the role of predictive coding, interoception, and the enteric nervous system as the processual and biological support for intuition and "gut feelings" in sports—the missing link that might explain outstanding expert performance based on action anticipation.

Keywords
Action observation network, Motor simulation, Action anticipation, Intention reading, Expertise, Neuroscience of sport, Mirror properties, Enteric nervous system, Interoception, Predictive coding

1 VISUAL ANTICIPATION AND INTENTION READING

As we observe Djokovic's return of serve, we might marvel on his technical abilities or on the harmonious composition of his movements that seem to anticipate the localization of the ball ahead of time, allowing for an adequate preparation of action, in order to successfully return the ball. Understanding the behavioral, neural, and cognitive skills that underpin perceptuomotor excellence in sport is a challenge for contemporary cognitive neuroscience and motor control fields. Some scientists have contended, in line with an information-processing model, that the ability to anticipate upcoming events is crucial for excellence, and that such anticipation derives from accessing crucial information ahead of time. Others argue, however, and in line with an ecological framework, that successful action anticipation is associated to the detection of affordances that occur online rather than to predicting what is going to happen in advance (for a comprehensive review, please see McMorris, 2004). Experimental evidence has provided support for both contentions and both early information pickup (e.g., Müller et al., 2006) and online fast-tracking, in a prospective information-processing framework (e.g., Bastin et al., 2006), seem to lead to an expert advantage in anticipating the actions of another, compared to novices.

Scientists have long debated not only on how expert athletes decide upon a course of action (i.e., what are the relevant features of the environment to be detected) but also on how the early information pickup occurs (i.e., what are the perceptual mechanisms governing the ability to rapidly acquire the relevant information). One of the most privileged sources of information is attained through visual perception. Some research suggests that the effective reading of gesture kinematics leads to a better anticipation of the action of another (Williams et al., 2002). Take soccer, for example, when defending a goal, in order to gain time and start moving in the right direction, it is crucial that the goalkeeper captures the movement information of the kicker before the penalty kick, so as to anticipate the direction of the ball in time and act upon it. Indeed, it has been shown that expert goalkeepers use more effective perceptual search strategies, with fewer fixations of longer duration, concentrating on specific salient body and kinematics cues, compared to novices—who tend to present a more scattered visual search strategy (Savelsbergh et al., 2002).

Recognition of the importance of perceptual strategies of experts has given rise to several eye-tracking and behavioral studies. These studies have mainly used edited stimuli by spatial (e.g., Mecheri et al., 2011; Panchuk and Vickers, 2009), temporal (e.g., Müller et al., 2015, 2016), or combined occlusion techniques (e.g., Causer et al., 2017), in order to determine the least amount of information needed for predicting the outcome of ball trajectories or of the motor actions of another. Processing the motion and content of objects is supported by two distinct anatomofunctional systems, the ventral and dorsal visual streams, whereby the ventral stream enables the conscious identification of objects (regardless of their spatial localization and motion) (the "what" stream), and the dorsal stream enables to unconsciously program, control, and carry out an action (regardless of object identification) (the "where" and "how" stream) (Goodale and Milner, 1992; Milner and Goodale, 2008; Ungerleider and Mishkin, 1982). The neuroanatomical circuitry of these two parallel streams, distributed across the parietal, temporal, and frontal lobes, and the existence of multiple dorsal and ventral streams are still being discussed (Binkofski and Buxbaum, 2013). Recent insights concerning ventral and dorsal pathways now seem to point that earlier evidence on the roles of these streams based on occlusion paradigms might have been biased. Because occlusion studies have mostly relied on perceptual reports based on explicit judgments, they have probably mainly exploited the workings of the ventral stream. Possibly, the data thus obtained has not resulted from the interplay between dorsal and ventral streams nor did it exploit the workings of the body-centered implicit movement control processed by the dorsal stream (Van der Kamp et al., 2008). Indeed, Montagne et al. (2008) have argued that there are many lab-made errors in anticipation that are not accounted for by contextualized expert action. Thus, these authors question if the studies that have been used to explore action anticipation actually tap into the same mechanisms that are assessed *in realistic contexts*. Crucially, Dicks et al. (2010) have shown that movement and gaze behaviors depend on the constraints of the experimental setting and so we should tread carefully, when generalizing experimental data, and strive to investigate in the most natural performance environments. As it seems, the available information concerning the perceptual mechanics of anticipation in expertise might be biased or, to say the least, only partially true.

Indeed, several studies report more effective visual search behaviors, attunement of information extraction, and optimal use of significant kinematic information sources in experts compared to novice individuals (e.g., Müller et al., 2006; Williams et al., 2002). However, if one considers that the fastest serves in tennis range from 250 to 260 km/h (i.e., tennis players need to hit the ball about 270 ms after the opponent's serve), could we conceive that elite athletes might have refined their perceptual capacities to a superhuman standard that enables them to track at high speed? One may wonder whether unconscious vision, albeit less flexible because it does not consider several conscious alternatives, might serve the purpose, as its best quality is its rapid speed, allowing for fast decisions (Ansorge et al., 2014). This perceptual unawareness framework nicely fits the intuitive model discussed ahead. Again, this hints at the importance of the dorsal stream, said to be fast and unconscious. Since it seems to be humanly impossible to "keep an eye on the ball," i.e., to centrally (i.e., via the fovea) process the information of a speeding ball up to contact, it is possible that this motion information is (at least partially) peripherally processed by experts. Research seems to point at this possibility as an fMRI study has shown that peripherally obtained information is sent to the central (foveal) region of the primary visual cortex. This surprising discovery points to the possibility that the foveal cortex receives feedback from peripheral vision (Williams et al., 2008). Together, these findings seem to support the possibility that experts are using peripheral visual pursuit strategies and that the input thus obtained is fed back to the center of resolution, allowing optimal anticipation of what should be their next move.

Increased ability to use unconscious signals and picking up relevant information from body movements may explain the superior perceptual skills of expert athletes, but how does one progress from sensory information to optimal action prediction and what are the neural mechanisms supporting predictive gazing behaviors and unconscious perception? Indeed, sampling the relevant information from the environment in order to organize one's behavior may still not be enough to allow for fast responses in sport. This is reflected in the evidence that reacting to environmental events is computationally incongruent with the time constraints of natural behavior and competitive interactions in particular. In turn, this contributes to support the idea that the brain builds models of sensory and motor events in order to allow for predictions about their unfolding. Studies are now supporting this view by providing evidence that specific brain structures entail the anatomical organization for implementing predictive processing (Bastos et al., 2012) and that humans' sensorimotor functions have developed in order to build predictive models of the hidden causes of incoming sensory information that may allow the implementation of adequate anticipatory behavioral patterns (Friston et al., 2006). In order to make sense of predictive perceptual processes, it has been suggested that the brain activity is structured in hierarchical generative models (Friston and Kiebel, 2009) that are in tune with predictive coding theories. These theories are particularly relevant for decision making in sports as discussed below. Independent of the neuroanatomical mappings of perceptual inferences, it seems that in ball sports, advance information prior to the ball release and view of the ball trajectory contribute to optimal ball catching performance (Stone et al., 2014). Expert anticipation may thus, in principle, be based on both (1) detecting early body cues and (2) "keeping an eye on the ball."

2 FROM AN ACTION OBSERVATION NETWORK TO PREDICTIVE EMBODIMENT

Excellence in interactive sports may not only be based on the ability for early pickup of relevant features in the movements of the others (Abernethy and Zawi, 2007; Abernethy et al., 2008), but even more so on the anticipation of their movements (and outcomes) in order to regulate one's own behavior (Aglioti et al., 2008; Farrow and Abernethy, 2003; Weissensteiner et al., 2008).

Proactive gaze has been a classical domain where predictive behaviors have been tested (Henderson, 2017). When observing someone grasping an object, for example, it has been shown that the gazing behavior of the observer anticipates the actual movements of the observed individuals, and that this proactive gaze is similar to when the individual is performing the movement himself (Flanagan and Johansson, 2003). Thus, when looking at someone else's actions, predictive eye movements seem to be based on the ability to internally simulate a sequence of movements as if the observer were the actor of the observed actions. In this respect, the richness of the motor copy that one possesses of an action may impact one's predictive abilities concerning the movements of others. At a behavioral level, it has been shown that the perceptual system is tuned to predictions about self-performed actions (i.e., individuals are better at predicting actions of which a motor copy belongs to them) (Knoblich and Flach, 2001). The evidence that perception is an active process (*active sensing*) supported the idea that perception–action coupling is a fundamental mechanism for predictive perception (see Feldman, 2016 for a critical review of the literature on the link between action and perception) and, in fact, contributed to the idea that individuals reactivate a (multimodal) internal simulation of observed actions in order to predict their fate.

In this perspective, perceptuomotor, multiple-duty cells (mirror neurons) may play an important function in action anticipation. Importantly, the human motor system is endowed with the necessary machinery to support predictive simulation for individual action control. Motor control theories propose that the predictive simulation of the sensory consequences of one's movement is realized through the coupling of forward and inverse models (Flanagan and Johansson, 2003; Wolpert and Flanagan, 2001; Wolpert and Kawato, 1998). If triggered by the observation of a movement, these predictive motor simulations, however, may be exapted for predicting the behavior of others, supporting interpersonal motor interactions (Candidi et al., 2015) and ultimately excellence in interactive sports. Neurophysiological evidence for such possibility comes from seminal animal studies (di Pellegrino et al., 1992; Fogassi et al., 2005) and evidence in human research (Chong et al., 2008; Kilner et al., 2009) reporting increased neural activity in motor (premotor and parietal) regions during passive action observation, an activity that is likely due to double-duty mirror neurons. More importantly, neurophysiological (EEG, TMS, and intracortical) studies in humans and monkeys highlighted the predictive nature of these motor activations during action observation (Kilner et al., 2004), during perception of implied movements (Urgesi et al., 2010), or when the final phase of a movement that can be predicted is occluded (Umiltà et al., 2001). Also, mirror properties are said to develop by means of sensorimotor learning (Catmur et al., 2007), thus explaining why stronger activation of the Mirror Neuron System (MNS) has been shown to occur in motor experts when observing movements belonging to the motor domain for which they are experts (Calvo-Merino et al., 2006).

Motor activations during action observation were initially reported using correlational approaches while having participants passively observing the actions and suggested that an action observation network (AON) existed in humans (Caspers et al., 2010). Following studies tried to understand whether these internal simulative mechanisms may serve a perceptual, predictive, function moving from the idea of passive simulation to an embodied form or predictive simulation (Grush, 2004).

The association between motor simulation and predictive perceptual advantages has been studied in elite basketball players by controlling the visual expertise of individuals (Aglioti et al., 2008). In Aglioti et al.'s (2008) study, elite players performed better than expert watchers (journalists or coaches) and novice individuals when predicting the fate of free shots at the basket. Moreover, while both players and expert watchers increased the excitability of their hand muscles during free shot observation, only athletes showed a time-specific, and fate-specific, motor facilitation that dissociated between in- and out-shots when the critical kinematic cues became available before hand-ball release, suggesting that fine motor simulative processes may underpin excellent perceptual anticipation. Interestingly, support in favor of an expertise-contingent visuomotor mechanism for error detection also comes from studies with musicians in which mute video-clips were used (Candidi et al., 2014; Panasiti et al., 2016).

In order to further describe the dependency of perceptual predictions on either ball trajectory or early kinematic cues, Tomeo et al. (2013) have studied the reactivity of the corticospinal system of expert soccer kickers, goalkeepers, and naïve subjects during a predictive trajectory discrimination task of soccer penalty kicks. In this study, the congruency between body kinematics and ball trajectory was fictitiously manipulated such that the ball could either follow the trajectory associated to the movement (congruent trials) or be directed in the other direction (incongruent trials). The authors showed that kickers' perceptual predictions were more susceptible to body kinematics rather than ball trajectory compared to goalkeepers whose perceptual decisions were also influenced by visual information concerning the ball. Furthermore, they also found that goalkeepers and naïve subjects showed, respectively, a reduced and increased leg motor activation for deceptive movements compared to nondeceptive ones. These results suggest that relying solely on movement kinematics may even have detrimental perceptual effects (as in the case of kickers) while being able to switch from motor simulation to visual cues (as goalkeepers did in this study) may be the most effective strategy in sports.

In order to assign any perceptual function to motor activations during action observation, however, the key evidence would be that motor activations during action observation are fed back to perceptual systems (high-order visual areas such as pSTS in the case of visual predictions concerning the movements of others) in order to facilitate perceptual functions and ultimately improve decisions about the fate of an observed movement ahead of its full realization. Brain stimulation (Candidi et al., 2008; D'Ausilio et al., 2009; Urgesi et al., 2007) and lesion (Aglioti and Pazzaglia, 2010; Moro et al., 2008; Nelissen et al., 2010; Pazzaglia et al., 2008a,b) studies provide causative evidence that activations in premotor and motor regions serve the transformation of visual, acoustic, and even olfactory (Aglioti and Pazzaglia, 2011) inputs in action representations. In a similar vein, if simulative activations interact with perceptual systems, inhibiting the activity of perceptual areas during action observation should be reflected in higher motor activations. This hypothesis has been studied by transiently reducing the activity of visual (Arfeller et al., 2013) and premotor (Avenanti et al., 2013) nodes of the AON and measuring the response of motor-related cortices during action observation. These studies converge in showing that reducing the reactivity of the visual system is paralleled by an increase in the activity of motor nodes which is interpreted as a compensatory mechanism where the increase in motor resources dedicated to the simulation of the observed action might facilitate the visual analysis of the movements (Avenanti et al., 2013). Direct evidence for a causal role of the left IFG for action prediction has recently been shown by transiently reducing/increasing the activity of this region via cathodal/anodal transcranial direct current stimulation (tDCS) and measuring the ability to predict the outcome of an observed hand movement (Avenanti et al., 2017).

In the case of sport excellence, the implication for this framework is that acquiring a refined (i.e., more accurate and multimodal) sensorimotor copy of the observed movements would endow the perceptual system with higher anticipatory skills. By altering the kinematic features of an observed penalty kick, Makris and Urgesi (2015) have supported this idea by asking expert soccer players (outfield players or goalkeepers) and novices to predict the direction of the ball after perceiving the initial phases of penalty kicks that contained or not incongruent body kinematics. The transient inhibition (repetitive transcranial magnetic stimulation, rTMS) of high-order visual areas (STS) disrupted performance in both experts and novices, especially in those with greater visual expertise (i.e., goalkeepers), while rTMS to premotor regions (dPM) impaired performance only in expert players (i.e., outfield players and goalkeepers).

Despite the importance of the AON in action prediction, we have previously shown (Abreu et al., 2012) that anticipating the consequences of the actions of others implies a functional reorganization, beyond the AON, that subtends a strong link between anticipation, error detection, body awareness, and motor expertise (see Fig. 1).

We have contended that experts use specific perceptual strategies and make use of refined brain, simulative, mechanisms to predict the outcome of the action of another. Although both embodiment and unconscious perception might explain fast and accurate predictions, some prediction performances are so outstanding that there seems to be more to expert prediction. Could intuition explain such optimal performances? Next we shall discuss this.

3 "GUT FEELING" IN SPORTS: FROM INTEROCEPTION TO INTUITION

Parkinson's disease is a neurodegenerative disorder that mainly affects the motor system. Despite many unknowns, this condition has been considered to be a disorder of the brain whereby the death of subcortical neurons leads to stiffness, tremors, and other motor difficulties (see, for example, Jankovic, 2008). Although we are discussing the expert mechanisms of motor anticipation, it is important to look at this disease model for it might bear important insights. Surprising new evidence suggests that Parkinson's disease might actually originate in the gut, and only later spread to the brain, as gut bacteria in mice have been found to regulate movement disorders owing to the signaling between the gut and the brain (Sampson et al., 2016).

Although the medical community has distanced itself from certain traditional medicine claims, the link between gut and behavior is not a new one and ancient Ayurvedic medicine already predicted gut health to be critical to overall health. These fresh new insights, however, now support a strong link between intestinal microbiota and neurodevelopment. Indeed, the gut is lined by millions of neurons that allow us to feel our bowls from the inside. This constitutes a whole other nervous system,

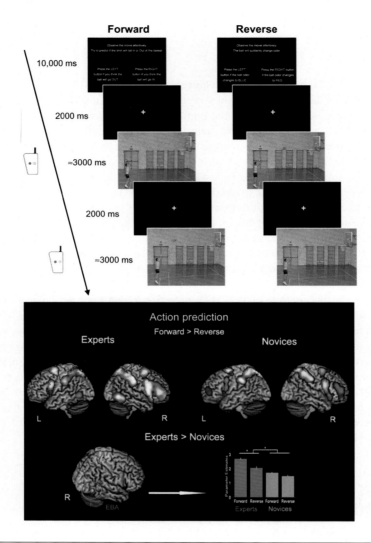

FIG. 1

Upper panel: the study design. Participants observed the free-shot movies during both an action prediction (forward, F) condition and a ball-color-change detection (reverse, R) control condition. F instructions: "please attend to the movies displayed and try to predict the outcome of the throws; use the left button press for predicted out shots, use the right button press for predicted in shots." R instructions: "please attend to the ball in the movies displayed. At a certain point of the ball trajectory, the ball will change color for a moment; use the left button press if the color changes to *red*, use the right button press if the color changes to *blue*." *Lower panel*: brain activations associated with action prediction (i.e., Forward > Reverse). Both expert and novice groups activated areas in the frontoparietal network associated with action observation and in the postcentral gyrus (somatosensory cortices) when predicting the outcome of an action. The between-groups comparison revealed the additional recruitment of an area within the posterior portion of the inferior temporal cortex, overlapping with the EBA, in experts but not in the novice group. Asterisks indicate significant comparisons ($P < 0.05$).

Adapted from Abreu, A.M., Macaluso, E., Azevedo, R., Cesari, P., Urgesi, C., Aglioti, S.M., 2012. Action anticipation beyond the action observation network: an fMRI study in expert basketball players. Eur. J. Neurosci., 35 (10), 1646–1654. doi:10.1111/j.1460-9568.2012.08104.x.

known as the enteric nervous system (ENS) (for review, please see Sasselli et al., 2012). But the complexity of the ENS seems to indicate that the gut is doing more than just helping with digestion and much information from the ENS is signaled to the brain, hinting that our emotional states might result, in part, from the messages sent from the ENS to the brain, and not the other way around. Moreover, the number of neurotransmitters in the ENS and the amount of serotonin found in the bowls seem to indicate that we can definitely "feel" with our gut (e.g., Hadhazy, 2010).

Can we then speak of "gut feelings" when discussing anticipatory reading of observed action in sports? We have previously suggested cue saliency, automatization, and early, finely tuned pattern-recognition processes to be implicated in expert decision making in sports, implicating that both gut and early information processing might be implied (Abreu, 2014). Indeed, it is possible that other than the information from the outside world, attained via our senses (exteroception), we make use of information from the musculoskeletal system (proprioception) and the viscera (interoception) to make inferences (Ondobaka et al., 2017). Together, this information allows the brain to generate models of prediction that are continuously updated due to the mismatch between predictions and actual sensory input given by corticocortical feedback connections following a predictive coding framework (Rao and Ballard, 1999). Such a predictive coding framework might also account for the interoceptive inferences of the MNS bringing a new wealth of understanding on how we might infer the intentions of another by minimizing prediction errors throughout the cortical hierarchy (Kilner et al., 2007).

It thus seems that interoceptive signals from the gut to the brain are part of the bidirectional communication system that might allow the correct functioning of higher cognitive functions, such as intuitive decision making (e.g., Mayer, 2011), seemingly important in elite sporting contexts. In line with this, a new embodied predictive interoception coding model highlights the importance of ascending visceral sensations in modulating limbic predictions of body states (Barrett and Simmons, 2015) and interoception's place in the formation of an embodied self is established (Seth and Friston, 2016).

Thus, it seems that there is a link between the ENS, interoception "gut feelings," and motor disorders. But how is interoception associated to expert anticipatory reading of observed actions in sports? Gu and FitzGerald (2014) seem to nicely link interoception with decision making by adding the concept of homeostasis. These authors argue that homeostasis arises because we seek familiar states and the minimization of surprise. Thus, the prediction-error minimization described earlier and attained by autonomic and somatic reflexes aids in the maintenance of homeostasis and such interoceptive state information in the insular cortex contextualizes choice behavior by transmitting this information to other decision-making nodes. In an attempt to quantify anticipation, a real-time quantification environment has been used to collect information (i.e., EEG, EMG, eye tracking, and motion data, among others) regarding a possible anticipatory profile in an array of several sporting modalities. Early data shows that anticipatory profiles are what really distinguish elite athletes from other sport practitioners (Nadin, 2015). While discussing awareness and anticipation, the author argues that such distinguishing expert anticipatory profile could be guided by "gut feeling." Indeed, it has been proposed that the reason expert athletes intuitively know how to tackle an opponent, kick a ball, or return a serve in a matter of fractions of seconds, is because they access previously stored experiences without conscious effort (Gigerenzer, 2007). In line with this, Voss and Paller (2009) suggest that an "implicit memory"-like neurocognitive mechanism might afford accurate discrimination of repeat stimuli from new stimuli without conscious knowledge. Apparently, then, "gut feelings" are not random hunches, but draw on previous experiences that have not been consciously acquired or retrieved by instructed awareness. On the other hand, these memories might not be explicit because they partly originate from visceral information sent by the vagus nerve to the brain. Accordingly, Reed et al. (2010) show that people have difficulty describing their gaze angle, when catching a ball, despite being proficient at catching it. The authors argue that when trying to fit explicit knowledge, to their unconscious actions, participants sometimes use incorrect descriptions. This is truly important to consider, as many sports studies rely on explicit descriptions given by athletes and novices alike. Importantly, it seems that unconscious perception does have a role in guiding successful motor action.

4 CONCLUSION

We have contended that the anticipatory mechanisms for motor prediction in sports might rely on skilled perceptual information pickup of relevant cues that might partially be supported by fast unconscious perceptual processing. Beyond the information pickup processes, we argue that the correct anticipation of the actions of another ultimately relies on proactive perception based on simulative processes that are supported by ones' own perceptual and motor experience. However, we find that in sports, some extraordinary performances are so exceptional that they cannot be explained by speed, consciousness, or even embodiment mechanisms. Here we suggest that intuition and "gut feelings" nicely link with a fast, simulative, and partially unconscious model of motor prediction and anticipation. Despite its' supernatural and mystical appearance, intuition is supported by interoceptive and predictive coding mechanisms and the ENS. Thus the study of interoception may ultimately open new doors for understanding expert anticipation in sports.

ACKNOWLEDGMENTS

The financial contribution of the Italian Ministry of University and Research (PRIN progetti di ricerca di rilevante interesse nazionale, Bando 2015, Prot. 20159CZFJK) is gratefully acknowledged by S.M.A. M.C. was supported by Sapienza University (Progetti Medi 2016).

REFERENCES

Abernethy, B., Zawi, K., 2007. Pickup of essential kinematics underpins expert perception of movement patterns. J. Mot. Behav. 39 (5), 353–367. http://dx.doi.org/10.3200/JMBR.39.5.353-368.

Abernethy, B., Zawi, K., Jackson, R.C., 2008. Expertise and attunement to kinematic constraints. Perception 37 (6), 931–948. http://dx.doi.org/10.1068/p5340.

Abreu, A.M., 2014. Action anticipation in sports: a particular case of expert decision-making. Trends Sport Sci. 1 (21), 197–203.

Abreu, A.M., Macaluso, E., Azevedo, R., Cesari, P., Urgesi, C., Aglioti, S.M., 2012. Action anticipation beyond the action observation network: an fMRI study in expert basketball players. Eur. J. Neurosci. 35 (10), 1646–1654. http://dx.doi.org/10.1111/j.1460-9568.2012.08104.x.

Aglioti, S.M., Pazzaglia, M., 2010. Representing actions through their sound. Exp. Brain Res. 206 (2), 141–151. http://dx.doi.org/10.1007/s00221-010-2344-x.

Aglioti, S.M., Pazzaglia, M., 2011. Sounds and scents in (social) action. Trends Cogn. Sci. 15 (2), 47–55. http://dx.doi.org/10.1016/j.tics.2010.12.003.

Aglioti, S.M., Cesari, P., Romani, M., Urgesi, C., 2008. Action anticipation and motor resonance in elite basketball players. Nat. Neurosci. 11 (9), 1109–1116. http://dx.doi.org/10.1038/nn.2182.

Ansorge, U., Kunde, W., Kiefer, M., 2014. Unconscious vision and executive control: how unconscious processing and conscious action control interact. Conscious. Cogn. 27, 268–287. http://dx.doi.org/10.1016/j.concog.2014.05.009.

Arfeller, C., Schwarzbach, J., Ubaldi, S., Ferrari, P., Barchiesi, G., Cattaneo, L., 2013. Whole-brain haemodynamic after-effects of 1-Hz magnetic stimulation of the posterior superior temporal cortex during action observation. Brain Topogr. 26 (2), 278–291. http://dx.doi.org/10.1007/s10548-012-0239-9.

Avenanti, A., Annella, L., Candidi, M., Urgesi, C., Aglioti, S.M., 2013. Compensatory plasticity in the action observation network: virtual lesions of STS enhance anticipatory simulation of seen actions. Cereb. Cortex 23 (3), 570–580. http://dx.doi.org/10.1093/cercor/bhs040.

Avenanti, A., Paracampo, R., Annella, L., Tidoni, E., Aglioti, S.M., 2017. Boosting and decreasing action prediction abilities through excitatory and inhibitory tDCS of inferior frontal cortex. Cereb. Cortex, 1–15. http://dx.doi.org/10.1093/cercor/bhx041. Epub ahead of print.

Barrett, L.F., Simmons, W.K., 2015. Interoceptive predictions in the brain. Nat. Rev. Neurosci. 16 (7), 419–429. http://dx.doi.org/10.1038/nrn3950.

Bastin, J., Craig, C., Montagne, G., 2006. Prospective strategies underlie the control of interceptive action. Hum. Mov. Sci. 25 (6), 718–732. http://dx.doi.org/10.1016/j.humov.2006.04.001.

Bastos, A.M., Usrey, W.M., Adams, R.A., Mangun, G.R., Fries, P., Friston, K.J., 2012. Canonical microcircuits for predictive coding. Neuron 76 (4), 695–711. http://dx.doi.org/10.1016/j.neuron.2012.10.038.

Binkofski, F., Buxbaum, L.J., 2013. Two action systems in the human brain. Brain Lang. 127 (2), 222–229. http://dx.doi.org/10.1016/j.bandl.2012.07.007.

Calvo-Merino, B., Grèzes, J., Glaser, D.E., Passingham, R.E., Haggard, P., 2006. Seeing or doing? Influence of visual and motor familiarity in action observation. Curr. Biol. 16 (19), 1905–1910. http://dx.doi.org/10.1016/j.cub.2006.07.065.

Candidi, M., Urgesi, C., Ionta, S., Aglioti, S.M., 2008. Virtual lesion of ventral premotor cortex impairs visual perception of biomechanically possible but not impossible actions. Soc. Neurosci. 3 (3–4), 388–400. http://dx.doi.org/10.1080/17470910701676269.

Candidi, M., Sacheli, L.M., Mega, I., Aglioti, S.M., 2014. Somatotopic mapping of piano fingering errors in sensorimotor experts: TMS studies in pianists and visually trained musically naives. Cereb. Cortex 24 (2), 435–443. http://dx.doi.org/10.1093/cercor/bhs325.

Candidi, M., Sacheli, L.M., Aglioti, S.M., 2015. From muscles synergies and individual goals to interpersonal synergies and shared goals: mirror neurons and interpersonal action hierarchies: comment on "Grasping synergies: a motor-control approach to the mirror neuron mechanism" by D'Ausilio et al. Phys. Life Rev. 12, 126–128. http://dx.doi.org/10.1016/j.plrev.2015.01.023.

Caspers, S., Zilles, K., Laird, A.R., Eickhoff, S.B., 2010. ALE meta-analysis of action observation and imitation in the human brain. Neuroimage 50 (3), 1148–1167. http://dx.doi.org/10.1016/j.neuroimage.2009.12.112.

Catmur, C., Walsh, V., Heyes, C., 2007. Sensorimotor learning configures the human mirror system. Curr. Biol. 17 (17), 1527–1531. http://dx.doi.org/10.1016/j.cub.2007.08.006.

Causer, J., Smeeton, N.J., Williams, A.M., 2017. Expertise differences in anticipatory judgements during a temporally and spatially occluded task. PLoS One 12 (2), e0171330. http://dx.doi.org/10.1371/journal.pone.0171330.

Chong, T.T., Cunnington, R., Williams, M.A., Kanwisher, N., Mattingley, J.B., 2008. fMRI adaptation reveals mirror neurons in human inferior parietal cortex. Curr. Biol. 18 (20), 1576–1580. http://dx.doi.org/10.1016/j.cub.2008.08.068.

D'Ausilio, A., Pulvermüller, F., Salmas, P., Bufalari, I., Begliomini, C., Fadiga, L., 2009. The motor somatotopy of speech perception. Curr. Biol. 19 (5), 381–385. http://dx.doi.org/10.1016/j.cub.2009.01.017.

Dicks, M., Button, C., Davids, K., 2010. Examination of gaze behaviors under in situ and video simulation task constraints reveals differences in information pickup for perception and action. Atten. Percept. Psychophys. 72 (3), 706–720. http://dx.doi.org/10.3758/APP.72.3.706.

di Pellegrino, G., Fadiga, L., Fogassi, L., Gallese, V., Rizzolatti, G., 1992. Understanding motor events: a neurophysiological study. Exp. Brain Res. 91 (1), 176–180. http://dx.doi.org/10.1007/BF00230027.

Farrow, D., Abernethy, B., 2003. Do expertise and the degree of perception—action coupling affect natural anticipatory performance? Perception 32 (9), 1127–1139. http://dx.doi.org/10.1068/p3323.

Feldman, A.G., 2016. Active sensing without efference copy: referent control of perception. J. Neurophysiol. 116 (3), 960–976. http://dx.doi.org/10.1152/jn.00016.2016.

Flanagan, J.R., Johansson, R.S., 2003. Action plans used in action observation. Nature 424 (6950), 769–771. http://dx.doi.org/10.1038/nature01861.

Fogassi, L., Ferrari, P.F., Gesierich, B., Rozzi, S., Chersi, F., Rizzolatti, G., 2005. Parietal lobe: from action organization to intention understanding. Science 308, 662–667. http://dx.doi.org/10.1126/science.1106138.

Friston, K., Kiebel, S., 2009. Cortical circuits for perceptual inference. Neural Netw. 22 (8), 1093–1104. http://dx.doi.org/10.1016/j.neunet.2009.07.023.

Friston, K., Kilner, J., Harrison, L., 2006. A free energy principle for the brain. J. Physiol. Paris 100 (1–3), 70–87. http://dx.doi.org/10.1016/j.jphysparis.2006.10.001.

Gigerenzer, G., 2007. Gut Feelings: The Intelligence of the Unconscious. Viking, New York, NY.

References

Goodale, M.A., Milner, A.D., 1992. Separate visual pathways for perception and action. Trends Neurosci. 15 (1), 20–25.

Grush, R., 2004. The emulation theory of representation: motor control, imagery, and perception. Behav. Brain Sci. 27 (3), 377–442.

Gu, X., FitzGerald, T.H., 2014. Interoceptive inference: homeostasis and decision-making. Trends Cogn. Sci. 18 (6), 269–270. http://dx.doi.org/10.1016/j.tics.2014.02.001.

Hadhazy, A., 2010. Think twice: how the gut's "second brain" influences mood and well-being. Sci. Am. February 12 Issue. Retrieved 12 April, 2017 from, http://www.scientificamerican.com/article.cfm?id=gut-second-brain.

Henderson, J.M., 2017. Gaze control as prediction. Trends Cogn. Sci. 21 (1), 15–23. http://dx.doi.org/10.1016/j.tics.2016.11.003.

Jankovic, J., 2008. Parkinson's disease: clinical features and diagnosis. J. Neurol. Neurosurg. Psychiatry 79 (4), 368–376. http://dx.doi.org/10.1136/jnnp.2007.131045.

Kilner, J.M., Vargas, C., Duval, S., Blakemore, S.J., Sirigu, A., 2004. Motor activation prior to observation of a predicted movement. Nat. Neurosci. 7, 1299–1301. http://dx.doi.org/10.1038/nn1355.

Kilner, J.M., Friston, K.J., Frith, C.D., 2007. Predictive coding: an account of the mirror neuron system. Cogn. Process. 8 (3), 159–166. http://dx.doi.org/10.1007/s10339-007-0170-2.

Kilner, J.M., Neal, A., Weiskopf, N., Friston, K.J., Frith, C.D., 2009. Evidence of mirror neurons in human inferior frontal gyrus. J. Neurosci. 29 (32), 10153–10159. http://dx.doi.org/10.1523/JNEUROSCI.2668-09.2009.

Knoblich, G., Flach, R., 2001. Predicting the effects of actions: interactions of perception and action. Psychol. Sci. 12 (6), 467–472. http://dx.doi.org/10.1111/1467-9280.00387.

Makris, S., Urgesi, C., 2015. Neural underpinnings of superior action prediction abilities in soccer players. Soc. Cogn. Affect. Neurosci. 10 (3), 342–351. http://dx.doi.org/10.1093/scan/nsu052.

Mayer, E.A., 2011. Gut feelings: the emerging biology of gut–brain communication. Nat. Rev. Neurosci. 12 (8). http://dx.doi.org/10.1038/nrn3071.

McMorris, T., 2004. Acquisition and Performance of Sports Skills. John Wiley & Sons, Ltd., Chichester, West Sussex.

Mecheri, S., Gillet, E., Thouvarecq, R., Leroy, D., 2011. Are visual cue masking and removal techniques equivalent for studying perceptual skills in sport? Perception 40 (4), 474–489. http://dx.doi.org/10.1068/p6828.

Milner, A.D., Goodale, M.A., 2008. Two visual systems re-viewed. Neuropsychologia 46 (3), 774–785. http://dx.doi.org/10.1016/j.neuropsychologia.2007.10.005.

Montagne, G., Bastin, J., Jacobs, D.M., 2008. What is visual anticipation and how much does it rely on the dorsal stream? Int. J. Sport Psychol. 39 (2), 149–156.

Moro, V., Urgesi, C., Pernigo, S., Lanteri, P., Pazzaglia, M., Aglioti, S.M., 2008. The neural basis of body form and body action agnosia. Neuron 60 (2), 235–246. http://dx.doi.org/10.1016/j.neuron.2008.09.022.

Müller, S., Abernethy, B., Farrow, D., 2006. How do world-class cricket batsmen anticipate a bowler's intention? Q. J. Exp. Psychol. 59 (12), 2162–2186. http://dx.doi.org/10.1080/02643290600576595.

Müller, S., McLaren, M., Appleby, B., Rosalie, S.M., 2015. Does expert perceptual anticipation transfer to a dissimilar domain? J. Exp. Psychol. Hum. Percept. Perform. 41 (3), 631–638. http://dx.doi.org/10.1037/xhp0000021.

Müller, S., Fadde, P.J., Harbaugh, A.G., 2016. Adaptability of expert visual anticipation in baseball batting. J. Sports Sci. 9, 1–9. http://dx.doi.org/10.1080/02640414.2016.1230225.

Nadin, M., 2015. Anticipation—the underlying science of sport. Report on research in progress. Int. J. Gen. Syst. 44 (4), 422–441. http://dx.doi.org/10.1080/03081079.2014.989224.

Nelissen, N., Pazzaglia, M., Vandenbulcke, M., Sunaert, S., Fannes, K., Dupont, P., Aglioti, S.M., Vandenberghe, R., 2010. Gesture discrimination in primary progressive aphasia: the intersection between gesture and language processing pathways. J. Neurosci. 30 (18), 6334–6341. http://dx.doi.org/10.1523/JNEUROSCI.0321-10.2010.

Ondobaka, S., Kilner, J., Friston, K., 2017. The role of interoceptive inference in theory of mind. Brain Cogn. 112, 64–68. http://dx.doi.org/10.1016/j.bandc.2015.08.002.

Panasiti, M.S., Pavone, E.F., Aglioti, S.M., 2016. Electrocortical signatures of detecting errors in the actions of others: an EEG study in pianists, non-pianist musicians and musically naïve people. Neuroscience 318, 104–113. http://dx.doi.org/10.1016/j.neuroscience.2016.01.023.

Panchuk, D., Vickers, J.N., 2009. Using spatial occlusion to explore the control strategies used in rapid interceptive actions: predictive or prospective control? J. Sports Sci. 27 (12), 1249–1260. http://dx.doi.org/10.1080/02640410903156449.

Pazzaglia, M., Pizzamiglio, L., Pes, E., Aglioti, S.M., 2008a. The sound of actions in apraxia. Curr. Biol. 18 (22), 1766–1772. http://dx.doi.org/10.1016/j.cub.2008.09.061. Epub 2008 Nov 13.

Pazzaglia, M., Smania, N., Corato, E., Aglioti, S.M., 2008b. Neural underpinnings of gesture discrimination in patients with limb apraxia. J. Neurosci. 28 (12), 3030–3041. http://dx.doi.org/10.1523/JNEUROSCI.5748-07.2008.

Rao, R.P., Ballard, D.H., 1999. Predictive coding in the visual cortex: a functional interpretation of some extra-classical receptive-field effects. Nat. Neurosci. 2 (1), 79–87. http://dx.doi.org/10.1038/4580.

Reed, N., McLeod, P., Dienes, Z., 2010. Implicit knowledge and motor skill: what people who know how to catch don't know. Conscious. Cogn. 19 (1), 63–76. http://dx.doi.org/10.1016/j.concog.2009.07.006.

Sampson, T.R., Debelius, J.W., Thron, T., Janssen, S., Shastri, G.G., Ilhan, Z.E., … Chesselet, M.F., 2016. Gut microbiota regulate motor deficits and neuroinflammation in a model of Parkinson's disease. Cell 167 (6), 1469–1480. http://dx.doi.org/10.1016/j.cell.2016.11.018.

Sasselli, V., Pachnis, V., Burns, A.J., 2012. The enteric nervous system. Dev. Biol. 366 (1), 64–73. http://dx.doi.org/10.1016/j.ydbio.2012.01.012.

Savelsbergh, G.J., Williams, A.M., Van der Kamp, J., Ward, P., 2002. Visual search, anticipation and expertise in soccer goalkeepers. J. Sports Sci. 20 (3), 279–287. http://dx.doi.org/10.1080/026404102317284826.

Seth, A.K., Friston, K.J., 2016. Active interoceptive inference and the emotional brain. Philos. Trans. R. Soc. B Biol. Sci. 371 (1708), 20160007. http://dx.doi.org/10.1098/rstb.2016.0007.

Stone, J.A., Panchuk, D., Davids, K., North, J.S., Maynard, I., 2014. Integrating advanced visual information with ball projection technology constrains dynamic interceptive actions. Procedia Eng. 72, 156–161. http://dx.doi.org/10.1016/j.proeng.2014.06.027.

Tomeo, E., Cesari, P., Aglioti, S.M., Urgesi, C., 2013. Fooling the kickers but not the goalkeepers: behavioral and neurophysiological correlates of fake action detection in soccer. Cereb. Cortex 23 (11), 2765–2778. http://dx.doi.org/10.1093/cercor/bhs279.

Umiltà, M.A., Kohler, E., Gallese, V., Fogassi, L., Fadiga, L., Keysers, C., Rizzolatti, G., 2001. I know what you are doing. A neurophysiological study. Neuron 31 (1), 155–165. http://dx.doi.org/10.1016/S0896-6273(01)00337-3.

Ungerleider, L.G., & Mishkin, M., 1982. Two cortical visual systems. In: Ingle, D.J., Goodale, M.A., & Mansfield, R.J.W. (Eds.), Analysis of Visual Behavior. MIT Press, Cambridge, pp. 549–586.

Urgesi, C., Candidi, M., Ionta, S., Aglioti, S.M., 2007. Representation of body identity and body actions in extrastriate body area and ventral premotor cortex. Nat. Neurosci. 10 (1), 30–31. http://dx.doi.org/10.1038/nn1815.

Urgesi, C., Maieron, M., Avenanti, A., Tidoni, E., Fabbro, F., Aglioti, S.M., 2010. Simulating the future of actions in the human cortico-spinal system. Cereb. Cortex 20 (11), 2511–2521. http://dx.doi.org/10.1093/cercor/bhp292.

Van der Kamp, J., Rivas, F., Van Doorn, H., Savelsbergh, G., 2008. Ventral and dorsal system contributions to visual anticipation in fast ball sports. Int. J. Sport Psychol. 39 (2), 100–130.

Voss, J.L., Paller, K.A., 2009. An electrophysiological signature of unconscious recognition memory. Nat. Neurosci. 12 (3), 349–355. http://dx.doi.org/10.1038/nn.2260.

Weissensteiner, J., Abernethy, B., Farrow, D., Müller, S., 2008. The development of anticipation: a cross-sectional examination of the practice experiences contributing to skill in cricket batting. J. Sport Exerc. Psychol. 30 (6), 663–684. http://dx.doi.org/10.1123/jsep.30.6.663.

Williams, A.M., Ward, P., Knowles, J.M., Smeeton, N.J., 2002. Anticipation skill in real world task: measurement, training, and transfer in tennis. J. Exp. Psychol. Appl. 8 (4), 259–270. http://dx.doi.org/10.1037/1076-898X.8.4.259.

Williams, M.A., Baker, C.I., De Beeck, H.P.O., Shim, W.M., Dang, S., Triantafyllou, C., Kanwisher, N., 2008. Feedback of visual object information to foveal retinotopic cortex. Nat. Neurosci. 11 (12), 1439–1445. http://dx.doi.org/10.1038/nn.2218.

Wolpert, D.M., Flanagan, J.R., 2001. Motor prediction. Curr. Biol. 11 (18), R729–32.

Wolpert, D.M., Kawato, M., 1998. Multiple paired forward and inverse models for motor control. Neural Netw. 11 (7–8), 1317–1329. http://dx.doi.org/10.1016/S0893-6080(98)00066-5.

CHAPTER

Exploring the applicability of the contextual interference effect in sports practice

5

Damian Farrow[*,†,‡,1], Tim Buszard[*,‡]

[*]Institute of Sport, Exercise and Active Living, Victoria University, Melbourne, VIC, Australia
[†]Skill Acquisition, Australian Institute of Sport, VIC, Australia
[‡]Game Insight Group, Tennis Australia, Richmond, VIC, Australia
[1]Corresponding author: Tel.: +61-2-9919-5001, e-mail address: damian.farrow@vu.edu.au

Abstract

This review will consider three key issues considered critical when determining the efficacy of the contextual interference effect when applied to sports practice. First, the issue of complexity is considered in relation to the amount of interference actually needed in the applied sports setting to create effective learning. Second, the traditional underpinning mechanism/s of contextual interference are discussed in relation to recent neurophysiological perspectives on their viability. A counter-position to these dominant theories is also presented drawing on an implicit learning framework. The final issue considers the typical measures of learning used within the contextual interference literature and scrutinizes them relative to the needs of bridging the apparent theory–practice divide. The concluding section then presents a model to measure the degree of contextual interference within the applied setting, which in turn offers both future research directions as well as guidelines for practitioners.

Keywords

Sport, Cognitive effort, Skill acquisition, Expertise, Practice

1 INTRODUCTION

Motor learning research has investigated numerous variables that influence the efficacy of skill practice. One of the most widely researched practice issues is the contextual interference effect (see Brady, 1998, 2008; Magill and Hall, 1990 for reviews). The contextual interference effect posits that when multiple skills (or skill variations) need to be practiced they are more effectively learned when there is interference present during practice. The interference is created by manipulating the scheduling of practice trials such that skills are learned in either a blocked or random fashion. Random scheduling involves the learner being required to switch between the skills "randomly" throughout practice, whereas blocked practice requires the learner to practice one skill for a block of repetitions before switching to the other skill. While practice performance is usually supressed during random practice relative to blocked practice, research has repeatedly demonstrated that retention or transfer performance is superior for random practice.

The majority of contextual interference research has emanated from controlled laboratory experiments, where usually unskilled participants complete a high volume of practice repetitions over a short time period in an effort to learn a relatively simple movement task. Such a context is very different to that of sport, where athletes spend day after day, year after year, practicing whole-body, multidegree of freedom movement skills to reach high levels of expertise. Despite the disparity between the two settings, recommendations from laboratory-based research are regularly generalized to practitioners in the field concerning optimal methods of practice scheduling (e.g., Schmidt, 1991). While there is clearly a place for such laboratory research, it is argued that the extant literature is at a point where there is a need to validate laboratory findings in the field. Indeed some more recent reviews of the literature suggest that the contextual interference effect cannot be simply generalized to the applied setting (Barreiros et al., 2007; Brady, 2008; Wulf and Shea, 2002).

This review will consider three key issues that are pivotal in this apparent discrepancy between the laboratory and field and ensuing theory and practice. First, the issue of skill complexity will be considered. The need to control unconstrained multiple degrees of freedom is suggested to place a significant load on the performer that may not be present in more simplified laboratory tasks. Consequently consideration is given to how much interference is actually needed in the applied setting to create effective learning relative to the expertise level of the performer. The second issue considers the purported underpinning mechanism/s of contextual interference. Traditional theoretical explanations have predominantly focused on the role of a limited capacity system, such as working memory, in either reconstructing (Lee and Magill, 1983, 1985) or elaborating a skill's representation (Shea and Zimny, 1983). Recently, these explanations have been subjected to neurophysiological measurement in an effort to determine which mechanism is more likely. A counter-position to these dominant theories is also proposed where a more implicit mode of learning is enacted (Rendell et al., 2011) which in turn has different implications for applied settings. The third issue will consider the measurement of learning. Retention and transfer testing have both been used to examine the impact of contextual interference on learning. When the applied setting is considered, an argument is made that the results of transfer testing need to take precedence over those accrued from retention testing. The final section presents a model for determining the amount of contextual interference present in a practice session and in turn offers some future research directions and guidelines for practitioners.

2 THE INFLUENCE OF SKILL COMPLEXITY

It is generally agreed that skill complexity influences the contextual interference effect (Guadagnoli and Lee, 2004; Magill and Hall, 1990; Wulf and Shea, 2002). When practicing simple motor skills, the contextual interference effect consistently emerges. Conversely, when practicing complex motor skills, the contextual interference effect is less apparent. Comprehensive reviews of the contextual interference effect exemplify this occurrence. Brady (1998, 2004) revealed that effect sizes for the contextual interference effect were much larger for simple motor skills (e.g., finger tapping tasks) compared to applied skills (e.g., tennis groundstrokes). Likewise, Barreiros et al. (2007) reported that only 11 of 27 experiments reviewed focusing on applied skills observed the contextual interference effect. The results of recent applied experiments are equally ambiguous. Contrary to the typical contextual interference effect, differences in practice performance between high and low contextual interference practice are often not observed (Cheong et al., 2012, 2016; Porter and Magill, 2010, see second experiment). With regards to skill retention and transfer, progressively increasing contextual interference from low to high was reported to elicit the greatest outcomes when learning to putt in golf (Porter and Magill, 2010, first experiment) and perform basketball passes (Porter and Magill, 2010, second experiment). However, in a study focusing on passing the ball in hockey, no differences emerged between groups of varying practice schedules (Cheong et al., 2016), including a group that progressively increased from low to high contextual interference (Cheong et al., 2012).

The inconsistent findings have been attributed to the complexities associated with applied skills. Skill complexity is the product of the degrees of freedom associated with the movement coupled with the demands of the environment. For example, for finger tapping tasks, the degrees of freedom in movement is low, and the environment would be considered closed, whereby there are little or no external stimuli that could either distract or demand a learners attention. This would therefore be considered a simple skill. Conversely, a tennis serve is a multijoint action that requires controlling many degrees of freedom. Moreover, typically the serve is performed in an open environment, whereby the player needs to make a decision of where to serve the ball based on the opponent's skill and location, and the match situation. This is therefore regarded as a complex skill. Guadagnoli and Lee's (2004) challenge point hypothesis also provides definitions of skill complexity. *Nominal task difficulty* is the term used to describe the level of difficulty for a task regardless of the performer's skill level or external influences, whereas *functional task difficulty* describes the level of difficulty for the performer under the specific conditions. Finger tapping tasks would be considered to be of low nominal and functional task difficulty. A tennis serve, however, would be of moderate nominal difficulty, while functional task difficulty would differ depending on the skill level of the performer and the environmental conditions (e.g., wind, temperature).

Skill complexity is thought to moderate the contextual interference effect as result of cognitive effort. Cognitive effort (in the context of contextual interference) refers to the amount of cognitive processing that occurs when planning to perform a motor skill. High cognitive effort is considered advantageous for learning. When practicing in conditions of high contextual interference, as opposed to low contextual interference, cognitive effort is typically higher (Patterson and Lee, 2008). This is especially evident when practicing simple motor skills. For instance, when performing a finger-tapping task, cognitive effort is greater when required to tap a different key on each trial, as this is more difficult, as opposed to tapping the same key repeatedly. When practicing a complex motor skill, however, cognitive effort is also likely to be high under conditions of low contextual interference. This is because the complexity of the skill demands greater cognitive processing to perform the skill. By way of example, executing a tennis forehand is often challenging for a novice performer and the heightened difficulty increases the amount of cognitive processing required to perform the skill.

Experimental evidence demonstrates that high contextual interference can place demands on cognitive processing that are too high to reap the benefits that typically are found to emerge from such practice in laboratory settings. In a drawing task that involved varying levels of task difficulty, high contextual interference practice was most beneficial for the simplest task, but this effect weakened as task difficulty increased (Albaret and Thon, 1998). Hence, task difficulty, or skill complexity, relative to the performer, appears to moderate the contextual interference effect. Certainly this argument reflects various accounts of learning, whereby learning is more robust when the task presents an optimal challenge to the performer (e.g., Challenge Point Framework, Guadagnoli and Lee, 2004; desirable difficulty proposition, Christina and Bjork, 1991). We suspect that the primary reason why the study of complex motor skills reveals ambiguous results with regards to the contextual interference effect is because the manipulation of contextual interference is unlikely to alter task difficulty for complex motor skills. Thus, the contextual interference effect, as derived from laboratory-based research, would not emerge for such skills.

Where does this leave scientists and practitioners? Let us consider the final outcome. The aim in motor learning is to (a) maximize performance improvements and (b) ensure that the improvements transfer to an environment where the skill will be performed. In sport, this is the competition environment. To achieve these goals, cognitive effort should be maximized, which can be attained by manipulating contextual interference. However, if the skill is relatively complex for the performer, the amount of contextual interference will likely have minimal impact on the amount of cognitive effort expended. We argue that in such situations, the most effective practice schedule will depend on the competition environment. If competition features high contextual interference, then random practice is more likely to produce skills that can be transferred to competition, and vice versa if competition features low contextual interference. This is consistent with Russell and Newell (2007) who argued that random practice teaches learners to flexibly switch between tasks minimizing the impact of the "switch cost" (see Monsell, 2003 for a review), while blocked practice is equally as effective as random practice when the task involves repeatedly performing the same skill. Given many sports demand rapid task switching, it may be that the purported benefits of random practice are simply one of specificity.

3 MECHANISMS AND PRACTICE

Explanations of the contextual interference effect have focused on cognitive processing during performance. The *forgetting-reconstruction hypothesis* (Lee and Magill, 1983, 1985) suggests that high contextual interference caused the performer to constantly forget task-specific information between practice trials, therein requiring the performer to (re)construct the action plan for every trial. This process is thought to improve the ability to develop action plans, which subsequently improves skill acquisition. A similar hypothesis—the *elaboration hypothesis* (Shea and Morgan, 1979; Shea and Zimny, 1983)—proposed that high contextual interference practice causes the performer to engage in a process of comparing and contrasting the skills being practiced, consequently leading to a more elaborate and distinctive representation of the motor skill in memory.

Common to both theories is the assumption that a limited capacity system, such as working memory, influences the learning process. A limited capacity causes the performer to either (a) forget information about previous performance of a particular skill under conditions of high contextual interference (forgetting-reconstruction hypothesis) or to (b) compare and contrast performance with previous trials (elaboration hypothesis). Clever experimental manipulations support the assertion that a limited capacity system influences the contextual interference effect. Reaction time to a probe that occurs immediately prior to movement, which is when action reconstruction is thought to occur, was slower for high contextual interference practice compared to low contextual interference practice (Li and Wright, 2000). The same trend was apparent when reacting to a probe that was inserted during the intertrial period, which is when elaboration processing is thought to occur. Slower reaction time is indicative of greater working memory involvement (hence cognitive processing).

Advances in neurophysiological measures offer support to the traditional accounts of the contextual interference effect. Inhibiting activation of the primary motor cortex during the intertrial period via transcranial magnetic stimulation negatively impacted higher contextual interference practice, but not lower contextual interference practice (Lin et al., 2008, 2009). Hence, the learning benefits associated with high contextual interference are dependent upon the neural activity that occurs during the intertrial period. It is assumed that heightened neural activity represents greater cognitive processing. Elsewhere higher contextual interference practice has correlated with increased excitability in the primary motor cortex (Cross et al., 2007; Lin et al., 2011). It must be emphasized, however, that these findings are based solely on simple motor skills, such as finger tapping tasks. The transferability of these findings to applied skills is therefore still questionable. Hence a challenge for neuroscientists is to continue to add complexity to the skills being examined in the laboratory context as well as further develop in situ measures of brain behavior so that the findings can be generalized across settings.

An alternate explanation of the contextual interference effect has been derived from the theory of implicit motor learning (Rendell et al., 2009). Implicit motor learning refers to the acquisition of a motor skill with minimal conscious awareness of the step-by-step processes underpinning the learnt skill, thereby reducing reliance on working memory (Masters, 1992; Masters and Poolton, 2012). Motor skills are often acquired implicitly when the performer does not attempt to consciously analyze their movement patterns (i.e., error detection and correction; Maxwell et al., 2003). Rendell et al. (2011) presented preliminary evidence to suggest that the retention benefits of random practice might be driven by mechanisms which are also present during implicit motor learning (the implicit learning hypothesis for random practice). The study employed a number of measures to assess the level of cognitive effort and implicit/explicit processes that occurred during blocked and random practice of two motor skills (kicking and handball in Australian Rules Football). The results showed that relative to blocked practice, random practice resulted in higher levels of cognitive activity. However, the results also suggested that random practice shared characteristics with implicit learning (i.e., superior secondary task transfer performance and less access to verbally based task knowledge). These findings were evident for random practice of the more complex of the two motor skills studied (i.e., kicking). The authors proposed that working memory resources may be so overwhelmed by the information required to switch tasks during random practice that participants are unable to test hypotheses relating to the movement solutions that are generated, or rehearse and store verbal, task-related rules and information (the implicit learning hypothesis—Masters, 1992). Thus, random practice may involve high levels of cognitive activity that are not directly allocated to developing movement solutions.

The findings from Rendell et al. (2011) contradict the two predominant theories of contextual interference within the current literature. The elaboration hypothesis (Shea and Zimny, 1983) suggests that high contextual interference is advantageous for learning because learners make a greater number of comparisons among the tasks they are learning and therefore appreciate the distinctiveness of the different tasks. In contrast, the implicit learning hypothesis implies that learning benefits of high contextual interference practice might be due to a *reduction* in the comparative processing between tasks and *less* conscious access to movement solutions underlying the tasks being learned.

The reconstruction hypothesis (Lee and Magill, 1983, 1985) conversely suggests that high contextual interference practice evokes superior learning because learners encounter a partial forgetting of one skill during the period that they are practicing other skills. The implicit learning hypothesis shares some commonality with the reconstruction hypothesis in so far as it agrees that high contextual interference learners tend to forget the movement solution for one skill as they switch to an alternative skill. The key difference between the two hypotheses is the explanation of the effect of "forgetting." The reconstruction hypothesis suggests that the advantage of high contextual interference emerges as a result of effortful processing that occurs when the learner reconstructs the forgotten skill, whereas the implicit learning hypothesis suggests that task switching instead encourages learners to place a greater reliance on passive or tacit knowledge to produce their movements.

A common thread in the elaboration and reconstruction theories is that both suggest that contextual interference is a function of increased cognitive effort stimulated by high contextual interference and diminished cognitive effort ensuing from low contextual interference (Brady, 1998; Li and Wright, 2000). The implicit learning hypothesis concurs that high contextual interference evokes higher levels of cognitive effort than low contextual interference but suggests that the cognitive processing caused by task switching may in fact prevent learners from actively processing their previous movement solutions. This would therefore reduce task-related processing. In this way, high contextual interference learners might be so consumed by the requirements of the random practice structure that they ultimately learn in a more implicit manner than low contextual interference learners who have more attention capacity available for conscious interpretation and planning of their movements. Clearly this is an opportunity for further research and subsequently provides a range of interesting challenges for coaches seeking to develop the skills of their athletes in the field.

4 THE IMPORTANCE OF DEMONSTRATING THE TRANSFER OF LEARNING

Consistent with motor learning approaches more broadly, the learning benefits of contextual interference have been inferred through the administration of retention and/or transfer testing. Retention tests assess the level of skill performance after a period of no practice on the task/s under the same conditions as encountered during practice. The results of such testing are used to infer how permanent any skill improvement has been. Alternately, transfer testing assesses skill performance in either different performance conditions to those experienced during practice or when the task itself is varied in some way.

Interestingly when transfer testing approaches are used to infer the robustness of the contextual interference effect in the applied setting, mixed findings are apparent. Barreiros et al. (2007) suggested transfer testing did not produce more robust support for contextual interference than retention testing. In contrast, Shewokis and Snow (1997) found moderate to high support. As highlighted by Barreiros et al. (2007) the difference in results may be simply due to a larger sample of research being included in their review as compared to Shewokis and Snow's. Irrespective of the test, it is suggested that skill transfer measures, particularly in the competition setting, are the "gold standard" measure of learning (Pinder et al., 2011; Shewokis and Snow, 1997). Measuring skill during competition allows for the assessment of whether the learning that emerged under specific task and environmental constraints during practice transferred to the competitive environment (Pinder et al., 2011). The use of such measures has been rare in the contextual interference literature. In one exception, Brady (1997) counted the number of strokes golfers accrued in an actual round of golf after practicing in either a blocked or random fashion. While the intent of this approach was good, the lack of experimental control coupled with a very broad metric of skill performance (simply the number of shots) limited the value of the results.

More recently, Cheong et al. (2016) considered the issue of transfer from not only a testing perspective but also during actual practice performance. They considered the importance of environmental factors by comparing more traditional random and blocked practice schedules with a game-based random practice approach. The logic behind this comparison was motivated by the observation that even the applied studies that have examined contextual interference effects have tended to practice and examine sport skills devoid of the game context in which they would be performed. That is, skills are often performed in combination such as dribble and shoot, and in open sports are performed in unpredictable settings where there are demands on the performers' perceptual–cognitive skills. While Cheong et al. (2016) findings were far from conclusive, the representativeness of the practice context and subsequent increased complexity of the skills are critical factors known to influence the generalizability of the contextual interference effect that must continue to be considered in future research.

5 FUTURE RESEARCH DIRECTIONS TO PRODUCE GUIDELINES FOR APPLIED PRACTICE

Clearly the overarching issue considered in this review is the need to shift the focus of contextual interference research from the laboratory to the field. There are clearly a number of challenges in achieving this aim which are selectively considered below.

It is interesting to revisit Newell and McDonald's (1992) reservations about the generalizability of the contextual interference effect as summarized in Brady's (1998) review of the contextual interference literature. A particular concern was the task specificity of the effect and the absence of a systematic application of the effect across all stages of skill expertise (see Brady, 1998, p. 296). Of particular interest to the current review was Newell and McDonald's disagreement with Magill's (1992) assertion that contextual interference could be employed to develop elite sports performance. Direct experimental evidence regarding the benefits of contextual interference when investigating elite or skilled athletes will always be difficult to acquire (see Hall et al., 1994 for an exception). Commonly reported problems when examining elites include the inability to randomize or purposefully split the group into different treatment conditions and limited control of confounding variables such as coach feedback (Sands et al., 2005). Consequently innovative approaches are needed to accurately tease out the impact of a specific practice approach on the skill transfer performance of the skilled athletes in competition.

One possible strategy to achieve this above goal is via the implementation of an observational tool to document the organization of practice coupled with competition-based transfer testing. To date, observations of the practice habits of skilled athletes have revealed several commonalities. Specific to the contextual interference effect, it seems that athletes and coaches more habitually structure practice to feature low contextual interference (e.g., Coughlan et al., 2014). We suspect that this tendency is driven by: (i) the immediacy of visible changes in motor performance that may (or may not) correspond to learning; (ii) practice reflecting the way in which coaches practiced as players (Reid et al., 2007); and (iii) a practical solution to the coaches observations that performers are already sufficiently challenged. Lower contextual interference practice is also thought to be easy to implement and conducive to remedial technical practice or confidence building. However, what is absent from such practice observation is a more detailed understanding of practice organization.

One approach to gaining a finer grained insight into the organization of practice used is to derive a statistical metric of between-skill and within-skill variability by recording the predictability of practice order. We have implemented this approach within a high-performance tennis training environment to classify the degree of practice variability without the need to complete a controlled intervention study. Table 1 describes the shot-by-shot information for a drill focusing on serving in tennis. The data provide information about between-skill and within-skill variability, which is used to inform the amount of contextual interference in practice (see Table 2). These values are visually represented in Fig. 1. Here, the contextual interference of multiple sessions can be plotted, and assessments can be made based on performance feedback.

Performance feedback provides an indication of task difficulty. In this example, performance was measured based on accuracy and movement. *Accuracy* referred to the percentage of serves that landed in the service box. *Movement* referred to a predetermined movement quality, in this case the percentage of serves whereby the player achieved a desired range of knee flexion, as determined by the coach. Performance in practice should be assessed relative to competition standards. This player's serving accuracy in competition was 72%; thus, the task in this practice session was seemingly at an appropriate level of difficulty. If accuracy in practice was significantly more or less than 72%, the coach may be prompted to manipulate task difficulty. Data regarding knee flexion in competition data were not available; however, 35.0% in practice is quite low. It would therefore be appropriate to modify the task in practice so that the desired knee flexion could be achieved on more serves.

Significantly, this metric allows for the objective measurement of contextual interference without the need for deliberate intervention within the training setting. Hence, researchers can assess the efficacy of the contextual interference effect with skilled performers in an applied setting over a substantial period of practice. From a practical perspective, we encourage coaches and athletes to assess between-skill and within-skill variability based on the contextual

Table 1 Shot-by-Shot Information for a Drill Focusing on Serving in Tennis

Shot No.	Shot Type (Between-Skill Variability)	Court Position (Within-Skill Variability)	Serve Direction (Within-Skill Variability)
1	Serve	Deuce	Wide
2	Serve	Deuce	Wide
3	Serve	Deuce	Wide
4	Serve	AD	Body
5	Backhand	—	—
6	Serve	Deuce	Wide
7	Serve	Deuce	Body
8	Backhand	—	—
9	Volley	—	—
10	Serve	AD	T
11	Serve	AD	Wide
12	Backhand	—	—
13	Serve	Deuce	Wide
14	Backhand	—	—
15	Serve	Deuce	T
16	Backhand	—	—
17	Backhand	—	—
18	Serve	AD	Wide
19	Forehand	—	—
20	Forehand	—	—
21	Forehand	—	—
22	Serve	Deuce	Wide
23	Volley	—	—
24	Serve	Deuce	T
25	Backhand	—	—
26	Backhand	—	—
27	Forehand	—	—
28	Serve	AD	Body
29	Forehand	—	—
30	Serve	AD	Wide
31	Backhand	—	—
32	Serve	AD	Body
33	Serve	AD	Body
34	Forehand	—	—

interference that is experienced in competition. In other words, practice should be structured so that it features similar between-skill and within-skill variability as competition. To ensure that practice is of appropriate difficulty for the performer, performance in practice should also be compared to competition

5 Future research directions to produce guidelines

Table 2 Calculating Contextual Interference for the Session Outlined in Table 1

Between-Skill Variability[a]	Within-Skill Variability[a]
$1-|\rho|$	$1-|\rho|$
$1-0.25$	$1-0.07$
0.75[b]	0.93[b]

[a]Contextual interference is considered to be the product of between-skill variability and within-skill variability. Estimates of between-skill and within-skill predictability were obtained from the correlation estimate from a first-order autoregression model $(1-|\rho|)$. This model assumes that the expected skill (drill) at time t is a function of the skill (drill) performed immediately beforehand $(t-1)$. When there is minimal variability in the sequence of skills (drills) performed in a session, the autocorrelation, ρ, in a sequence of skill (drills) will be high. A value of between 0 and 1 is returned, with 1 representing the highest possible variability. The outcome of the between-skill model was an indicator for a service skill activity (1 if serve skill was used, 0 otherwise). For the within-skill model, we subset the analyses to all service skills, and the outcome was a six-category indicator describing the court side (Ad/Deuce) and serve type (Wide/Body/Down the T) used.

[b]These values should be assessed relative competition. Based on 85 men's matches from the 2017 Australian Open, the mean between-skill variability during service games was 0.82, and the mean within-skill variability during service games was 0.90. Thus, the between-skill variability during this practice drill was slightly less than competition, but within-skill variability was very similar to competition.

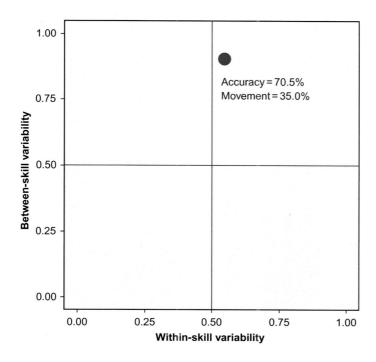

FIG. 1

Visually representing the degree of contextual interference (between-skill and within-skill variability) for the session described in Table 1.

standards. Significantly better or poorer performance is an indication that task difficulty needs to be altered.

The other most pressing issue for the application of contextual interference in sports settings is the issue of progression. The typical recommendation of researchers has been that a progressive approach to the use of contextual interference is warranted (e.g., Landin and Hebert, 1997; Magill and Hall, 1990). In contrast, Russell and Newell (2007) using laboratory-based data, argued that if skill transfer is to occur in an open environment, then random practice can provide an advantage over a progressive combination of blocked then random practice. More recently, Porter and Magill (2010) demonstrated that skill retention and in some instances transfer was superior when sports skills were practiced in a manner that progressively increased the amount of contextual interference. However, the testing conditions used by Porter and Magill could not be considered as an open environment and hence limit their generalizability. Clearly, it would be useful to further verify Russell and Newell's claims in the applied setting.

One alternative practice approach that has received comparatively little investigation relative to contextual interference is the concept of Win Shift Lose Stay (Simon et al., 2008). In this practice approach, a shift to practicing a different task variation or skill is based on an individual's capacity to achieve a predetermined level of success. For example, while practicing basketball shots, if the ball goes in for a "swish" (i.e., does not touch the ring on the way in) the player moves to a different shooting location. However, if the shot is missed or hits the ring before going in, the shot has to be repeated from the same location. While the supporting evidence for this approach is inconclusive (e.g., Simon et al., 2008), the practical benefits of such an approach in determining an appropriate degree of practice progression for a performer warrants further examination. Interestingly, the popular physical education or basketball shooting game of "Round the World" has been played by children for many years and is predicated on exactly the same principles as Win Shift Lose Stay.

6 CONCLUDING REMARKS

In summary, a lingering question concerning the establishment of practical guidelines is whether high contextual interference approaches are somewhat redundant for highly complex skills? In short—no, but we believe that the discussion should focus less on contextual interference per se and more on identifying important task and environmental constraints that are present in the transfer context. In sport, this refers to the competition environment, as the sole purpose of practice is to augment competition performance. We challenge researchers from the neurosciences to apply their methods where reasonably possible in a manner in which the findings can be used to support the development of more informed guidelines regarding the organization of practice for the practitioner.

ACKNOWLEDGMENTS

The authors would like to acknowledge Megan Rendell's and Rich Masters' input into this chapter.

REFERENCES

Albaret, J.M., Thon, B., 1998. Differential effects of task complexity on contextual interference in a drawing task. Acta Psychol. (Amst), 100 (1), 9–24. http://dx.doi.org/10.1016/S0001-6918(98)00022-5.

Barreiros, J., Figueiredo, T., Godinho, M., 2007. The contextual interference effect in applied settings. Eur. Phys. Educ. Rev. 13 (2), 195–208. http://dx.doi.org/10.1177/1356336X07076876.

Brady, F., 1997. Contextual interference and teaching golf skills. Percept. Mot. Skills 84, 347–350. http://dx.doi.org/10.2466/pms.1997.84.1.347.

Brady, F., 1998. A theoretical and empirical review of the contextual interference effect and the learning of motor skills. Quest, 50, 266–293. http://dx.doi.org/10.1080/00336297.1998.10484285.

Brady, F., 2004. Contextual interference: A meta-analytic study. Percept. Mot. Skills 99 (1), 116–126.

Brady, F., 2008. The contextual interference effect and sport skills. Percept. Mot. Skills 106, 461–472. http://dx.doi.org/10.2466/PMS.106.2.461-472.

Cheong, J.P.G., Lay, B., Grove, J.R., Medic, N., Razman, R., 2012. Practicing field hockey skills along the contextual interference continuum: A comparison of five practice schedules. J. Sports Sci. Med. 11 (2), 304–311.

Cheong, J.P., Lay, B., Razman, R., 2016. Investigating the contextual interference effect using combination sports skills in open and closed skill environments. J. Sports Sci. Med. 15 (1), 167–175.

Christina, R.W., Bjork, R.A., 1991. Optimizing long-term retention and transfer. In: Druckman, D., Bjork, R.A. (Eds.), In the Mind's Eye: Enhancing Human Performance. National Academy Press, Washington, DC, pp. 23–56.

Coughlan, E.K., Williams, A.M., McRobert, A.P., Ford, P.R., 2014. How experts practice: A novel test of deliberate practice theory. J. Exp. Psychol. Learn. Mem. Cogn. 40 (2), 449. http://dx.doi.org/10.1037/a0034302.

Cross, E.S., Schmitt, P.J., Grafton, S.T., 2007. Neural substrates of contextual interference during motor learning support a model of active preparation. J. Cogn. Neurosci. 19 (11), 1854–1871. http://dx.doi.org/10.1162/jocn.2007.19.11.1854.

Guadagnoli, M.A., Lee, T.D., 2004. Challenge point: A framework for conceptualising the effects of various practice conditions in motor learning. J. Mot. Behav., 36, 212–224. http://dx.doi.org/10.3200/JMBR.36.2.212-224.

Hall, K.G., Domingues, D.A., Cavasos, R., 1994. Contextual interference effects with skilled baseball players. Percept. Mot. Skills 78, 834–841.

Landin, D., Hebert, E.P., 1997. A comparison of three practice schedules along the contextual interference continuum. Res. Q. Exerc. Sport 68, 357–361.

Lee, T.D., Magill, R.A., 1983. The locus of contextual interference in motor-skill acquisition. J. Exp. Psychol. Learn. Mem. Cogn., 9, 730–746. http://dx.doi.org/10.1037/0278-7393.9.4.730.

Lee, T.D., Magill, R.A., 1985. Can forgetting facilitate skill acquisition? In: Goodman, D., Wilberg, R.B., Franks, I.M. (Eds.), Differing Perspectives in Motor Learning, Memory, and Control. Elsevier, Amsterdam, pp. 3–22.

Li, Y., Wright, D.L., 2000. An assessment of the attention demands during random- and blocked-practice schedules. Q. J. Exp. Psychol., 53 (A)(2), 591–606. http://dx.doi.org/10.1080/713755890.

Lin, C-H., Fisher, B.E., Winstein, C.J., Wu, A.D., Gordon, J., 2008. Contextual interference effect: Elaborative-processing or forgetting-reconstruction? A post-hoc analysis of TMS-induced effects on motor learning. J. Mot. Behav., 40 (6), 578–86. http://dx.doi.org/10.3200/JMBR.40.6.578-586.

Lin, C.H., Fisher, B.E., Wu, A.D., Ko, Y.A., Lee, L.Y., Winstein, C.J., 2009. Neural correlate of the contextual interference effect in motor learning: A kinematic analysis. J. Mot. Behav., 41 (3), 232–242. http://dx.doi.org/10.3200/JMBR.41.3.232-242.

Lin, C.H.J., Knowlton, B.J., Chiang, M.C., Iacoboni, M., Udompholkul, P., Wu, A.D., 2011. Brain–behavior correlates of optimizing learning through interleaved practice. Neuroimage, 56 (3), 1758–1772. http://dx.doi.org/10.1016/j.neuroimage.2011.02.066.

Magill, R.A., 1992. Practice schedule considerations for enhancing human performance in sport. In: Christina, R., Eckert, H. (Eds.), Enhancing Human Performance in Sport: New Concepts and Developments. Human Kinetics, Champaign, IL, pp. 38–50. The American Academy of Physical Education Papers, No. 25.

Magill, R.A., Hall, K.G., 1990. A review of the contextual interference effect in motor skill acquisition. Hum. Mov. Sci., 9, 241-289. http://dx.doi.org/10.1016/0167-9457(90)90005-X.

Masters, R.S.W., 1992. Knowledge, knerves and know-how: The role of explicit versus implicit knowledge in the breakdown of a complex motor skill under pressure. Br. J. Psychol. 83, 343–358. http://dx.doi.org/10.1111/j.2044-8295.1992.tb02446.x.

Masters, R.S.W., Poolton, J.M., 2012. Advances in implicit motor learning. In: Hodges, N.J., Williams, A.M. (Eds.), Skill Acquisition in Sport: Research, Theory and Practice, second ed. Routledge, London, UK, pp. 59–75.

Maxwell, J.P., Masters, R.S.W., Eves, F.F., 2003. The role of working memory in motor learning and performance. Conscious. Cogn., 12, 376-402. http://dx.doi.org/10.1016/S1053-8100(03)00005-9.

Monsell, S., 2003. Task switching. Trends Cogn. Sci., 7 (3), 134-140. http://dx.doi.org/10.1016/S1364-6613(03)00028-7.

Newell, K.M., McDonald, P.V., 1992. Practice: A search for solutions in enhancing human performance in sport. Quest 25, 51–59.

Patterson, J.T., Lee, T.D., 2008. Organizing practice: the interaction of repetition and cognitive effort for skilled performance. In: Farrow, D., Baker, J., MacMahon, C. (Eds.), Developing Sport Expertise: Researchers and Coaches Put Theory Into Practice. Routledge, Abingdon, pp. 119–134.

Pinder, R.A., Davids, K.W., Renshaw, I., Araújo, D., 2011. Representative learning design and functionality of research and practice in sport. J. Sport Exerc. Psychol., 33, 146-55. http://dx.doi.org/10.1123/jsep.33.1.146.

Porter, J.M., Magill, R.A., 2010. Systematically increasing contextual interference is beneficial for learning sport skills. J. Sports Sci., 28 (12), 1277-1285. http://dx.doi.org/10.1080/02640414.2010.502946.

Reid, M., Crespo, M., Lay, B., Berry, J., 2007. Skill acquisition in tennis: Research and current practice. J. Sci. Med. Sport, 10, 1-10. http://dx.doi.org/10.1016/j.jsams.2006.05.011.

Rendell, M.A., Masters, R.S.W., Farrow, D., 2009. The paradoxical role of cognitive effort in contextual interference and implicit motor learning. Int. J. Sport Psychol. 40, 636–647.

Rendell, M.A., Masters, R.S.W., Farrow, D., Morris, T., 2011. An implicit basis for the retention benefits of random practice. J. Mot. Behav. 43, 1–13. http://dx.doi.org/10.1080/00222895.2010.530304.

Russell, D.M., Newell, K.M., 2007. How persistent and general is the contextual interference effect? Res. Q. Exerc. Sport, 78 (4), 318-327. http://dx.doi.org/10.1080/02701367.2007.10599429.

Sands, W.A., McNeal, J.R., Stone, M.H., 2005. Plaudits and pitfalls in studying elite athletes. Percept. Mot. Skills, 100, 22-24. http://dx.doi.org/10.2466/pms.100.1.22-24.

Schmidt, R.A., 1991. Motor Learning and Performance. From Principles to Practice. Human Kinetics, Champaign, IL.

Shea, J.B., Morgan, R.L., 1979. Contextual interference effects on the acquisition, retention, and transfer of a motor skill. J. Exp. Psychol. Hum. Learn. Mem., 5 (2), 179-187. http://dx.doi.org/10.1037/0278-7393.5.2.179.

Shea, J.B., Zimny, S.T., 1983. Context effects in memory and learning movement information. In: Magill, R.A. (Ed.), Memory and Control of Action. North-Holland, Amsterdam, pp. 345–366.

Shewokis, P.A., Snow, J., 1997. Is the contextual interference effect generalizable to non-laboratory tasks? Res. Q. Exerc. Sport 68, A-64.

Simon, D.A., Lee, T.D., Cullen, J.D., 2008. Win-shift, lose-stay: Contingent switching and contextual interference in motor learning. Percept. Mot. Skills 107, 407–418.

Wulf, G., Shea, C.H., 2002. Principles derived from the study of simple skills do not generalize to complex skill learning. Psychon. Bull. Rev. 9 (2), 185–211. http://dx.doi.org/10.3758/BF03196276.

CHAPTER

Sport, time pressure, and cognitive performance

6

Chia N. Chiu*, Chiao-Yun Chen[†], Neil G. Muggleton*,[‡,§,¶,1]

*Institute of Cognitive Neuroscience, National Central University, Jhongli, Taiwan
[†]National Chung Cheng University, Chiayi, Taiwan
[‡]Brain Research Center, College of Health Science and Technology, National Central University, Jhongli, Taiwan
[§]Institute of Cognitive Neuroscience, University College London, London, United Kingdom
[¶]Goldsmiths, University of London, New Cross, London, United Kingdom
[1]Corresponding author: Tel.: +886 03 4227151x65201; Fax: +886 03 4263502,
e-mail address: neil.muggleton@gmail.com

Abstract

Sport participation, fitness, and expertise have been associated with a range of cognitive benefits in a range of populations but both the factors that confer such benefits and the nature of the resulting changes are relatively unclear. Additionally, the interactions between time pressure and cognitive performance for these groups is little studied. Using a flanker task, which measures the ability to selectively process information, and with different time limits for responding, we investigated the differences in performance for participants in (1) an unpredictable, open-skill sport (volleyball), (2) an exercise group engaged in predictable, closed-skill sports (running, swimming), and (3) nonsporting controls. Analysis by means of a drift diffusion analysis of response times was used to characterize the nature of any differences. Volleyball players were more accurate than controls and the exercise group, particularly for shorter time limits for responding, as well as tending to respond more quickly. Drift diffusion model analysis suggested that better performance by the volleyball group was due to factors such as stimulus encoding or motor programming and execution rather than decision making. Trends in the pattern of data seen also suggest less noisy cognitive processing (rather than greater efficiency) and should be further investigated.

Keywords

Sport, Cognition, Time pressure, Drift diffusion model, Flanker task

1 INTRODUCTION

Sporting expertise has been associated with improved cognitive performance in an increasing number of studies (e.g., Yarrow et al., 2009). This seems to be the case when sport or exercise is employed as an "intervention," typically in elderly populations (e.g., Kramer et al., 1999; Langlois et al., 2012 but see also, for example, Komulainen et al., 2010) and, importantly, also when comparing groups who have chosen to engage in a sport or fitness training voluntarily where the individuals are more often younger participants (e.g., Wang et al., 2013a,b). Additionally, higher fitness has been associated with better cognitive performance (Chaddock et al., 2011; Voss et al., 2011) and school performance (Castelli et al., 2007) in children. Despite this, there is uncertainty about the mechanisms underlying the reported effects, either in terms of what exactly about sport or exercise is causing the improvements seen or the neural mechanisms underlying any changes seen. For example, in terms of the aspects of sport or exercise resulting in cognitive modulation, an investigation by Küster et al. (2016) found a stronger effect related to an active lifestyle, rather than due to the training intervention, in a study of older adults. In terms of the nature of the brain-related changes that might account for cognitive changes seen, a study by Chaddock-Heyman et al. (2016) found a relationship between aerobic fitness and hippocampal perfusion in children aged 7–9 years, suggested to reflect better microcirculation and cerebral vasculature and postulated to affect memory task performance.

Here, we aimed to better characterize the cognitive differences, if any, associated with exercise or sporting participation by means of a more detailed analysis of the responses made on a task than is typically performed. While performance on most cognitive tasks is measured using response times, accuracy, or a combination of the two, this can be problematic in terms of either which of the two measures will be affected or there being a contribution of a speed-accuracy trade-off to performance changes (e.g., faster but less accurate or slower but more accurate responses), making results difficult to interpret (see Voss et al., 2013 for a brief overview). One way this can be addressed is through use of analysis based on the drift diffusion model (DDM) (Ratcliff, 1978; Ratcliff and McKoon, 2008; Ratcliff and Rouder, 1998; Ratcliff et al., 1999). This allows greater information to be derived from response time information collected from binary choice tasks and the model of the decision process assumes there is continuous accumulation of information until a decision threshold for one of the two responses in reached (Ratcliff, 1978). Importantly, it is possible to interpret parameters obtained from this analysis in terms of cognitive processes, potentially allowing for better interpretation of the reasons underlying any differences in performance seen. Typically, three of the main measures obtained through diffusion model analysis are: (1) drift rate (the rate at which information is accumulated toward making a decision); (2) decision threshold (essentially indicating how conservative or liberal the criterion for making a decision is); and (3) nondecision time (the duration of nondecision-related components, typically information encoding, and motor processes). A slower response could therefore be due to a lower drift rate, a higher decision threshold, or a longer nondecision time, which, as can be seen from the nature of the measures, could each be a result of different underlying causes. For example, drift rate can be a measure of the speed of cognitive or perceptual processing of stimuli (Schmiedek et al., 2007), threshold separation can be affected by age, with increased age associated with more conservative responding and leading to slower performance (e.g., Ratcliff et al., 2000), and affected by instructions emphasizing speed or accuracy (Voss et al., 2004), and nondecision time can also be higher in older participants (Ratcliff et al., 2000).

We carried out such an analysis of effects on cognitive performance on a flanker task (Eriksen and Eriksen, 1974), which can test the ability to suppress responses to inappropriate information, in conjunction with an investigation of the effects of time pressure on responding. Time pressure can be an important factor affecting decision making and can be manipulated with relative ease in cognitive tasks and, based on the challenge–hindrance framework, is defined as challenge stress which can positively affect performance (LePine et al., 2005). Consequently, this leads to the prediction that both flanker task performance will be better for sporting groups tested and that this would interact with the effects of time pressure, with the sports groups showing less of a cost when the time to make a response was particularly short.

Performance was compared for controls (normal, nonsporting participants), those who took part in regular exercise (running, swimming), and those who performed a skill-based sport (volleyball).

The study aimed to investigate several questions. First, does exercise or sporting participation improve performance on a flanker task, indicative of better ability to suppress inappropriate information? Second, is this affected differently for those involved in closed-skill sports (such as running or swimming), who typically have higher fitness but little skill loading in their sport, and those involved in open-skill sports, where fitness and responding to a dynamic and variable environment is required? Third, is there an effect of a time limit being placed on responding on the task and is this affected by sport/exercise? Finally, can appropriate analysis of performance data offer insight into the mechanisms underlying any performance differences seen. This led to the following specific hypotheses: (1) sporting participation would be associated with improved task performance, increased fitness alone would result in an intermediate improvement and participation in an open-skill sport would result in the largest improvement; (2) time pressure would be associated with increased task difficulty, hence expected to be associated with worse accuracy but reduced response times. Reduced accuracy was less expected for the sporting groups (and again, with intermediate performance for the fit group), hence differences in performance for the groups were expected to be largest for the shortest time limit; and (3) improved performance in the sporting group would be due to (a) a reduced nondecision time (i.e., better motor responding or information encoding) and/or (b) an increased drift rate (due to, for example, more rapid cognitive processing) and this would again be clearer for the short time limit than the longer time limit.

2 METHODS
2.1 STUDY DESIGN
A between-subjects design was used with repeated measures of flanker task accuracy and response times (for each of three different time limits for making a response) as within-subject factors. Response times were used in diffusion model analysis to allow assessment of further relevant within-subject parameters. Each participant performed blocks of the flanker task (detailed later) with a fixed response time limit (of which they were informed) for each block, with block orders (and hence the order of presentation of the time limits) being randomized across participants.

2.2 ETHICAL APPROVAL AND CONSENT
All participants provided informed consent prior to taking part in the study. The study was approved by the local ethics committee.

2.3 PARTICIPANTS
Thirty-one participants were recruited from National Central University, Taiwan. Twelve of these were in the exercise group, who regularly engaged in running or swimming, 11 were members of the university volleyball teams, and 8 were controls. For each, height, weight, and gender were recorded, and fitness was evaluated using a standard measure, the Progressive Aerobic Cardiovascular Endurance Run (PACER) test (Léger and Lambert, 1982), to give the VO2Max fitness measure (see below for details). Participant attributes are summarized in Table 1.

Table 1 Characteristics of the Participants

Group	Control	Exercise	Volleyball
Count	8	12	11
Gender	3 male, 5 female	6 male, 6 female	6 male, 5 female
Age (years)	21.75 ± 0.70	21.50 ± 0.58	23.36 ± 0.53
VO2Max (mL kg^{-1} min^{-1})	29.07 ± 0.51	35.69 ± 0.78	37.58 ± 1.25
Height (cm)	164.13 ± 3.37	164.67 ± 2.14	167.86 ± 2.23
Weight (kg)	61.50 ± 5.92	57.67 ± 2.79	62.73 ± 2.88
BMI (kg m^{-2})	22.53 ± 1.50	21.31 ± 0.54	22.25 ± 0.58

All values are means ± standard error of the mean.

2.4 TEST PROCEDURES
2.4.1 PACER test
This test requires participants to run a 20 m distance repeatedly, with a time limit for each 20 m repeat. Following a START instruction, participants were required to run over the 20 m distance, aiming to do so before a BEEP sounded. This beep indicated that the participant must change direction and run the 20 m back to their starting point. At the end of each minute a TRIPLE BEEP (rather than a single beep) was used as the signal, indicating both a change of direction and that the required pace was increased (i.e., the time between beeps was shortened). Failure to cover the distance resulted in a "miss" and two misses meant the end-point of the test had been reached for an individual. The number of times the distance was covered before failure allowed estimation of the VO2Max of the participant. This is a measure of maximal oxygen uptake and is an indicator of aerobic physical fitness.

2.4.2 Flanker task
The task is illustrated in Fig. 1.

Stimuli were presented on a monitor at a distance of 57 cm from the participant in a quiet, darkened room. Following presentation of a fixation "+" symbol for 500 ms, a flanker stimulus was presented. The stimuli used were composed of five "<" and ">" symbols, with the participant required to indicate the direction of the center (i.e., the third) symbol. Four arrangements were used: <<<<<; >>>>>; <<><<; and >><>>. Each of these occurred with equal frequency, such that there were an equal number of trials with the "<" and ">" targets and an equal number with congruent and incongruent flanking elements. Participants were required to make a button press indicating the direction of the target element, using their first finger to indicate "left" and their second finger to indicate "right," as quickly and accurately

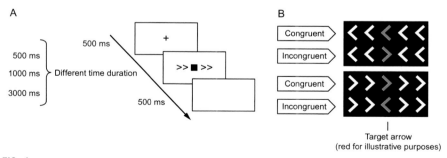

FIG. 1

The flanker task. (A) The timeline of a typical trial, with a 500 ms fixation followed by presentation of the stimulus for 500, 1000, or 3000 ms and a 500 ms intertrial interval. The *square* indicates the location of the element to which a response (indicating the direction it was facing) was required. (B) The four possible stimulus arrangements with the target element highlighted in *red*.

as possible. In different blocks of trials the time limits for responding were 500, 1000, and 3000 ms, with participants informed prior to starting a block what the time limit was. Presentation of the stimulus was terminated should the time limit be reached and responses that were too slow were treated as errors with a "Too slow" message displayed on the screen. A 500 ms blank screen followed each trial. One hundred and sixty trials were presented for each time limit and the order that these were presented in was randomized across participants.

2.5 DATA PROCESSING AND ANALYSIS

Response times and response accuracy were collected for all trials. Task accuracy was initially analyzed using a multivariate mixed-design analysis of variance with within-subject factors of TIME allowed to make a response (500, 1000, or 3000 ms) and CONGRUENCY of the trial (congruent or incongruent). A between-subjects factor was used for the GROUP to which participants belonged (control, exercise, or volleyball). The same approach was also used to analyze response times and each of the parameters obtained from the DDM analysis.

Parameters representing the output of the drift diffusion analysis were obtained using the response times for the trials in conjunction with the fast diffusion model (Voss and Voss, 2007, 2008) and employing software made available for such analysis (available from: http://www.psychologie.uni-heidelberg.de/ae/meth/fast-dm/index.html) following labeling of responses in terms of both trial type (congruent or incongruent) and time limit (500, 1000, or 3000 ms). This allowed for analysis of the effects of the group to which participants belonged on behavior in conjunction with analysis of the effects of time limits on performance.

3 RESULTS
3.1 PARTICIPANT PROFILES

Univariate analysis of variance showed a significant effect of GROUP for VO2Max scores; $F(2) = 9.060$, $P = 0.001$. Post hoc Bonferroni-corrected comparisons showed the control group scores for this measure were significantly lower than both the exercise group (mean difference $= 6.6188$, standard error $= 2.02549$, $P = 0.009$) and the volleyball group (mean difference $= 8.5156$, standard error $= 2.06199$, $P = 0.001$). There was no significant difference between the exercise and the volleyball groups (mean difference $= 1.8968$, standard error $= 1.85237$, $P = 0.944$). Other parameters did not differ significantly between groups.

3.2 ACCURACY AND RESPONSE TIMES
3.2.1 Accuracy

Data illustrating overall accuracy for each group is shown in Fig. 2A and for each time limit, as well as for congruent and incongruent trials, in Fig. 3A.

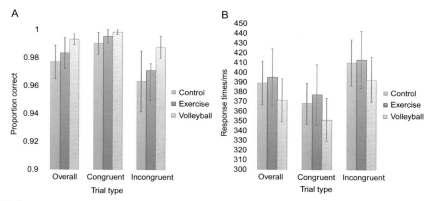

FIG. 2

(A) Overall accuracy data for each group for all trials as well as for congruent and incongruent trials separately and (B) overall response times for each group. *Error bars* represent standard deviations.

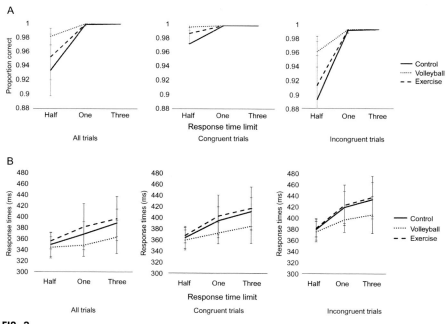

FIG. 3

(A) Accuracy and (B) response time data for all trials, congruent trials, and incongruent trials for each time limit for responding for the three groups tested.

For CONGRUENCY, a significant main effect was seen ($F(1) = 36.572$, $P < 0.001$). Congruent and incongruent trials accuracies were compared using a paired t-test, which showed significantly better performance on congruent than incongruent trials ($t = 5.752$, $P < 0.001$).

There was a significant main effect of GROUP ($F(2) = 7.337$, $P = 0.003$). This was analyzed using a univariate ANOVA with a factor of group and post hoc comparisons of means with Bonferroni correction. This showed that there was no significant difference between the control and exercise groups (mean difference $= 0.064$, standard error $= 0.0043$, $P = 0.440$) but the volleyball group performed significantly better than controls (mean difference $= 0.0163$, standard error $= 0.00438$, $P = 0.003$). The comparison of the volleyball and exercise groups approached significance (mean difference $= 0.0099$, standard error $= 0.00393$, $P = 0.053$).

There was a significant main effect of TIME ($F(2) = 73.956$, $P < 0.001$). This effect was investigated using (corrected) paired sample t-tests. These showed significantly worse performance for the 500 ms condition compared to the 1000 ms condition ($t = 7.013$, $P < 0.001$) and compared to the 3000 ms condition ($t = 7.008$, $P < 0.001$), but no difference between the 1000 and 3000 ms conditions ($t = 1.438$, $P = 0.161$).

In terms of interactions, a significant TIME × GROUP interaction was seen ($F(4) = 7.166$, $P < 0.001$). Analysis for each time restriction showed a significant difference between groups for the 500 ms condition ($F(2) = 7.245$, $P = 0.003$), with significantly better performance for the volleyball group compared to the control group (mean difference $= 0.0482$, standard error $= 0.01391$, $P = 0.003$) and better performance approaching significance for the volleyball group compared to the exercise group (mean difference $= 0.0292$, standard error $= 0.01169$, $P = 0.056$) (comparisons all Bonferroni corrected). No significant differences were seen for any comparisons for the 1000 ms or the 3000 ms conditions.

For the TIME × CONGRUENCY interaction a significant effect was seen ($F(2) = 35.6$, $P < 0.001$), with better accuracy for congruent than incongruent trials. This was significant for the 500 ms condition ($t = 5.731$, $P < 0.001$), but there was no significant difference for the 1000 ms condition ($t = 1.438$, $P = 0.161$) or the 3000 ms condition (where accuracy was 100% for congruent and incongruent trials for all participants).

Neither the CONGRUENCY × GROUP ($F(2) = 2.1222$, $P = 0.122$) nor the TIME × CONGRUENCY × GROUP interaction ($F(4) = 2.110$, $P = 0.092$) were significant.

3.2.2 Response times

Overall response time data are shown in Fig. 2B and for congruency and time limits in Fig. 3B.

There was no main effect of GROUP, although this did approach significance ($F(2) = 2.606$, $P = 0.092$), with fastest performance for the volleyball group.

There was a main effect of TIME ($F(2) = 71.133$, $P < 0.001$). Analysis showed that this was a consequence of fastest performance for the 500 ms condition, intermediate performance for the 1000 ms condition, and slowest performance for the 3000 ms condition, with all comparisons being significant (500 vs 1000: $t = -8.088$, $P < 0.001$; 500 vs 3000: $t = -9.044$, $P < 0.001$; 1000 vs 3000: $t = 5.843$, $P < 0.001$).

There was a main effect of CONGRUENCY ($F(1) = 377.075$, $P < 0.001$), with congruent trials being performed with lower response times than incongruent trials ($t = 19.663$, $P < 0.001$).

For the significant TIME × GROUP interaction ($F(4) = 3.108$, $P = 0.022$), there were no significant differences between groups for the 500 ms condition but the difference between the volleyball and the exercise groups approached significance for the 1000 ms (mean difference = 29.6820, standard error = 12.22641, $P = 0.066$) and 3000 ms conditions (mean difference = 32.0596, standard error = 13.92432, $P = 0.087$), with faster performance in the volleyball group in both cases.

For the TIME × CONGRUENCY interaction ($F(2) = 22.905$, $P < 0.001$), the congruent trials were faster than the incongruent trials for each time limit (500 ms: $t = 14.916$, $P < 0.001$; 1000 ms: $t = 15.161$, $P < 0.001$; 3000 ms: $t = 17.393$, $P < 0.001$). Similarly, for both the congruent and the incongruent trials, responses were fastest in the 500 ms condition, intermediate in the 1000 ms condition, and slowest in the 3000 ms condition (all $P < 0.001$). Hence, this interaction seems likely to reflect the sizes of the differences being larger for the incongruent trials than for the congruent trials for the different time conditions.

3.2.3 Accuracy and response time summary

Accuracy was worse for the shortest time limit, better for congruent trials than incongruent trials, and better at the shortest time limit for the volleyball group.

Congruent trials were performed faster than incongruent trials. Trials with a shorter time limit were performed faster than those with a longer time limit. There was a tendency for the volleyball group to respond faster, in particular for the 1000 and 3000 ms limit conditions.

3.3 DRIFT DIFFUSION ANALYSIS

3.3.1 Threshold separation (A)

Threshold separations (A) for each group and for different time limits, as well as for congruent and incongruent trials, are shown in Fig. 4.

There was a main effect of TIME ($F(2) = 16.430$, $P < 0.001$), reflecting a lower threshold for the 500 ms conditions, intermediate for the 1000 ms condition, and highest (i.e., most conservative) for the 3000 ms condition. For both congruent and incongruent trials, this was significant for the 500 ms vs the 3000 ms conditions ($t = 2.806$, $P = 0.009$ and $t = 6.263$, $P < 0.001$, respectively) and the 500 and 1000 ms conditions also differed for the incongruent trials ($t = 5.014$, $P < 0.001$).

The main effect of GROUP approached significance ($F(2) = 3.028$, $P = 0.064$), with post hoc analysis suggesting that this was due to a difference between the

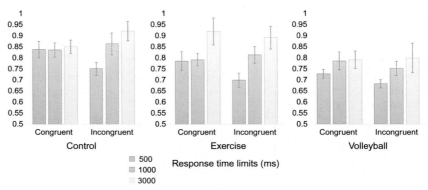

FIG. 4

Threshold separations calculated for each group for congruent and incongruent trials as well as for each time limit for responding for the three groups tested.

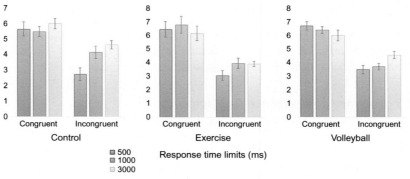

FIG. 5

Drift rates for each group for congruent and incongruent trials as well as for each time limit for responding for the three groups tested.

volleyball group and the control group, which also approached significance (mean difference $= 0.0875$, standard error $= 0.03839$, $P = 0.076$).

There was a significant interaction between CONGRUENCY and TIME ($F(2) = 4.639$, $P = 0.014$). This was found to reflect a significant difference between threshold separation for congruent and incongruent trials for the 500 ms condition ($t = 3.377$, $P = 0.002$), with a higher separation for congruent trials, but no difference for either the 1000 ms or 3000 ms conditions.

3.3.2 Drift rate (V)

The drift rates obtained are illustrated in Fig. 5.

There was a main effect of CONGRUENCY ($F(1) = 134.877$, $P < 0.001$), with a higher drift rate for congruent than incongruent trials ($t = 13.330$, $P < 0.001$).

There was also a significant main effect of TIME ($F(2) = 3.285$, $P = 0.045$), as well as a CONGRUENCY × TIME interaction ($F(2) = 16.257$, $P < 0.001$). This was found to reflect significant differences between drift rates for 500 ms vs 1000 ms, 500 ms vs 3000 ms, and 1000 ms vs 3000 ms conditions for incongruent trials only ($t = 3.701$, $P = 0.001$; $t = 6.227$, $P < 0.001$; $t = 2.286$, $P = 0.029$, respectively). For these conditions, the drift rate was lowest for the 500 ms condition, intermediate for the 1000 ms condition, and highest for the 3000 ms condition.

3.3.3 Nondecision time (T)

Nondecision times for each group and time limit are shown in Fig. 6.

There was a main effect of CONGRUENCY ($F(1) = 134.877$, $P < 0.001$), with a significantly higher nondecision time for incongruent than congruent trials ($t = 10.642$, $P < 0.001$).

There was a significant effect of TIME ($F(2) = 24.512$, $P < 0.001$) with the nondecision time lowest for the 500 ms condition, intermediate for the 1000 ms condition, and longest for the 3000 ms condition.

Finally, there was a significant TIME × GROUP interaction ($F(4) = 3.297$, $P = 0.017$). For the control group a one-way ANOVA showed a significant effect of TIME ($F(2) = 4.840$, $P = 0.019$) with post hoc tests showing a significant difference for the 500 ms condition compared to the 3000 ms condition only (mean difference = 0.333, standard error = 0.0107, $P = 0.016$). For the exercise group, there was no significant effect of TIME, although it approached significance ($F(2) = 2.552$, $P = 0.093$), and for the volleyball group, there was no significant effect of TIME ($F(2) = 1.469$, $P = 0.246$).

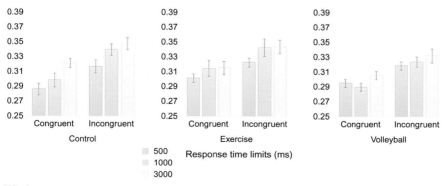

FIG. 6

Nondecision times for each group for congruent and incongruent trials as well as for each time limit for responding for the three groups tested.

3.3.4 Drift diffusion parameter summary

Analysis of the components of the DDM showed no significant differences in drift rates. There was a trend toward a difference in threshold separation, with this measure being lower for the volleyball group (usually indicative of a less conservative criterion). There was a significant interaction for nondecision time, with this being equivalent for groups for the 500 ms condition, but increasing significantly with time for the control group, approaching such a pattern for the exercise group, but with no significant effect of time for the volleyball group.

4 DISCUSSION

As predicted, sporting participation, and more specifically, playing volleyball, was associated with better performance on the flanker task, primarily in terms of accuracy on the task but also with a trend toward faster responding. Additionally, and also as predicted, time pressure was associated with reduced accuracy on the task. Response times also showed the expected reduction with the shorter time limits. The pattern of the effects on accuracy for the sporting groups were also broadly as predicted, with more accurate performance for the volleyball group for the shortest time limit compared to controls, with seemingly intermediate performance for the exercise group. The response time data differed slightly from predictions, with a trend toward faster responding for the volleyball group for the 1000 ms (intermediate) and 3000 ms (longest time limit) conditions. This may be a result of the high accuracy levels seen for these two time limits such that any improvement could only be seen in response times.

Thus, these data are broadly consistent with the sort of pattern seen in a number of previous studies, with involvement in a sport requiring responses in an unpredictable, dynamically changing and externally paced environment (open-skill sports such as tennis, football, and basketball) producing the largest effects on performance and involvement in sports with a more predictable environment (closed-skill sports such as running or swimming) producing an intermediate level of benefit (see Wang et al., 2013b for an example). It is noteworthy though, that the benefits of sporting performance seen here were general with respect to the flanker task. In other words, improvement was not seen to interact with congruency of the trials and so suggesting that such improvement was not necessarily related to specific cognitive aspects of task performance, such as being less affected by incongruent distractor elements, a point of particular relevance when considering the outcomes of the DDM analysis.

It was hypothesized that improved performance in the sporting group would possibly be due to a reduced nondecision time and/or an increased drift rate, with the former potentially being due to more rapid information processing or faster motor preparation with the latter potentially indicative of more rapid cognitive processing. Interestingly, the pattern of data obtained showed no significant effects on drift rate as a result of being involved in sporting participation but there was a significant change in nondecision time, with this increasing with the increased time allowed for responding for the control group, not changing significantly for the volleyball group, and (again) seemingly having an intermediate pattern for the exercise group. In addition, there was a trend toward a reduced threshold separation for the volleyball group (usually indicative of a *less* conservative criterion).

This is particularly interesting in the context of possible changes in neural efficiency in athletes (Del Percio et al., 2008; Nakata et al., 2010; Yarrow et al., 2009). Del Percio et al. (2008) investigated this idea in fencing athletes and karate athletes compared to controls and suggested their results did not fully fit with such a neural efficiency hypothesis. Furthermore, a study by Wang et al. (2015) using electroencephalographic (EEG) recording and a visuospatial working memory task found a pattern of results where badminton players were faster at the task with no reduction in accuracy. Interestingly the EEG analysis suggested that this might be due to better response preparation or better encoding and retrieval of information, rather than an improvement in neural efficiency.

The data here seem to be consistent with the pattern seen by Wang et al. (2015) with the differences in performance seemingly not indicative of a change in efficiency of cortical processing, but either related to more efficient stimulus encoding and/or more rapid response generation. In other words, while there was better accuracy in some conditions for the volleyball group in particular, this does not necessarily seem to be due to better cognitive processing.

One caveat on this observation is the trend toward a reduced threshold separation seen for the volleyball group compared to the control group. The most relevant conclusion for this is that further investigation would be sensible as it seems that a reduced threshold separation, which would usually be expected to be associated with *worse* performance, here is associated with better performance. One possibility, should this be a reliable effect, is that the drift process toward making a response on the task is essentially "cleaner" (for example, less variable or noisy), meaning individuals showing this pattern of responding can maintain good (or improved) performance by having a reduced threshold separation but no associated increase in error rates.

5 CONCLUSIONS

Performance on a flanker task showed the typical pattern of reduced accuracy and slower responding for trials where the target was accompanied by incongruent flankers. Increased time pressure for responding generally caused responses to be made more quickly but with an increase in error rates. Of the three groups tested, a volleyball group, an exercise group, and a control group, performance on the task was generally best for the volleyball group who also performed better with shorter time limits for responding. The data obtained is also consistent with performance being intermediate between the control and volleyball groups for the exercise group. More detailed analysis of responding, employing a DDM, suggests the volleyball group superiority was a consequence of more rapid motor responding or stimulus encoding, rather than more rapid drift toward the decision being made.

ACKNOWLEDGMENTS
This work was funded by the Ministry of Science and Technology, Taiwan (grant numbers: 104-2420-H-008-001-MY2, 105-2410-H-008-023-MY2, and 104-2410-H-194-034-MY2).

REFERENCES

Castelli, D.M., Hillman, C.H., Buck, S.M., Erwin, H.E., 2007. Physical fitness and academic achievement in third- and fifth-grade students. J. Sport Exerc. Psychol. 29, 239–252.

Chaddock, L., Hillman, C.H., Buck, S.M., Cohen, N.J., 2011. Aerobic fitness and executive control of relational memory in preadolescent children. Med. Sci. Sports Exerc. 43, 344–349.

Chaddock-Heyman, L., Erickson, K.I., Chappell, M.A., Johnson, C.L., Kienzler, C., Knecht, A., Drollette, E.S., Raine, L.B., Scudder, M.R., Kao, S.-C., Hillman, C.H., Kramer, A.F., 2016. Aerobic fitness is associated with greater hippocampal cerebral blood flow in children. Dev. Cogn. Neurosci. 20, 52–58.

Del Percio, C., Rossini, P.M., Marzano, N., Iacoboni, M., Infarinato, F., Aschieri, P., et al., 2008. Is there a "neural efficiency" in athletes? A high-resolution EEG study. Neuroimage 42, 1544–1553.

Eriksen, B.A., Eriksen, C.W., 1974. Effects of noise letters upon the identification of a target letter in a nonsearch task. Atten. Percept. Psychophys. 16 (1), 143–149.

Komulainen, P., Kivipelto, M., Lakka, T.A., Savonen, K., Hassinen, M., Kiviniemi, V., et al., 2010. Exercise, fitness and cognition—a randomised controlled trial in older individuals: the DR's EXTRA study. Eur. Geriatr. Med. 1 (5), 266–272.

Kramer, A., Hahn, S., Cohen, N., Banich, M., McAuley, E., Harriosn, C., et al., 1999. Ageing, fitness and neurocognitive function. Nature 400, 418–419.

Küster, O.C., Fissler, P., Laptinskaya, D., Thurm, F., Scharpf, A., Woll, A., Kolassa, S., Kramer, A.F., Elbert, T., von Arnim, C.A.F., Kolassa, I.-T., 2016. Cognitive change is more positively associated with an active lifestyle than with training interventions in older adults at risk of dementia: a controlled interventional clinical trial. BMC Psychiatry 16, 315.

Langlois, F., Vu, T.T.M., Chassé, K., Dupuis, G., Kergoat, M.-J., Bherer, L., 2012. Benefits of physical exercise training on cognition and quality of life in frail older adults. J. Gerontol. B Psychol. Sci. Soc. Sci. 68 (3), 400–404.

Léger, L.A., Lambert, J., 1982. A maximal multistage 20-m shuttle run test to predict VO_2 max. Eur. J. Appl. Physiol. Occup. Physiol. 49 (1), 1–12.

LePine, J.A., Podsakoff, N.P., LePine, M.A., 2005. A meta-analytic test of the challenge stressor–hindrance stressor framework: an explanation for inconsistent relationships among stressors and performance. Acad. Manage. J. 48 (5), 764–775.

Nakata, H., Yoshie, M., Miura, A., Kudo, K., 2010. Characteristics of the athletes' brain: evidence from neurophysiology and neuroimaging. Brain Res. Rev. 62, 197–211.

Ratcliff, R., 1978. A theory of memory retrieval. Psychol. Rev. 85, 59–108.

Ratcliff, R., McKoon, G., 2008. The diffusion decision model: theory and data for two-choice decision tasks. Neural Comput. 20, 873–922.

Ratcliff, R., Rouder, J.N., 1998. Modeling response times for two-choice decisions. Psychol. Sci. 9, 347–356.

Ratcliff, R., Van Zandt, T., McKoon, G., 1999. Connectionist and diffusion models of reaction time. Psychol. Rev. 106, 261–300.

Ratcliff, R., Spieler, D., McKoon, G., 2000. Explicitly modeling the effects of aging on response time. Psychon. Bull. Rev. 7, 1–25.

Schmiedek, F., Oberauer, K., Wilhelm, O., Süß, H.-M., Wittmann, W.W., 2007. Individual differences in components of reaction time distributions and their relations to working memory and intelligence. J. Exp. Psychol. Gen. 136, 414–429.

Voss, A., Voss, J., 2007. Fast-dm: A free program for efficient diffusion model analysis. Behav. Res. Meth. 39, 767–775.

Voss, A., Voss, J., 2008. A fast numerical algorithm for the estimation of diffusion-model parameters. J. Math. Psychol. 52, 1–9.

Voss, A., Rothermund, K., Voss, J., 2004. Interpreting the parameters of the diffusion model: an empirical validation. Mem. Cognit. 32, 1206–1220.

Voss, M.W., Chaddock, L., Kim, J.S., VanPatter, M., Pontifex, M.B., Raine, L.B., Kramer, A.F., 2011. Aerobic fitness is associated with greater efficiency of the network underlying cognitive control in preadolescent children. Neuroscience 199, 166–176.

Voss, A., Nagler, M., Lerche, V., 2013. Diffusion models in experimental psychology—a practical introduction. Exp. Psychol. 60 (6), 385–402.

Wang, C.H., Chang, C.C., Liang, Y.M., Chiu, W.S., Tseng, P., Hung, D.L., et al., 2013a. Open vs. closed sports and the modulation of inhibitory control. PLoS One 8, e55773.

Wang, C.H., Chang, C.C., Liang, Y.M., Shih, C.M., Muggleton, N.G., Juan, C.H., 2013b. Temporal preparation in athletes: a comparison of tennis players and swimmers with sedentary controls. J. Mot. Behav. 45, 55–63.

Wang, C.-H., Tsai, C.L., Tu, K.-C., Muggleton, N.G., Juan, C.-H., Liang, W.-K., 2015. Modulation of brain oscillations during fundamental visuo-spatial processing: a comparison between female collegiate badminton players and sedentary controls. Psychol. Sport Exerc. 16, 121–129.

Yarrow, K., Brown, P., Krakauer, J.W., 2009. Inside the brain of an elite athlete: the neural processes that support high achievement in sports. Nat. Rev. Neurosci. 10, 585–596.

CHAPTER 7

Effectiveness of above real-time training on decision-making in elite football: A dose–response investigation

Javid J. Farahani[1], Amir H. Javadi, Barry V. O'Neill, Vincent Walsh

Institute of Cognitive Neuroscience, University College London, London, United Kingdom
[1]*Corresponding author: Tel.: 20-7679-1177, e-mail address: javid@londoncognition.com*

Abstract

We examined the effects of video-based training in elite footballers' decision-making by presenting videos with training and testing scenarios at above real-time speeds. We also examined different training protocols to establish how much training is beneficial. We found that above real-time training improved accuracy and response time in football decision-making. In terms of scheduling, we found that the benefits were short lasting and did not last beyond 2 weeks.

Keywords

Decision-making, Sport expertise, Above real-time training, Football

1 INTRODUCTION

An understanding of expert sports performance has eluded fans, media, coaches, athletes, and scientists alike, and each of these groups is in agreement that the others do not understand. The popularity of the Olympic Games, FIFA World Cup, Wimbledon, and other major sports events is driven, in part, by the quality of expert performance in action as well as the intrinsic appeal of competition (Janelle and Hillman, 2003).

Many individuals around the world yearn to reach elite levels in different sports, but few succeed in doing so. As such, an important question arises: how do elite athletes differ from nonexperts in their performance and learning? In order to answer this question, significant research has been carried out throughout the last three decades, and expertise research has now become a well-established domain in sports science and cognitive psychology (e.g., Arroyo-Figueroa et al., 2006; Ericsson, 1996; Starkes and Ericsson, 2003).

During the process of investigating and understanding how elite performers function in a given domain, Ericsson and Smith (1991) proposed the expert performance approach (EPA) as a theoretical framework with three stages for the study of expertise. The first stage in this approach is to assess expert performance through laboratory or field testing. The expert–novice paradigm is used in the first stage to identify expert performers where field or laboratory testing is employed to elicit the differences in expert and novice performance. The second stage is to use process-tracing methods (e.g., eye movement recordings, film occlusion, and verbal reports) in the design of a representative task to investigate the mechanisms that impact the expert performance. At this stage, the goal is to understand how experts perform better than novices in the specific skill. The final stage of the EPA is to examine the acquisition of the identified characteristics of expertise. Retrospective training history profiling and training interventions can come into play at this stage.

Sport science researchers have been using the EPA to understand expertise within the sport with a significant amount of literature generated. Results to date show that elite athletes benefit from superior perceptual–cognitive skills, such as anticipation (Abernethy, 1994; Muller et al., 2006; Williams et al., 2003), pattern recall and recognition (Abernethy et al., 2005; Baker et al., 2003; Gilis et al., 2008), and decision-making (Abernethy, 1996; Lorains et al., 2013a,b; Starkes and Lindley, 1994). A comprehensive review on the expert–novice paradigm shows that the superior cognitive knowledge enables elite players to extract the most meaningful information from the environment, effectively committing it to memory, and when a player needs to perform a specific skill, this information can be retrieved to facilitate performance in similar scenarios (Williams and Davids, 1998).

In a very early attempt to study expertise, Fitts and Posner (1967) noted that elite athletes perform with a higher level of automaticity than novices. The concept of automaticity is not only applied in physical skills but has also been used to investigate cognitive skills in elite performance (e.g., Beilock et al., 2002, 2004). Speeded tasks may lead to processing efficiency as the nature of the task urges elite athletes to perform at a higher level of automaticity (Lorains et al., 2013a,b). The use of speeded manipulations has previously been applied in elite sports studies. For instance, in a study on elite handball athletes, Johnson and Raab (2003) suggested that time pressure may increase the level of accuracy in decision-making, rather than damaging it. In their study, participants were asked to make the decision as fast as possible, and they found that accuracy is higher when athletes chose the first option they generated. These results suggest that time pressure may force athletes to perform more automatically which is of benefit to the quality of their decisions. In another attempt, Hepler and Feltz (2012) found that when university basketball players perform more instinctively and with a higher level of automaticity by choosing the first option they generate, the quality of their decisions improves.

Given the results mentioned earlier there are limited intervention studies that have sought to apply this information in order to improve decision-making skills in situational or strategic sports such as football (Gabbett et al., 2007; Lorains et al., 2013a,b; Milazzo and Fournier, 2015; Schweizer et al., 2011).

Early studies in this area include those by Thiffault (1974, 1980) which were some of the first decision-making training interventions to enhance athletes' level of performance. In these studies, decision-making skills among elite ice hockey players were assessed using pictures. Athletes were asked where they would place the ball if they were in possession. This method proved to be effective in comparing decision-making behavior between elite and novice athletes, but it was far from creating a training environment that made the athletes feel like they were making decisions in a real-match situation.

As technology advanced, using videos within sport became more common in the 1990s, and studies began employing video stimuli alongside static images. In one such study, Starkes and Lindley (1994) pioneers in designing video-based decision-making tasks, trained basketball players through both static images and videos. More significant performance improvements were observed in video-based training methods than by using static images.

According to the authors, the major benefits of video-based training revolve around the practical applications in elite sport—coaches can control some scenarios for specific needs, involve injured players, and avoid increasing the physical load on athletes—claims that are plausible but yet to be empirically demonstrated. Although this study demonstrated promise, Araujo et al. (2006) and Farrow and Abernethy (2003) discussed the lack of "life-like" environments for making decisions. In order to create a more life-like environment for decision-making in sports, Hays and Singer (1989) discussed the importance of fidelity. Fidelity could be used to describe physical or psychological tasks, referring to the similarity of simulation task to the real-world one. It can also be used to describe how participants perceive the simulation environment compared to the life-like one (Stoffregen et al., 2003) and has been investigated in different domains outside of the sports literature. For example, Olmos et al. (2012) discussed the importance of fidelity in improving the learning level among musicians via computer-based simulations. In another study by Kolf (1973) in aviation, results showed that training in a faster environment than real time is perceived as the real experience by pilots. This study was the first reported use of "above real-time training" (ARTT) in aviation as cited in Guckenberger et al. (1993), but there are no full documents available for the study.

ARTT refers to a training paradigm that places participants in a simulated environment that functions at faster-than-normal time, i.e., a video played at faster-than-normal speed. Schneider (1985) discussed the importance of automaticity in expertise and claimed that "critical high-performance skills that are practiced at least in part in an above real time environment could lead to a faster acquisition of automaticity patterns of performance, less opportunity for memory decay, and a sustained level of motivation during training." Later on, Guckenberger et al. (1993) conducted a study using ARTT and in their report, they discussed two interconnected experiments. In the first study, 25 novice male subjects performed three tank gunnery tasks on a table-top simulator under varying levels of time acceleration (i.e., 1.0× (normal speed), 1.6× (1.6 times normal speed), 2.0× (twice normal speed), sequential, and mixed). They were then transferred to a standard 1.0× condition for testing. Every accelerated condition or combination of conditions produced better training and transfer than the standard real-time or 1.0× condition. They also found that the most effective method for presentation of stimuli was the presentation of trials at 1.0×, 1.6×, and 2.0× in a random order during training. Overall, the best ARTT group accomplished a score approximately 50% higher and trained in 25% less time compared to the real-time control group.

While the use of ARTT in training high skills performance in aviation has shown promise and despite the potential benefit in athletes, only one study has investigated the impact of this training method in athletic settings. Lorains et al. (2013a,b) conducted the first reported study in a sporting context in Australian Rules Football by investigating the effects of speeded video (between 0.75, 1.0, 1.25, 1.5, 1.75, and 2.0) on decision-making performance of elite, semi-elite, and novice participants. The group reported that elite athletes made more accurate decisions under faster video speeds, and participants rated speeds of 1.25 and 1.5 as most gamelike. This effect is thought to be driven by a more automatic decision-making process under increased time constraints in experts compared to novices (Lorains et al., 2013a,b).

Lorains et al. (2013a,b) elaborated on this original investigation by looking at training decision-making skills in Australian Rules footballers, using the ARTT method. Specifically, pre- and postmeasures were employed following 5 weeks of video-based ARTT among 45 elite Australian Rules football players who were divided into three randomly assigned groups. There were 16 players in the fast-speed training group, 15 in the normal speed one, and 14 in the control group. Additionally, two retention tests were carried out at 2 and 10 weeks posttraining in order to assess the level of memory decay of the decision-making skills. Additionally, and arguably most important in the world of elite sport, a transfer test of decision-making was included in an effort to evaluate the impact of video-based training on match performance.

The authors reported no noteworthy difference during the pretest compared with other groups who had achieved a similar level of accuracy. However, in the posttest conducted after 5 weeks training and the retention test (only 2 weeks and not 10 weeks after), the results demonstrated that the fast video group outperformed their normal speed and control group counterparts.

Employing process-tracing methods, such as verbal reports, biomechanical profiling, temporal and spatial occlusion techniques, and visual search behavior, were the next step to understand elite performance (Williams and Ericsson, 2005). Henderson (2011) describes the use of eye movement recording in opening the door to the real-world visual processes. An advantage of this technique is access to a real-time snapshot of visual behavior. In the expertise literature, it has been reported that elite athletes make more accurate decisions while using fewer fixations of longer duration (North et al., 2009; Vaeyens et al., 2007; Williams et al., 1994).

In a follow-up study, Lorains et al. (2014) investigated the effect of video-based training in above real time and normal speed on the visual search behavior of elite

Australian football athletes. In their experiment, eye movement data were collected pre- and posttest from three groups of participants, two training, and one control group without any training. Based on previous literature in expertise studies such as Sailer et al. (2005), they hypothesized that visual search behavior will be more efficient due to training. Improvement in visual search efficiency was measured by a decrease in the number of fixations and higher fixation duration. Lorain et al. reported that regardless of video speed, the fixation duration became longer at posttest and retention compared to the control group. Furthermore, the above real-time training group spent a longer duration fixating on the best option after the retention test, compared to other groups. While no significant differences were found in the average number of fixations, there was a trend for the fast video group to use fewer fixations following training compared to the normal speed and control group. It has been suggested that athletes developed their ability to pay attention to the relevant information on the screen and therefore they perform better. Change in fixation locations suggests that training using above real-time methods resulted in a more efficient visual search strategy and improved response time.

Effects of ARTT have never been investigated in elite football (soccer) performance. The primary objective of this research is to investigate the effect of ARTT on decision-making skills (as measured by decision accuracy) in professional academy footballers. A secondary objective is to investigate the effect of four different duration training schedules on decision-making (as measured by decision accuracy) in professional academy footballers. We predict that based on previous research in applying ARTT on training decision-making, elite footballers' performance on video-based task will be improved as the result of training. This prediction is in line with the previous research by Lorains et al. (2013a,b) in applying ARTT in Australian football.

2 METHODS AND MATERIALS
2.1 PARTICIPANTS

73 male participants (range 17–21 years old, mean 19 years old) took part in five groups (1 Session/Week $n=15$, 2 Sessions/Week $n=14$, 5 Sessions/Week $n=15$, Random Sessions $n=14$, Control group $n=15$). They were all based at a major football academy with at least 3 years of playing football at a professional level. All participants were naïve to the study. All participants had normal or corrected-to-normal vision. All participants gave their written informed consent in accordance with the Declaration of Helsinki and the guidelines approved by the Ethical Committee of University College London (UCL).

2.2 EXPERIMENTAL DESIGN

The study comprised of five experimental groups. Groups differed in the timing of the sessions. Table 1 shows the protocol of each experimental group.

Table 1 Structure of the Five Experimental Groups

Group	Week								
	1	2	3	4	5	6	7	8	9
1 Session/Week	Pre-T					Post-T			Ret-T
	T1	T2	T3	T4	T4				
2 Sessions/Week	Pre-T			Post-T			Ret-T		
	T1	T2–T3	T4–T5						
5 Sessions/Week	Pre-T	Post-T			Ret-T				
	T1–T5								
Random Sessions	Pre-T					Post-T			Ret-T
	T1	T2–T5 at random							
Control	Pre-T	Post-T		Post-T	Post-T	Post-T			
	No training								

Training (T1–T5 sessions) and testing sessions (Pre-T, Post-T, and Ret-T sessions) consisted of 20 trials. Testing sessions began with three extra warm-up trials. *Pre-T,* pretraining test session; *Post-T,* posttraining test; *Ret-T,* retention-test session; *T1–T5,* training sessions 1–5; *W,* week.

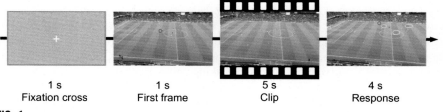

FIG. 1

Procedure of one trial. Playback speed was 1.0× for testing sessions (5 s of presentation) and 1.5× for training sessions (5 s presentation of 7.5 s videos).

In each session participants were asked to perform a computer task. The computer task consisted of presentation of a screen highlighting the location of the ball, a short video clip followed by a screen containing three possible options. The participants' task was to indicate the best option. Options were ordered with one indicating the best choice (value 1), one indicating an intermediate choice (value 2), and one indicating the worst choice (value 3). Each video clip was watched and rated by three UEFA (Union of European Football Associations) A license coaches independently. The video clips were extracted from real competitions by a camera placed in the middle of the field. The duration of the video clips was 5 s. During the training sessions feedback was provided for 4 s after each response (T1–T5 as shown in Table 1), but no feedback was provided during testing sessions (Pre-T, Post-T, and Ret-T as shown in Table 1). Trials were separated by a fixation cross on the screen for 1 s. Participants were asked to respond as quickly and as accurately as possible. Fig. 1 shows sample screens of one trail.

The experiment was developed in MATLAB® (v2013b, MathWorks, USA) using Psychtoolbox (v3). A tablet computer with 13 in. touch screen (Lenovo Yoga) was used for presentation of stimuli and participants were required to tap on the screen to indicate their responses.

2.3 STATISTICAL ANALYSIS

Choice and response time were recorded for analysis. Response *accuracy* was calculated for each choice based on the following: (3—*option's value*) × 50 (%). Option's value represents a number between 1 and 3 with 1 indicating the best option and 3 indicating the worst option. Mean accuracy and median response time were calculated for all testing sessions. To account for individual *differences Accuracy Difference and Response Time Difference were* calculated:

Accuracy Difference = Accuracy Posttraining/Retention test − Pretraining

Response Time Difference = Response Time Pretraining − Posttraining/Retention test

To investigate the baseline activity of the participants in different groups, two one-way ANOVAs were run on Group (1 Session/Week, 2 Sessions/Week, 5 Sessions/Week, Random Sessions, and Control) with accuracy and response time as independent variables. Furthermore, to ensure a stable performance in the Control group, two repeated measures ANOVAs (rANOVA) were run on accuracy and response time of all five sessions.

Two series of analysis were run to (1) investigate effects of different protocols of training on performance and (2) effectiveness of training compared to the Control group. Two 4 × 2 mixed factor ANOVA were run to investigate the differential effects of training protocols on accuracy difference and response time difference with Group (1 Session/Week, 2 Sessions/Week, 5 Sessions/Week, and Random Sessions) as between-subject factor and Session (Posttraining Difference/Retention-test Difference) as within-subject factor. Finally, to investigate the effects of training, two independent sample *t*-tests were run to compare performance of the training groups with those of Control group.

Post hoc paired samples *t*-tests were used to study the difference between conditions. For further scrutiny of the data we report post hoc tests regardless of the results of the ANOVAs. False discovery rate (FDR) correction is used for correction of multiple comparisons.

3 RESULTS

To investigate whether all five groups had a similar baseline level in the Pretraining session, we ran two one-way ANOVAs on accuracy and response time. This ANOVA showed nonsignificant main effects of Group ($F_{(4,68)} = 2.105$, $P = 0.090$, $\eta_p^2 = 0.110$) for accuracy and for response time ($F_{(4,68)} = 1.145$, $P = 0.343$, $\eta_p^2 = 0.063$), which shows that participants began the study with similar level of performance. To investigate changes in the Control group from Pretraining to the four Posttraining sessions, we ran two repeated measures ANOVAs with Session as within-subject factor. These rANOVAs showed nonsignificant main effects of Session for accuracy ($F_{(3,42)}=1.042$, $P=0.384$, $\eta_p^2=0.069$) and response time ($F_{(3,42)}=0.223$, $P=0.880$, $\eta_p^2=0.016$).

A 4 × 2 mixed factor ANOVA on accuracy difference showed a trend toward significant main effect of Session ($F_{(1,54)}=3.368$, $P=0.072$, $\eta_p^2=0.059$), a nonsignificant main effect of Group ($F_{(3,54)}=2.011$, $P=0.123$, $\eta_p^2=0.101$), and a nonsignificant interaction ($F_{(3,54)}=1.864$, $P=0.147$, $\eta_p^2=0.094$). A similar mixed factor ANOVA on response time difference showed a significant main effect of Session ($F_{(1,54)}=20.847$, $P<0.001$, $\eta_p^2=0.279$), a significant main effect of Group ($F_{(3,54)}=49.597$, $P<0.001$, $\eta_p^2=0.479$), and a significant interaction ($F_{(3,54)}=2.863$, $P=0.045$, $\eta_p^2=0.137$). To explore the results further we ran planned post hoc one sample t-tests on both accuracy and response time Posttraining and Retention-test Differences to investigate whether there was an improvement in both Posttraining and Retention-test sessions compared to baseline. This test showed a nonsignificant improvement in accuracy for the 2 Sessions/Week group during Posttraining session and a nonsignificant improvement for the Random Sessions group during Retention-test session. Other comparisons became significant. Response times showed a significant improvement for both Posttraining and Retention-test sessions for all the groups (see Tables 2 and 3).

Paired sample t-tests were also run to look at the changes between Posttraining and Retention-test sessions for both accuracy and response time. These tests showed a significant difference for the 1 Session/Week group for accuracy and 5 Sessions/Week group for response time (see Table 4 and Fig. 2). This shows that the acquired still was perishable.

Performance of different groups was also compared for Posttraining and Retention-test sessions. Neither of the comparisons was significantly different (see Table 5).

To investigate the effects of training in training groups (Table 6) (1 Session/Week, 2 Sessions/Week, 5 Sessions/Week, and Random Sessions) compared

Table 2 One Sample t-Test on Accuracy Difference for Posttraining and Retention-Test Sessions

Group	DoF	Posttraining Difference		Retention-Test Difference	
		t	P	t	P
1 Session/Week	14	4.819	<0.001[a]	3.803	0.002[a]
2 Sessions/Week	13	0.935	0.367	2.656	0.020[b]
5 Sessions/Week	14	3.481	0.004[a]	3.401	0.004[a]
Random Sessions	13	3.414	0.005[a]	1.696	0.114

DoF represents degrees of freedom.
[a]Significant difference false discovery rate (FDR) corrected $\alpha<0.01$.
[b]Significant difference FDR corrected $\alpha<0.05$.

Table 3 One Sample *t*-Test on Response Time Difference for Posttraining and Retention-Test Sessions

Group	DoF	Posttraining Difference		Retention-Test Difference	
		t	P	t	P
1 Session/Week	14	4.512	<0.001[a]	3.297	0.005[a]
2 Sessions/Week	13	2.210	0.046[b]	2.243	0.043[b]
5 Sessions/Week	14	4.672	<0.001[a]	3.494	0.004[a]
Random Sessions	13	4.275	0.001[a]	3.374	0.005[a]

DoF represents degrees of freedom.
[a]Significant difference false discovery rate (FDR) corrected $\alpha < 0.01$.
[b]Significant difference FDR corrected $\alpha < 0.05$.

Table 4 Accuracy Difference and Response Time Difference Refers to the Difference Between Posttraining Difference and Retention-Test Difference for Accuracy and Response Time, Respectively

Group	DoF	Accuracy Difference		Response Time Difference	
		t	P	t	P
1 Session/Week	14	3.216	0.006[a]	1.827	0.089
2 Sessions/Week	13	0.518	0.613	0.599	0.560
5 Sessions/Week	14	1.260	0.228	5.634	<0.001[b]
Random Sessions	13	1.585	0.137	1.801	0.095

[a]Significant difference false discovery rate (FDR) corrected $\alpha < 0.05$.
[b]Significant difference FDR corrected $\alpha < 0.01$.

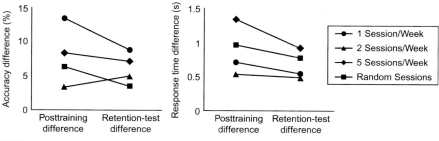

FIG. 2

Performance difference for training groups for Posttraining and Retention-test sessions. Higher values in the *right panel* show faster responses. Error bars are not displayed for clarity.

to the Control group, two independent sample *t*-tests were run on accuracy and response time measures. These tests showed a nonsignificant difference for the 2 Sessions/Week group compared to the Control group and significant differences for other groups (see Table 7 and Fig. 3).

Table 5 Two Independent Sample t-Tests on Accuracy Difference for Posttraining and Retention-Test Sessions Between Training Groups

Group	DoF	Posttraining Difference		Retention-Test Difference	
		t	P	t	P
1 vs 2 Sessions/Week	27	2.228	0.034	1.294	0.207
1 vs 5 Sessions/Week	28	1.381	0.178	0.534	0.598
1 vs Random Session/Week	27	2.102	0.045	1.700	0.101
2 vs 5 Sessions/Week	27	1.168	0.253	0.783	0.441
2 vs Random session/Week	26	0.726	0.475	0.515	0.611
5 vs Random Session/Week	27	0.671	0.508	1.232	0.228

DoF represents degrees of freedom. None of the comparisons became significantly different using false discovery rate (FDR) correction $\alpha < 0.05$.

Table 6 Two Independent Sample t-Tests on Response Time Difference for Posttraining and Retention-Test Sessions Between Training Groups

Group	DoF	Posttraining Difference		Retention-Test Difference	
		t	P	t	P
1 vs 2 Sessions/Week	27	0.603	0.552	0.233	0.817
1 vs 5 Sessions/Week	28	1.927	0.064	1.208	0.237
1 vs Random Session/Week	27	0.942	0.355	0.822	0.418
2 vs 5 Sessions/Week	27	2.119	0.043	1.279	0.212
2 vs Random Sessions/Week	26	1.291	0.208	0.931	0.630
5 vs Random Session/Week	27	1.014	0.320	0.414	0.682

DoF represents degrees of freedom. None of the comparisons became significantly different using false discovery rate (FDR) correction $\alpha < 0.05$.

Table 7 Comparison of Four Training Groups With Control Group for Accuracy and Response Time Difference for Posttraining Session

Group	DoF	Accuracy Difference		Response Time Difference	
		t	P	t	P
1 Session/Week	28	4.347	<0.001[a]	3.834	0.001[a]
2 Sessions/Week	27	0.872	0.391	1.957	0.061
5 Sessions/Week	28	2.886	0.007[a]	4.387	<0.001[a]
Random Sessions	27	3.837	0.001[a]	4.297	<0.001[a]

DoF represents degrees of freedom.
[a]Significant difference false discovery rate (FDR) corrected $\alpha < 0.05$.

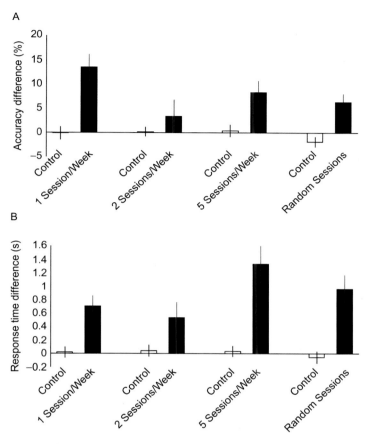

FIG. 3

Performance differences between training and control groups for Posttraining sessions for accuracy difference (A) and response time difference (B). Higher values in panel (B) show faster responses. Error bars represent one s.e.m.

4 DISCUSSION

To our knowledge, this is the first experiment with elite footballers specifically applying above real-time training with different schedules. The aim of this study was first to test the effect of video-based training in above real time on elite footballers decision-making quality and the second objective was to investigate the difference between different training protocols. The pretest showed no significant difference between groups in terms of decision-making accuracy and response time which means they all started at an equivalent level. However, all the training groups' results showed greater overall improvement both in accuracy and response time between pre- and posttest and retention. This is in line with previous research by Lorains et al. (2013a,b) on applying ARTT to Australian football. There is no significant change in the performance of participants in the control group, which is an indicator of effect of training for the training groups.

Comparing results of performance from posttest to retention shows that training needs to be continued as a 2-week break time between posttest and retention resulted in decrease in accuracy and longer response time. This effect was significant for 1 Session/Week in terms of accuracy and 5 Sessions/Week in terms of response time. However, comparing participants' performances between pretest and retention suggests that accuracy improvements for 1, 2, and 5 Sessions/Week schedule is improved. Also, all participants in four group performances in terms of response time are improved.

In terms of finding the best schedule for training between the four options, there is no significant difference in terms of accuracy and response time. However, there is a trend for 1 and 5 Sessions/Week for accuracy and 5 sessions per week for response time.

Video-based training has been shown to be effective in improving the decision-making quality of pilots (Guckenberger et al., 1993) and elite Australian football athletes (Lorains et al., 2013a,b). The criticism was always about the level of similarity and fidelity between the computer task and on pitch performance. However, Lorains et al. (2013a,b) showed a higher level of fidelity by applying ARTT to the video-based training in comparison with normal speed video and using this method has been shown effective in training athletes.

A challenge to the current research is to measure the transfer of learning on the task to the pitch as there is no measure of transfer in the current study. We are currently implementing this. We are also pursuing the effects of training and testing at different times of day and different schedule of training.

In conclusion, our findings show that video-based training above real time results in an improvement in decision-making accuracy and response times of elite footballers; however, in order to retain the skill, continued training is necessary.

ACKNOWLEDGMENTS

We are grateful to Southampton FC, Fulham, FC, Watford FC, Brentford FC, and Crystal Palace FC for their assistance with this project.

REFERENCES

Abernethy, B., 1994. The nature of expertise in sport. In: Serpa, S., Alves, J., Pataco, V. (Eds.), International Perspective on Sport and Exercise Psychology. Fitness Information Technology, Morgantown, WV, pp. 57–68.

Abernethy, B., 1996. Training the visual perceptual skills of athletes: insights from the study of motor expertise. Am. J. Sports Med. 24, S89–S92.

References

Abernethy, B., Baker, J., Cote, J., 2005. Transfer of pattern recall skills may contribute to the development of sport expertise. Appl. Cogn. Psychol. 19 (6), 705–718.

Araujo, D., Davids, K., Hristovski, R., 2006. The ecological dynamics of decision making in sport. Psychol. Sport Exerc. 7, 653–676.

Arroyo-Figueroa, G., Hernandez, Y., Sucar, E., 2006. Intelligent environment for training of power systems operators. Lect. Notes Artif. Intell. 4251, 943–950.

Baker, J., Cote, J., Abernethy, B., 2003. Sport-specific practice and the development of expert decision-making in team ball sports. J. Appl. Sport Psychol. 15, 12–25.

Beilock, S., Carr, T., MacMahon, C., Starkes, J., 2002. When paying attention becomes counterproductive: impact of divided versus skill-focused attention on novice and experienced performance of sensorimotor skills. J. Exp. Psychol. Appl. 8 (1), 6–16.

Beilock, S., Bertenthal, B., McCoy, A., Carr, T., 2004. Haste does not always make waste: expertise, direction of attention, and speed versus accuracy in performing sensorimotor skills. Psychon. Bull. Rev. 11 (2), 373–379.

Ericsson, K.A., 1996. The Road to Excellence: The Acquisition of Expert Performance in the Arts and Sciences, Sports, and Games. Erlbaum, Mahwah, NJ.

Ericsson, K.A., Smith, J., 1991. Toward a General Theory of Expertise: Prospects and Limits. Cambridge University Press, Cambridge.

Farrow, D., Abernethy, B., 2003. Do expertise and the degree of perception—action coupling affect natural anticipatory performance? Perception 32 (9), 1127–1139.

Fitts, P.M., Posner, M.I., 1967. Human Performance. Brooks/Cole, Monterey, CA.

Gabbett, T., Rubinoff, M., Thorburn, L., Farrow, D., 2007. Testing and training anticipation skills in softball fielders. Int. J. Sports Sci. Coach. 2 (1), 15–24.

Gilis, B., Helsen, W., Catteeuw, P., Wagemans, J., 2008. Offside decisions by expert assistant referees in association football: perception and recall of spatial positions in complex dynamic events. J. Exp. Psychol. Appl. 14, 21–35.

Guckenberger, D., Ullano, K.C., Lane, N.E., 1993. Training High Performance Skills Using Above Real-Time Training. National Aeronautics and Space Administration.

Hays, R., Singer, M., 1989. Simulation Fidelity in Training System Design: Bridging the Gap Between Reality and Training. Springer-Verlag, New York.

Henderson, J., 2011. Eye Movements and Scene Perception. In: Liversedge, S., Gilchrist, I., Everling, S. (Eds.), The Oxford Handbook of Eye Movements. Oxford University Press, Oxford, England.

Hepler, T., Feltz, D., 2012. Take the first heuristic, self-efficacy, and decision-making in sport. J. Exp. Psychol. Appl. 18 (2), 154–166.

Janelle, C.M., Hillman, C.H., 2003. Expert performance in sport: current perspectives and critical issues. In: Starkes, J.L., Ericsson, K.A. (Eds.), Expert Performance in Sports. Advances in Research on Sport Expertise. Human Kinetics, Champaign, IL, pp. 19–49.

Johnson, J., Raab, M., 2003. Take the first: option-generation and resulting choices. Organ. Behav. Hum. Decis. Process 91 (2), 215–229.

Kolf, J., 1973. Documentation of a Simulator Study of an Altered Time Base. Unpublished manuscript, NASA Dryden Flight Research Facility, Edwards, CA.

Lorains, M., Ball, K., MacMahon, C., 2013a. Expertise differences in a video decision-making task: speed influences on performance. Psychol. Sport Exerc. 14, 293–297.

Lorains, M., Ball, K., MacMahon, C., 2013b. An above real time training intervention for sport decision making. Psychol. Sport Exerc. 14, 670–674.

Lorains, M., Panchuk, D., Ball, K., MacMahon, C., 2014. The effect of an above real time decision-making intervention on visual search behaviour. Int. J. Sports Sci. Coach. 9 (6), 1383–1392.

Milazzo, N., Fournier, J., 2015. Effect of individual implicit video-based perceptual training program on high-skilled karatekas' decision making. Mov. Sport Sci. Sci. Mot. (88), 13–19.

Muller, S., Abernethy, B., Farrow, D., 2006. How do world-class cricket batsmen anticipate a bowler's intention? Q. J. Exp. Psychol. 59, 2162–2186.

North, J.S., Williams, A.M., Hodges, N., Ward, P., Ericsson, K.A., 2009. Perceiving patterns in dynamic action sequences: investigating the processes underpinning stimulus recognition and anticipation skill. Appl. Cogn. Psychol. 23 (6), 878–894.

Olmos, A., Bouillot, N., Knight, T., Mabire, N., Redel, J., Cooperstock, J., 2012. A high fidelity orchestra simulator for individual musicians' practice. Comput. Music J. 36 (2), 55–73.

Sailer, U., Flanagan, J.R., Johansson, R.S., 2005. Eye–hand coordination during learning of a novel visuomotor task. J. Neurosci. 25 (39), 8833–8842.

Schneider, K.C., 1985. Uninformed response rates in survey research: new evidence. J. Bus. Res. 13 (2), 153–162.

Schweizer, G., Plessner, H., Kahlert, D., Brand, R., 2011. A video-based training method for improving soccer referees' intuitive decision-making skills. J. Appl. Sport Psychol. 23 (4), 429–442.

Starkes, J.L., Ericsson, K.A., 2003. Expert Performance in Sports. Advances in Research on Sport Expertise. Human Kinetics, Champaign, IL.

Starkes, J.L., Lindley, S., 1994. Can we hasten expertise by video simulation? Quest 46, 211–222.

Stoffregen, T.A., Bardy, B.G., Smart, L.J., Pagulayan, R.J., 2003. On the nature and evaluation of fidelity in virtual environments. In: Hettinger, L.J., Haas, M. (Eds.), Virtual and Adaptive Environments: Applications, Implications, and Human Performance Issues. CRC Press, Mahwah, pp. 111–128.

Thiffault, C., 1974. In: Tachistoscopic training and its effect upon visual perceptual speed of ice hockey players. Proceedings of the Canadian Association of Sport Sciences, Edmonton, Alberta.

Thiffault, C., 1980. Construction et validation d'une measure de la rapidite de la pensee tactique des joueurs de hockey sur glace. In: Nadeau, C.H., Halliwell, W.R., Newell, K.M., Roberts, G.C. (Eds.), Psychology of Motor Behaviour and Sport. Human Kinetics, Champaign, IL, pp. 643–649.

Vaeyens, R., Lenoir, M., Williams, A.M., Mazyn, L., Philippaerts, R.M., 2007. The effects of task constraints on visual search behavior and decision-making skill in youth soccer players. J. Sport Exerc. Psychol. 29 (2), 147–169.

Williams, A.M., Davids, K., 1998. Visual search strategy, selective attention, and expertise in soccer. Res. Q. Exerc. Sport 69 (2), 111–128.

Williams, A.M., Ericsson, K.A., 2005. Perceptual-cognitive expertise in sport: some considerations when applying the expert performance approach. Hum. Mov. Sci. 24, 283–307.

Williams, A.M., Davids, K., Burwitz, L., Williams, J.G., 1994. Visual search strategies in experienced and inexperienced soccer players. Res. Q. Exerc. Sport 65 (2), 127–135.

Williams, A.M., Ward, P., Knowles, J., Smeeton, N., 2003. Anticipation skill in a real-world task: measurement, training and transfer in tennis. J. Exp. Psychol. Appl. 8 (4), 259–270.

FURTHER READING

Anderson, J.R., 1987. Methodologies for studying human knowledge. Behav. Brain Sci. 10, 467–505.

Anderson, A.F., Kludt, R., Bavelier, D., 2011. Verbal versus visual working memory skills in action video game players. Poster presented at the Psychonomics Soc. Meet., Seattle.

Bavelier, D., Green, C.S., Han, D.H., Renshaw, P.F., Merzenich, M.M., Gentile, D.A., 2011. Brains on video games. Nat. Rev. Neurosci. 12 (12), 763–768.

Blundell, N.L., 1984. Critical visual-perceptual attributes of championship level tennis players. In: Howell, M., Wilson, B. (Eds.), Proceedings of the VII Commonwealth and International Conference on Sport, Physical Education, Recreation and Dance, University of Queensland, BrisbaneKinesiological Science7, pp. 51–59.

Blundell, N.L., 1985. The contribution of vision to the learning and performance of sports skills: part 1: the role of selected visual parameters. Aust. J. Sci. Med. Sport 17 (3), 3–11.

Boot, W.R., Kramer, A.F., Simons, D.J., Fabiani, M., Gratton, G., 2008. The effects of video game playing on attention, memory, and executive control. Acta Psychol. (Amst) 129, 387–398.

Broadbent, D.P., Causer, J., Williams, A.M., Ford, P.R., 2015. Perceptual-cognitive skill training and its transfer to expert performance in the field: future research directions. Eur. J. Sport Sci. 15 (4), 322–331.

Casanova, F., Oliveira, J., Williams, M., 2009. Expertise and perceptual-cognitive performance in soccer: a review. Rev. Port. Cien. Desp. 9 (1), 115–122.

Chase, W.G., Simon, H.A., 1973. The mind's eye in chess. In: Chase, W.G. (Ed.), Visual Information Processing. Academic, New York, pp. 215–281.

Dye, M.W.G., Bavelier, D., 2010. Differential development of visual attention skills in school-age children. Vision Res. 50, 452–459.

Ericsson, K.A., Kintsch, W., 1995. Long-term working memory. Psychol. Rev. 102, 211–245.

Ericsson, K.A., Lehmann, A.C., 1996. Expert and exceptional performance: evidence of maximal adaptation to task constraints. Annu. Rev. Psychol. 47, 273–305.

Farrow, D., Abernethy, B., Jackson, R.C., 2005. Probing expert anticipation with the temporal occlusion paradigm: experimental investigations of some methodological issues. Motor Control 9, 332–351.

Galton, F., 1869. Heredity Genius. Macmillian, New York.

Glockner, A., Betsch, T., 2012. Decisions beyond boundaries: when more information is processed faster than less. Acta Psychol. (Amst) 139, 532–542.

Godden, D., Baddeley, A., 1975. Context dependent memory in two natural environments. Br. J. Psychol. 66, 325–331.

Green, C.S., Bavelier, D., 2006. Effects of action video game playing on the spatial distribution of visual selective attention. J. Exp. Psychol. Hum. Percept. Perform. 32, 1465–1478.

Greenfield, P.M., 2009. Technology and informal education: what is taught, what is learned. Science 323, 69–71.

Harlow, H.F., 1949. The formation of learning sets. Psychol. Rev. 56, 51–65.

Healy, A.F., Bourne Jr., L.E., 2012. Training Cognition: Optimizing Efficiency, Durability, and Generalizability. Psychology Press, New York, NY.

Hillman, C.H., Erickson, K.I., Kramer, A.F., 2008. Be smart, exercise your heart: exercise effects on brain and cognition. Nat. Rev. Neurosci. 9, 58–65.

Kemp, C., Goodman, N.D., Tenenbaum, J.B., 2010. Learning to learn causal models. Cognit. Sci. 23, 159.

Lejuez, C.W., Read, J.P., Kahler, C.W., Richards, J.B., Ramsey, S.E., Stuart, G.L., Brown, R.A., 2002. Evaluation of a behavioral measure of risk taking: the Balloon Analogue Risk Task (BART). J. Exp. Psychol. Appl. 8 (2), 75.

Lutz, A., Slagter, H.A., Dunne, J.D., Davidson, R.J., 2008. Attention regulation and monitering in meditation. Trends Cogn. Sci. 12, 163–169.

MacMahon, C., McPherson, S., 2009. Knowledge base as a mechanism for perceptual cognitive tasks: skill is in the details. Int. J. Sport Psychol. 40 (4), 565.

MacMahon, C., Starkes, J., Deakin, J., 2009. Differences in processing speed of game information in basketball players, coaches and referees. Int. J. Sport Psychol. 40 (3), 403–424.

Renner, M.J., Rosenzweig, M.R., 1987. Enriched and Impoverished Environments Effects on Brain and Behavior Recent Research in Psychology. Springer-Verlag, New York.

Revien, L., 1987. Eyerobics. Visual Skills Inc, Great Neck, NY.

Sanderson, F.H., 1981. Visual acuity and sports performance. In: Cockerill, I.M., MacGillivary, W.W. (Eds.), Vision in Sport. Stanley Thornes, Cheltenham, England, pp. 64–79.

Shapiro, D.C., Schmidt, R.A., 1982. The schema theory: recent evidence and developmental implications. In: Kelso, J.A.S., Clark, J.E. (Eds.), The Development of Movement Control and Co-ordination. Wiley, New York, pp. 113–150.

Vicenzi, D.A., Wise, J.A., Mouloua, M., Hancock, P.A., 2008. Human Factors in Simulation and Training. CRC Press, Boca Raton, FL, USA.

Ward, P., Farrow, D., Harris, K., Williams, M., Eccles, D., Ericsson, K.A., 2008. Training perceptual-cognitive skills: can sport psychology research inform military decision training? Mil. Psychol. 20 (1), 71–102.

Williams, A.M., Hodges, N.J., North, J., Barton, G., 2006. Perceiving patterns of play in dynamic sport tasks: investigating the essential information underlying skilled performance. Perception 35, 317–332.

Williams, A.M., Ford, P.R., Eccles, D.W., Ward, P., 2011. Perceptual-cognitive expertise in sport and its acquisition: implications for applied cognitive psychology. Appl. Cogn. Psychol. 25 (3), 432–442.

Wilson, M.R., Vine, S.J., Bright, E., Masters, R.S., Defriend, D., McGrath, J.S., 2011. Gaze training enhances laparoscopic technical skill acquisition and multi-tasking performance: a randomized, controlled study. Surg. Endosc. 25 (12), 3731–3739.

CHAPTER

Can athletes benefit from difficulty? A systematic review of growth following adversity in competitive sport

Karen Howells*, Mustafa Sarkar[†,1], David Fletcher[‡]

*School of Education, Childhood, Youth and Sport, The Open University, Milton Keynes, United Kingdom
[†]School of Science and Technology, Nottingham Trent University, Nottingham, United Kingdom
[‡]School of Sport, Exercise and Health Sciences, Loughborough University, Loughborough, United Kingdom
[1]Corresponding author: Tel.: +44-11-5848-6359, e-mail address: mustafa.sarkar@ntu.ac.uk

Abstract

Research points to the notion that athletes have the potential to benefit from difficulty. This phenomenon—otherwise known as growth following adversity—has attracted increasing attention from sport psychology scholars. In this paper, we systematically review and synthesize the findings of studies in this area to better understand: (a) how growth has been conceptualized in competitive sport, (b) the theory underpinning the study of growth in sport performers, (c) the nature of research conducted in this area, and (d) the adversity- and growth-related experiences of competitive athletes. Following the application of inclusion criteria and methodological quality assessment, 17 studies were deemed suitable for inclusion in the systematic review. The findings of these studies are reviewed and synthesized in relation to study characteristics (viz. growth terminology, theoretical underpinning, study design, participant details, and data analysis), quality appraisal, adversity-related experiences (viz. negative events and experiences, and response to negative events and experiences), and growth-related experiences (viz. mechanisms of growth and indicators of growth). To facilitate understanding of growth following adversity in competitive sport, we address the definitions and theories that have informed the body of research, discuss the associated findings related to the adversity- and growth-related experiences of competitive athletes, and outline avenues for future research. It is hoped that this review and synthesis will facilitate understanding and inform practice in this area.

Keywords

Adversarial growth, Athlete, Perceived benefits, Posttraumatic growth, Performance, Psychology, Sport, Stress, Stress-related growth, Trauma

1 INTRODUCTION

The notion that athletes have the potential to benefit from difficulty has received increasing attention in recent years. In their study of Olympic champions, for example, Fletcher and Sarkar (2012) reported that experiencing demanding situations is an important factor in the development of optimal sport performance: "Exposure to stressors was an essential feature of … Olympic champions. Indeed, most of the participants argued that if they had not experienced certain types of stressors at specific times, including highly demanding adversities … they would not have won their gold medals" (p. 672). Similarly, after claiming that "talent needs trauma" (p. 907), Collins and MacNamara (2012) speculated that "overcoming early life challenge is a precursor to high-level achievement" (p. 908) and that "there is a disproportionally high incidence of early trauma … in the life histories of elite. The knowledge and skills the athletes accrued from 'life' traumas … certainly appears to affect their subsequent development and performance in sport" (p. 909). Most recently, Hardy et al. (2017) provided evidence that early life (i.e., nonsport) adversity is essential and that later career (sport or nonsport) adversity often acts as a developmental catalyst in Olympic and/or World champions. What this research and other scholars point to is the intriguing possibility that sport performers can turn the most trying times around and develop as a result; a phenomenon referred to in the psychosocial literature as growth following adversity (cf. Gucciardi, 2017; Howle and Eklund, 2017; Sarkar and Fletcher, 2017).

Over the past couple of decades, a growing body of psychosocial research has shown that individuals can grow following adversity to the extent that they report development beyond their pretrauma functioning (see, for a review, Linley and Joseph, 2004). According to Tedeschi and Calhoun (2004), this growth entails an increased appreciation for life, more meaningful relationships, an increased sense of personal strength, a change in priorities, and/or a richer existential and spiritual awareness. Various terms have been used to conceptualize this phenomenon, including *perceived benefits* (Affleck et al., 1987), *positive changes in outlook* (Joseph et al., 1993), *posttraumatic growth* (PTG; Tedeschi and Calhoun, 1996), *stress-related growth* (SRG; Park et al., 1996), and *adversarial growth* (Linley and Joseph, 2004). Furthermore, scholars have proposed several theoretical explanations of growth (cf. Joseph and Linley, 2006; Zoellner and Maercker, 2006a), including a functional descriptive model of PTG (FDM; Calhoun et al., 2010; Calhoun and Tedeschi, 1998; Tedeschi and Calhoun, 1995, 2004), an organismic valuing theory of growth through adversity (OVT; Joseph and Linley, 2005), an affective-cognitive processing model of PTG (ACPM; Joseph et al., 2012), and the Janus-faced model of self-perceived PTG (Maercker and Zoellner, 2004; Zoellner and Maercker, 2006a). These theoretical models are summarized in Table 1. Collectively, this body of work indicates that (constructive) growth occurs as a result of a person's struggle to deal with a traumatic experience, the consequential "shattered self" (cf. Janoff-Bulman, 1989), and the subsequent interaction of a variety of person and situational factors.

Table 1 Summary of Growth Theoretical Models

Model	Reference(s)	Key Characteristics
Functional Descriptive Model of Posttraumatic Growth (FDM of PTG)	Tedeschi and Calhoun (1995, 2004), Calhoun and Tedeschi (1998), and Calhoun et al. (2010)	Cognitive processing begins when a negative event or experience challenges an individual's assumptive world. The early response to the event and the subsequent distress involve automatic cognitive processing in the form of intrusive thoughts and images which tend to be negative in nature. Over a period of time, this process, which the authors posit requires the presence of a considerable level of distress, leads to disengagement from previous goals and assumptions. Whether growth occurs is, in part, dependent on an individual's personality characteristics
Organismic Valuing Theory of Growth (OVT)	Joseph and Linley (2005)	Individuals are intrinsically motivated toward growth and the authors posited that a traumatic event causes a shattering of beliefs that instigates the individual to question their beliefs and seek meaning in their experience. There are four key elements of the model: a drive for completion which involves the cognitive-emotional integration of new trauma-related material; an emphasis on the differences between the cognitive processes of assimilation (i.e., incorporating the new information into your current worldview) and accommodation (i.e., the development of a new worldview to include new information); an appreciation of the differences between meaning as comprehensibility (i.e., why something happened) vs meaning as significance (i.e., what it means for the individual); and an acknowledgment of the hedonic (i.e., subjective well-being—SWB) and the eudemonic (i.e., psychological well-being—PWB) theoretical traditions

Continued

Table 1 Summary of Growth Theoretical Models—cont'd

Model	Reference(s)	Key Characteristics
Affective Cognitive Processing Model of Posttraumatic Growth (ACPM of PTG)	Joseph et al. (2012)	The model suggests that when an event occurs that has negative or traumatic aspects, event cognitions occur that are at both conscious (i.e., ruminative brooding or reflective pondering) and unconscious (i.e., the presence of intrusive thoughts and dreams) levels. The subsequent cognitive appraisal of the event will in turn impact upon the individual's emotional state, which will impact on what, and how, coping strategies are engaged. This may then impact upon further cognitive appraisal. This cycle of feedback continues indefinitely as the individual seeks meaning in their experience. This cyclical process is influenced by the social environment, the satisfaction of basic psychological needs (viz. autonomy, competence, and relatedness), and levels of personality. Discrepancies are resolved through (negative and positive) accommodation or assimilation in accordance with the propositions of OVT
Janus-Faced model of Growth	Maercker and Zoellner (2004) and Zoellner and Maercker (2006a)	The model is based on the assumption that following adverse events, individuals engage in a process of self-deception in an attempt to convince themselves of positive outcomes. The authors proposed a two-component or Janus-faced model which comprises a functional (or constructive) side to growth and an illusory side to growth. The functional, constructive, self-transcending side is analogous with the FDM. However, the illusory or dysfunctional side involves self-deception and is associated with denial, avoidance, wishful thinking, self-consolidation, and palliation which may occur following adversity

As noted at the outset of this paper, research points to the notion that athletes have the potential to benefit from difficulty (Fletcher and Sarkar, 2012; Hardy et al., 2017). This phenomenon—otherwise known as growth following adversity (cf. Gucciardi, 2017; Howle and Eklund, 2017; Sarkar and Fletcher, 2017)—has attracted increasing attention from sport psychology scholars. Over the past few years, researchers have specifically explored how athletes grow from adversity (Tamminen and Neely, 2016). In this paper, we review and synthesize the findings of these studies to better understand: (a) how growth has been conceptualized in competitive sport, (b) the theory underpinning the study of growth in sport performers, (c) the nature of research conducted in this area, and (d) the adversity- and growth-related experiences of competitive athletes. Reviewing these areas is needed because growth-related research in sport has become somewhat fragmented, drawing on similar but subtly different concepts and theories, and reporting a wide range of athletes' experiences. Synthesis will facilitate understanding and inform practice in this area.

2 METHOD

The methods of reviewing research are diverse. One way of differentiating between approaches is based on the process of literature collection which can be highly subjective (e.g., a narrative review) or criteria-based (e.g., a systematic review). Although narrative reviews can provide a critical discussion of pertinent issues within a topic area, they typically do not adopt a study protocol or transparent procedures. In contrast, systematic reviews involve a rigorous approach that reduces reporter bias and random error (Cook et al., 1997). Other strengths of a systematic review include the articulation of a research question(s), a thorough search for evidence, a criteria-based selection of relevant research, an evaluation of study quality, and evidence-based inferences (Collins and Fauser, 2005). For these reasons, a systematic rather than narrative approach was deemed appropriate for this review.

Depending on the type of research included, the analysis may involve metaanalytic techniques (i.e., for quantitative research) or metasynthesis techniques (i.e., for qualitative research), or a combination of these or similar techniques (i.e., for varied or mixed methods research) (Hong et al., 2017). Furthermore, when systematically reviewing literature, Collins and Fauser (2005) argued that "the choices made … should be explicit, transparent, clearly stated and reproducible by interested readers" (p. 103). Accordingly, guidance for conducting systematic reviews informed the review process (cf. Harris et al., 2014; Van Tulder et al., 2003). Van Tulder et al. (2003) identified five stages of a systematic review involving: a literature search, inclusion criteria, methodological quality assessment, data extraction, and data analysis, and emphasized that a minimum of two reviewers should be involved throughout the review process (see also Harris et al., 2014).

2.1 SEARCH STRATEGY

The search strategy adopted three main approaches to gather research on growth following adversity in competitive sport. First, in December 2016, an online search of the research literature was conducted via the following electronic databases: Applied Social Sciences Index and Abstracts, Medline, Physical Education Index, PsycARTICLES, PsycINFO, SportDISCUS, and Zetoc. The search protocol involved a search of titles, abstracts, and full papers using the terms: "adversarial growth," "adversity-related," "athlete," "athletic," "competitive," "elite," "growth-related," "perceived benefits," "performance," "performer," "positive adaptation," "positive by-products," "positive changes in outlook," "positive outcomes," "posttraumatic growth," "post-traumatic growth," "sport," "stress-related growth," and "thriving." The second strategy involved conducting a manual search of the pertinent journals.[a] Once the manual search was complete, the third strategy involved searching reference lists of the full papers that were collected and that met the inclusion criteria.

2.2 INCLUSION CRITERIA

In this review, studies were required to meet the following inclusion criteria: (a) comprise original empirical data, (b) constitute a full text peer-reviewed journal article, (c) be published in the English language, (d) involve competitive sport performers, and (e) include a research question, aim(s), or objective(s) that explicitly made reference to growth or growth-related terminology.

2.3 SIFTING OF PAPERS

The documents that were potentially appropriate for the review were evaluated by title, abstract, and then full text (see Fig. 1). At each stage of appraisal, papers were excluded from the sifting process if any of the inclusion criteria were not satisfied. To illustrate, papers were excluded if they were a book chapter (e.g., Tamminen and Neely, 2016), were comprised solely of an abstract (e.g., Galli et al., 2013), were not in the English language (e.g., Almeida et al., 2014), did not sample competitive sport performers (e.g., Smith et al., in press), or did not specify or include growth in the research question, aim(s), or objective(s) (e.g., Sabiston et al., 2007).

---

[a]Hand-searched journals include: *European Journal of Sport Science* (2001–2016); *International Journal of Applied Sport Sciences* (2000–2016); *International Journal of Sport and Exercise Psychology* (1970–2016); *International Journal of Sport Psychology* (1970–2016); *Journal of Applied Sport Psychology* (1989–2016); *Journal of Clinical Sport Psychology* (2007–2016); *Journal of Sport Psychology* (1979–1987); *Journal of Sport and Exercise Psychology* (1988–2016); *Journal of Sport Behavior* (1978–2016); *Journal of Sport Psychology in Action* (2010–2016); *Journal of Sports Sciences* (1983–2016); *Psychology of Sport and Exercise* (2000–2016); *Research Quarterly in Sport and Exercise* (2000–2016); *Scandinavian Journal of Medicine and Science in Sports* (1991–2016); *Sport, Exercise, and Performance Psychology* (2012–2016); and *The Sport Psychologist* (1984–2016).

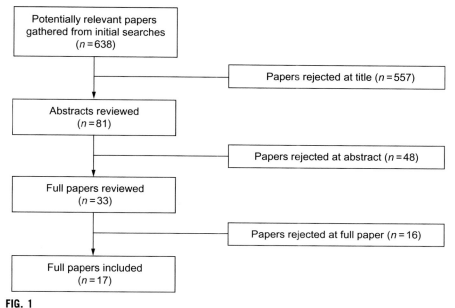

FIG. 1
PRISMA diagram of the stages of the systematic review.

2.4 METHODOLOGICAL RIGOR

To enhance methodological rigor, and in line with systematic review guidelines (e.g., Harris et al., 2014; Hong et al., 2017; Van Tulder et al., 2003) and contemporary systematic reviews in the sport literature (e.g., Forsdyke et al., 2016; Gledhill et al., 2017), a peer review team was formed to minimize bias and error. The team comprised of the first author and two researchers from different institutions (the second and third authors). The full text evaluation of eligibility was reviewed and discussed via peer debriefing and by playing "devil's advocate" (cf. Forsdyke et al., 2016; Gledhill et al., 2017). Although there were some disagreements about the eligibility of particular articles, agreement was reached through a process of constructive debate.

Further, given the heterogeneity of the included study designs, all studies were appraised for reporting quality based on the standards of the Mixed Methods Appraisal Tool (MMAT; Pluye and Hong, 2014; Pluye et al., 2011). The MMAT consists of 19 quality criteria for appraising quantitative, qualitative, and mixed methods studies. The MMAT has been recognized as a valid and reliable tool for appraising mixed methods research (Souto et al., 2015) and has been used recently in systematic reviews in the sport literature (viz. Forsdyke et al., 2016; Gledhill et al., 2017). All 17 studies were subject to two general screening questions and were considered of an appropriate quality for further appraisal. Those studies that utilized qualitative methods, including those that used multiple methods where all of the methods utilized were qualitative in nature, were assessed using Section 1 of the MMAT (i.e., the qualitative criteria 1.1–1.4). Those studies that utilized quantitative (descriptive) methods were assessed using Section 4 of the MMAT (i.e., the quantitative descriptive criteria 4.1–4.4). Sections 2 and 3 of the MMAT were not used because these criteria were only relevant for quantitative randomized controlled trials (2.1–2.4) and nonrandomized controlled trials (3.1–3.4). Those studies that employed both quantitative and qualitative methods (i.e., mixed methods) were appraised with both qualitative (i.e., 1.1–1.4) and quantitative (i.e., 4.1–4.4) criteria as well as the mixed methods component (i.e., the mixed methods criteria 5.1–5.3). All of the items were rated as "yes," "no," or "cannot tell" (referring to a lack of evidence), with one point awarded for each "yes" response. Initially, scores ranged from 0 to 4, with mixed method studies only able to score as high as their lowest score for each study design (cf. Smith et al., 2016). However, due to an excessively low score on the (sole) study that utilized a mixed methods design, an approach that allowed consideration of all responses (which resulted in a percentage quality score) was also utilized (cf. Pace et al., 2012).

2.5 DATA EXTRACTION

Following the application of the inclusion criteria and the methodological quality assessment, 17 studies were deemed to be suitable to include in the systematic review. Fig. 1 presents a summary of the data extraction process using a Preferred Reporting Items for Systematic Reviews and Meta-Analyses (PRISMA) flow diagram.

2.6 DATA ANALYSIS

The first phase of data analysis was indwelling (cf. Swann et al., 2012) where the peer review team read the full text of each study and became immersed in the findings and inferences. A summary of the included studies was then entered into a table. Convergent thematic analysis (CTA; Centre for Reviews and Dissemination; CRD, 2009; see also Hong et al., 2017) was subsequently used to synthesize data from different empirical findings and the assessment of methodological quality. A CTA consists of identifying the main or recurring themes from a body of research and is typically used for detecting, grouping, and summarizing findings from studies (Pope et al., 2007).

3 RESULTS

We review and synthesize the findings of the 17 studies deemed suitable for inclusion in relation to study characteristics, quality appraisal, adversity-related experiences, and growth-related experiences. In addition, Table 2 provides a summary of the following aspects of each study: the growth terminology used;

Table 2 Summary of Studies Included in the Review

Author(s) (Year)	Research Question(s)/Aim(s)/Objective(s)	Growth Terminology	(Growth) Framework/Model	Participants and Sport	Adversity	Data Collection	Data Analysis	Main Findings
Udry et al. (1997)	To examine: • The psychological responses of injured athletes' to season-ending injuries • The long-term benefits perceived they obtained from their injuries	Positive outcomes and benefits	None	21 participants 11 male and 10 female Mean age = 23.9 years Elite US ski team members	Season-ending injuries	Semistructured interviews	Inductive data analysis	Four general dimensions of growth in reaction to being injured were identified: • Injury-relevant information processing/awareness • Emotional upheaval/reactive behavior • Positive outlook/coping attempts • Other Four dimensions of injury benefits were identified: • Personal growth • Psychologically based enhancements • Physical-technical development • Also identified one participant with no growth
McDonough et al. (2011)	To qualitatively explore the development of social relationships, social support, and posttraumatic growth (PTG) among breast cancer survivors participating in a dragon boat program	PTG	None	17 participants 17 female Mean age = 51.24 years (SD = 11.09) Dragon boating	Breast cancer (diagnosed 4.06 years before their enrolment on dragon boating program)	Semistructured interviews	Interpretative phenomenological analysis (IPA)	Four profiles were identified. Each profile was discussed in terms of developing social relationships and support, providing support to others, physicality and athleticism, negative interactions and experiences. • Developing a feisty spirit of survivorship • I do not want it to be just about me • It's not about the pink, it's about the paddling • Hard to get close
Wadey et al. (2011)	To examine the antecedents and mechanisms underlying the perceived benefits following sport injury	Perceived benefits	Wiese-Bjornstal et al.'s (1998) integrated model of response to sport injury	10 participants 10 male Mean age = 21.7 years (SD = 1.8) 3 team sports (rugby union, soccer, and basketball)	Lower limb injuries	Semistructured interviews	Causal networks	There were a number of perceived benefits, which came from several antecedents across three temporal phases: • injury onset (e.g., emotional response) • rehabilitation (e.g., free time) • return to competition sport (e.g., reflective practice) The participants also reported mechanisms through which they derived their perceived benefits, including self-disclosure to others and mobilizing one's social support network

Continued

Table 2 Summary of Studies Included in the Review—cont'd

Author(s) (Year)	Research Question(s)/Aim(s)/Objective(s)	Growth Terminology	(Growth) Framework/Model	Participants and Sport	Adversity	Data Collection	Data Analysis	Main Findings
Galli and Reel (2012a)	To purposefully select participants who reported high stress-related growth on a self-report measure for further in-depth qualitative study. Three main questions: What are the athletes' experiences of potentially stressful events in sport? In what ways does growth manifest as a result of sport-related stress? What personal, environmental, and social mechanisms assist athletes' growth?	SRG	Functional descriptive model (FDM); Organismic valuing theory (OVT), Galli and Vealey's (2008) conceptual model of resilience	11 participants 8 female, 3 male Mean age = 20.82 years (SD = 1.67) Multiple Division 1 NCAA sports	Stressors (viz. chronic depression, stress fracture, changing sports, no scholarship, car accident, expectations, balancing school and sport, torn rotator cuff, overall demands of sport, lack of time, poor performance)	Posttraumatic growth inventory (PTGI) and semistructured interviews	General inductive approach	Four dimensions were identified: • Personal and sociocultural context • Disruption (struggles and working through) • Social support • Positive psychosocial outcomes Based on these dimensions, a conceptual model of SRG was developed that involved athletes' struggles and attempts to work through sport stressors led them to perceive growth in the form of: a new life philosophy, self-changes, and interpersonal changes. Social support was critical in facilitating athletes' attempts to work through and make meaning from their stressor. The entire SRG process was framed by athletes' life context, including personal characteristics and sociocultural conditions
Galli and Reel (2012b)	To examine adversarial growth in a sample of Division 1 NCAA athletes. Three main questions: Do Division I student-athletes report growth in response to their most difficult intercollegiate sport adversity, and if so, which subdomains of growth are most apparent? Are there differences in growth by type of adversity? Are there gender differences between male and female college athletes in overall growth, or any of the subdomains of growth?	Adversarial growth	Seven vectors of Development in College Students (Chickering and Reisser, 1993)	214 participants 150 female, 64 male Mean age = 20.79 years (SD = 1.55) Multiple Division 1 NCAA sports	Six types of stressors: • Mental and physical stress of sport • Sport injury • Time demands • Interpersonal issues • Personal struggles • Other	PTGI	Independent samples t-tests, one-way ANOVAs, and one-way MANOVAs	Athletes reported positive change at low to moderate levels from adversity, with the most improvement in personal strength. Female athletes reported greater spiritual growth, as well as more of a change in their ability to relate to others than males. Of the three types of adversities analyzed (i.e., time demands, injury, and the mental and physical stress of sport), athletes who reported time demands as their most difficult adversity exhibited more appreciation for life than athletes who cited the mental and physical stress of sport

Study	Aim	Growth concept	Theoretical framework	Participants	Type of adversity	Data collection	Data analysis	Findings
Day (2013)	To explore Paralympic athletes lived experiences of becoming physically active after disability and the role this has in the development of PTG	PTG	OVT	7 participants 4 male, 3 female Age unknown Elite Paralympic athletes from a range of sports	Acquired disability	Life history interviews and sport participation observation	Holistic content analysis	Three main themes were identified: • Recognizing possibility by acknowledging limitations • Responsibility for choice and consequences • Reestablishing and enhancing meaning
Tamminen et al. (2013)	To explore experiences of adversity and to examine perceptions of growth following adversity among elite female athletes	Adversarial growth	OVT	5 participants 5 female Age 18–23 years Multiple sports at elite	Adversities: • Performance slumps • Coach conflicts • Bullying • Eating disorders • Sexual abuse • Injury	Semistructured interviews	IPA	The participants experienced isolation/withdrawal, emotional disruption, questioning identity as an athlete, and understanding experiences within a context of perceived expectations following adversity. As participants sought and found meaning in their experiences, they identified opportunities for growth associated with social support and also as they realized the role of sport in their lives. Aspects of growth include realizing strength, gaining perspective of their problems, and gaining a desire to help others. Athletes' experiences with adversity were seen as part of an ongoing journey through elite sport
Wadey et al. (2013)	To examine coaches' perceptions of athletes SRG following sport injury	SRG	Wiese-Bjornstal et al.'s (1998) integrated model of response to sport injury	8 participants 8 male Mean age = 45.8 years (SD = 11.2) Coaches from club to international in multiple sports	Injury	Semistructured (life world) interviews	Deductive and inductive content analysis	Four dimensions of SRG identified by the coaches: • Personal growth (e.g., beliefs) • Psychological growth (e.g., sporting qualities) • Social growth (e.g., social support) • Physical growth (e.g., strength) Coaches also identified behavioral indicators
Crawford et al. (2014)	Does participation in ParaSport following acquired spinal cord injury (SCI) influence peoples' perceptions of PTG? What specific dimensions of PTG, if any, do ParaSport athletes report experiencing?	PTG	None	12 participants Gender unknown Age 24–55 years (M = 40.67, SD = 9.96) Multiple ParaSports	Acquired SCI	Survey and semistructured interviews	Phenomenological approach	Five general dimensions of growth emerged from the data • Injury relevant processing • Appreciation for life • Reactive behavior as a result of attempted integration into ParaSport • Relating to others • Health and well-being Participants reported increased physical functioning and independence related to their involvement in sport. Emotional and psychological gains were also associated with ParaSport including reestablishment of self-identity, improved clarity, and perception of life, changed priorities, greater confidence, and enhanced social relationships

Continued

Table 2 Summary of Studies Included in the Review—cont'd

Author(s) (Year)	Research Question(s)/Aim(s)/Objective(s)	Growth Terminology	(Growth) Framework/Model	Participants and Sport	Adversity	Data Collection	Data Analysis	Main Findings
Howells and Fletcher (2015)	To explore the adversity- and growth-related experiences of swimmers at the highest competitive level	Adversity- and growth-related experiences	Affective cognitive processing model of growth (ACPM) with reference to others	7 participants 4 male, 3 female Age (at Olympic swims) 14–41 years ($M = 23.39$, $SD = 6.04$) Olympic champion swimmers	Adversities characterized as: • Developmental stressors • External stressors • Embodied states • Psychological states • Externalized behaviors	Autobiographies	Narrative analysis	Adversity was perceived as traumatic. There was an initial attempt to maintain normality through the development of an emotional and embodied relationship with water. This involved nondisclosure and the development of multiple identities. As these strategies proved to be maladaptive and exposed the swimmers to further adversity, the swimmers sought meaning in their experiences and looked to others for support. Growth was identifiable through: superior performance, enhanced relationships, spiritual awareness, and prosocial behavior. Swimmers were vulnerable to exposure to further adversities—swimmers sought meaning in experiences and looking to others for support
Salim et al. (2015)	To examine the relationship between hardiness, coping, and perceived SRG in a sport injury context	Perceived SRG	OVT	206 participants 148 male, 58 female Mean age = 22.23 years ($SD = 6.50$) Team and individual sports, from recreational to elite	Injury	Dispositional Resilience Scale (DRS), Stress-Related Growth Scale (SRGS), and the Brief COPE Scale	Independent sample t-tests, one way ANOVAs, Pearson product-moment correlations, and bootstrapping for mediation analysis	Key findings: • There was a significant positive relationship between hardiness and perceived SRG The study identified two coping strategies found to mediate this relationship—emotional support and positive reframing
Sarkar et al. (2015)	Despite the recognition that experiencing adversity can have beneficial outcomes for human growth, little is known about the adversities that the world's best athletes encounter, and the perceived role that these experiences play in their psychological and performance development. These issues were investigated in this study	Adversity-related growth	FDM	10 participants 6 male, 4 female Age 33–70 years ($M = 47.60$ years, $SD = 12.06$) Olympic Gold Medalists from multiple sports	Adversity-related experiences: • Repeated nonselection • Significant sporting failure • Serious injury • Political unrest • Death of a family member	Semistructured interviews	Inductive thematic analysis	The findings indicate that the participants encountered a range of sport- and nonsport adversities that they considered were essential for winning their gold medals. The participants described the role that these experiences played in their psychological and performance development, specifically focusing on their resultant trauma, motivation, and learning

Study	Aim	Concept	Theory	Participants	Acquired disability	Data collection	Analysis	Key findings
Day and Wadey (2016)	A focus on recovery and growth after trauma to: (a) illuminate the process of working through trauma (b) identify alternative narrative types beyond positive accommodation	Positive accommodation and assimilation	OVT	2 participants 2 male Age mid-30s and mid-40s High level athletes with an acquired disability Nonspecified sports		Narrative interviews	Narrative analysis	Two narrative types identified: • Assimilation • Positive accommodation The narrative of positive accommodation demonstrated how sport provided mastery experiences, enhanced relationships, corporeal understanding, and enhanced life philosophies. The alternative narrative of assimilation was associated with resilience to trauma
Howells and Fletcher (2016)	To adopt a critical stance on the veridicality of growth by exploring Olympic swimmers' experiences of constructive and illusory growth	Adversarial growth (illusory vs constructive)	Janus-faced model of growth	4 participants 2 male, 2 female, Age (at time of Olympic participation) 17–27 years Olympic swimmers	Adversities included illness, injury, coach conflict, relationship breakdown, and performance slumps	Semistructured interviews and timelining	IPA	Key findings: • Some outcomes are illustrative of illusory growth, others more constructive • Earlier phases of growth are more likely to be characterized by illusory aspects. As time progresses, more constructive aspects are likely to be identified
Salim et al. (2016)	To explain how injured athletes high in hardiness experienced SRG and why athletes low in hardiness are less likely to derive such benefits	SRG	None	20 participants 13 male, 7 female Mean age = 23.7 years (SD = 6.4) Multiple sports from recreational to national	Injury	Semistructured interviews	Composite sequence analysis	Findings revealed that athletes high in hardiness experienced stress-related growth from having an emotional outlet, which enabled them to reframe their injury and experience positive affect (PA). In contrast, athletes low in hardiness had no emotional outlet, which led to suboptimal outcomes. The study identified four broad and sequential themes: • Emotional trauma • Emotional outlet • Subsequent responses • Resultant outcomes
Wadey et al. (2016)	To explore the relationship between needs satisfaction, SRG, and subjective well-being among injured athletes	SRG	OVT (and self-determination theory—SDT)	520 participants 316 male, 204 female Age range 18–59 years (M = 23.3 years, SD = 6.5) 39 different sports represented	Injury	NSS, PANAS, SRGS	Pearson product-moment correlations and structural equation modeling	Key findings: • Positive correlations were found between need satisfaction variables and SRG • Positive relationship between competence, relatedness, and positive affect (PA) • Moderate positive correlation between PA and SRG • The effect of competence and relatedness on PA was mediated by SRG
Roy-Davis et al. (2017)	To develop a context specific (i.e., substantive) and grounded explanatory theory that explores and explains the relationship between sport injury and SIRG	SIRG	None	37 participants 23 male, 14 female Age range 19–39 years (M = 27.3 years, SD = 5.4) Variety of sports (team and individual), recreational to elite	Injury	Semistructured interviews	Grounded theory	Production of a new term Sport Injury-Related Growth (SIRG). Internal (personality, coping styles, knowledge, and prior experience and perceived social support) and external (cultural scripts, physical resources, time, and received social support) factors enable athletes to transform their injury into an opportunity for growth and development

the (growth) framework or model (if any) that informed the study; the research question, aim(s), or objective(s); demographic data; the data collection method(s); the data analysis procedures; and the main findings.

3.1 STUDY CHARACTERISTICS
3.1.1 Growth terminology
The studies conceptualized growth using a variety of terms. Four studies referred to PTG (viz. Crawford et al., 2014; Day, 2013; Day and Wadey, 2016; McDonough et al., 2011); four studies referred to SRG (viz. Galli and Reel, 2012a; Salim et al., 2016; Wadey et al., 2013, 2016); one study referred to perceived SRG (viz. Salim et al., 2015); three studies referred to adversarial growth (viz. Galli and Reel, 2012b; Howells and Fletcher, 2016; Tamminen et al., 2013); one study referred to positive outcomes and benefits (viz. Udry et al., 1997); one study referred to perceived benefits (viz. Wadey et al., 2011); one study referred to adversity- and growth-related experiences (viz. Howells and Fletcher, 2015); one study referred to adversity-related growth (viz. Sarkar et al., 2015); and one study referred to Sport Injury-Related Growth (SIRG; viz. Roy-Davis et al., 2017).

3.1.2 Theoretical underpinning
The studies tended to be underpinned by theoretical models although they were not necessarily confined to models of growth and there was no dominant theory or model utilized. Four studies were informed by OVT (viz. Day, 2013; Day and Wadey, 2016; Salim et al., 2015; Tamminen et al., 2013); one study was informed by ACPM (viz. Howells and Fletcher, 2015); one study was informed by FDM (viz. Sarkar et al., 2015); one study was informed by the Janus-faced model of growth (viz. Howells and Fletcher, 2016); two studies were informed by multiple models [viz. Galli and Reel, 2012a: OVT, FDM, Galli and Vealey's (2008) conceptual model of resilience; Wadey et al., 2016: OVT and Self Determination Theory (SDT; Deci and Ryan, 1985)]; and four studies were informed by models not related to growth (viz. Galli and Reel, 2012b; Udry et al., 1997; Wadey et al., 2011, 2013). Specifically, two studies were informed by Wiese-Bjornstal et al.'s (1998) integrated model of response to sport injury (viz. Wadey et al., 2011, 2013): one study was informed by Chickering and Reisser's (1993) seven vectors of development in college students (viz. Galli and Reel, 2012b) and one study (viz. Udry et al., 1997) was informed by a number of stage and adjustment theories (e.g., Heil, 1993; Kubler-Ross, 1969). Lastly, four out of the 17 studies were not informed by any model or theory (viz. Crawford et al., 2014; McDonough et al., 2011; Roy-Davis et al., 2017; Salim et al., 2016).

3.1.3 Study design
The majority of studies employed qualitative designs with eight utilizing (semistructured) interviews (viz. McDonough et al., 2011; Roy-Davis et al., 2017; Salim et al., 2016; Sarkar et al., 2015; Tamminen et al., 2013; Udry et al., 1997; Wadey et al., 2011, 2013), one using narrative interviews (viz. Day and Wadey, 2016), and one

using autobiographies (viz. Howells and Fletcher, 2015). Three studies employed multiple qualitative methods; namely, one used semistructured interviews and timelining (viz. Howells and Fletcher, 2016), one used life history interviews and sport participation observation (viz. Day, 2013), and one utilized a (predominantly qualitative) survey with interviews (viz. Crawford et al., 2014). One study employed mixed methods involving interviews and the measurement of SRG using the Posttraumatic Growth Inventory (PTGI; Tedeschi and Calhoun, 1996; viz. Galli and Reel, 2012a). Lastly, three studies adopted a wholly quantitative approach employing questionnaires (viz. Galli and Reel, 2012b; Salim et al., 2015; Wadey et al., 2016). Specifically, one study utilized the PTGI to measure adversarial growth (Galli and Reel, 2012b); one study (viz. Salim et al., 2015) measured hardiness, perceived SRG, and coping utilizing the Dispositional Resilience Scale (DRS; Bartone et al., 1989), Stress-Related Growth Scale (SRGS; Park et al., 1996), and the Brief COPE (Carver, 1997), respectively; and one study (viz. Wadey et al., 2016) measured psychological needs satisfaction, subjective well-being, and perceived SRG utilizing the Needs Satisfaction Survey (NSS; Podlog et al., 2010), the Positive and Negative Affect Scale (PANAS; Watson et al., 1988), and the SRGS (Park et al., 1996), respectively.

3.1.4 Participant details
Demographic data of each study sample is presented in Table 2.

3.1.4.1 Sample size
The sample sizes for each study ranged from 2 to 520 participants.[b] Collectively, across the 17 studies, this equated to 1111 participants sampled in total. Ten studies had 12 or fewer participants, four studies had between 13 and 50 participants, and three studies had 51 participants or more.

3.1.4.2 Age and gender
Thirteen of the 17 studies provided details about the mean age of the participants. The mean age of all participants (excluding participants from Day, 2013; Day and Wadey, 2016; Howells and Fletcher, 2016; Tamminen et al., 2013 who did not provide mean ages of their participants) was 30.18 years (SD=11.52). Of the 1111 participants, across all 17 studies, 485 (43.7%) were female, 614 (55.2%) were male, and 12 (1.1%) were unknown (viz. Crawford et al., 2014). The participants of five studies were either exclusively male (viz. Day and Wadey, 2016; Wadey et al., 2011, 2013) or female (viz. McDonough et al., 2011; Tamminen et al., 2013).

---

[b]This includes the seven swimmers who wrote eight autobiographies in Howells and Fletcher's (2015) study of Olympic swimming champions. For the purposes of this review, the sample size was counted as seven.

3.1.4.3 Sport
Participants represented 48 (team and individual) sports. The most common sports sampled were swimming, track and field, basketball, soccer/association football, volleyball, tennis, and rugby union. However, the level of detail about which participants were engaged in which sports varied across the studies. Three studies provided the detail about the sport but did not allocate participant numbers to those sports (viz. Crawford et al., 2014; Roy-Davis et al., 2017; Sarkar et al., 2015), seven studies investigated growth in multiple sports and allocated participants to those sports (viz. Galli and Reel, 2012a,b; Salim et al., 2016; Tamminen et al., 2013; Wadey et al., 2011, 2013, 2016), four studies sampled participants in one sport only (viz. Howells and Fletcher, 2015, 2016: swimming; McDonough et al., 2011: dragon boating; Udry et al., 1997: skiing), and three studies did not indicate the participants' sports (viz. Day, 2013; Day and Wadey, 2016; Salim et al., 2015).

3.1.4.4 Sport modality and standard
All of the studies investigated growth in competitive sport but there was considerable variation in the modality and standard of sport that the participants competed in. Five studies sampled able-bodied elite athletes (viz. Howells and Fletcher, 2015, 2016; Sarkar et al., 2015; Tamminen et al., 2013; Udry et al., 1997), three studies sampled disabled elite athletes (viz. Crawford et al., 2014; Day, 2013; Day and Wadey, 2016), two studies sampled Division I NCAA college level athletes (viz. Galli and Reel, 2012a,b), one study sampled club level athletes (viz. McDonough et al., 2011), and six studies sampled a variety of levels ranging from club to international level (viz. Roy-Davis et al., 2017; Salim et al., 2015, 2016; Wadey et al., 2011, 2013, 2016).

3.1.5 Data analysis
The studies utilized a variety of methods of analysis. Analysis of the qualitative data comprised of a phenomenological approach (viz. Crawford et al., 2014); holistic content analysis (viz. Day, 2013); general inductive approach (viz. Galli and Reel, 2012a); narrative analysis (viz. Day and Wadey, 2016; Howells and Fletcher, 2015); interpretative phenomenological analysis (IPA) (viz. Howells and Fletcher, 2016; McDonough et al., 2011; Tamminen et al., 2013); grounded theory (viz. Roy-Davis et al., 2017); composite sequence analysis (viz. Salim et al., 2016); inductive thematic analysis (viz. Sarkar et al., 2015); inductive data analysis (viz. Udry et al., 1997); causal networks (viz. Wadey et al., 2011); and deductive and inductive content analysis (viz. Wadey et al., 2013). Analysis of the quantitative data comprised of independent-samples t-tests and one-way ANOVAs (viz. Galli and Reel, 2012b; Salim et al., 2015), one-way MANOVAs (viz. Galli and Reel, 2012b), Pearson product–moment correlations (viz. Salim et al., 2015; Wadey et al., 2016), bootstrapping for mediation analysis (viz. Salim et al., 2015), and structural equation modeling (viz. Wadey et al., 2016).

3.2 QUALITY APPRAISAL

Specific details of the MMAT criteria are provided in Table 3 and the assessment of each study against the (relevant) criteria is reported in Table 4. All of the studies which were appraised using the qualitative criteria were judged to be of excellent or high quality with eight papers scoring 4 out of 4, and four papers scoring 3 out of 4. The quantitative papers scored lower with two studies scoring 2 out of 4, and one study scoring 3 of 4. The study that utilized mixed methods was rated low in quality scoring 2 out of 4 due to low scores on the quantitative assessment (2 out of 4). All of the studies were also given a percentage quality score based on all assessable criteria (cf. Pace et al., 2012). Two studies scored 50%, one study scored 73%, six studies scored 75%, and eight studies scored 100% (see Table 4).

A number of issues were identified through the quality appraisal. Specifically, several qualitative studies (viz. Crawford et al., 2014; McDonough et al., 2011; Sarkar et al., 2015; Udry et al., 1997; Wadey et al., 2013) provided little or no information about the role of the researchers, how the researchers influenced the participants, or how the researchers' epistemological stance impacted on the interpretation of the data. With regards to the quantitative studies, none of the studies (viz. Galli and Reel, 2012b; Salim et al., 2015; Wadey et al., 2016) reported on response rate and one study (viz. Galli and Reel, 2012b) did not provide any details about sampling.

3.3 ADVERSITY-RELATED EXPERIENCES

The adversity-related experiences reported in the 17 studies are divided and synthesized in the following two sub-sections: negative events and experiences, and response to negative events and experiences.

3.3.1 Negative events and experiences

Participants experienced multiple negative events and experiences (see Table 5) that varied from being construed as stressors, adversities, or traumas. Eleven studies investigated growth in relation to one specific negative event or experience (viz. acquired spinal cord injury: Crawford et al., 2014; acquired disability: Day, 2013; Day and Wadey, 2016; breast cancer: McDonough et al., 2011; injury: Roy-Davis et al., 2017; Salim et al., 2015, 2016; Udry et al., 1997; Wadey et al., 2011, 2013, 2016). Of the nine studies that investigated growth following injury, five studies reported that the injuries were sustained in training or competition (viz. Salim et al., 2015, Roy-Davis et al., 2017; Udry et al., 1997; Wadey et al., 2011, 2016). Despite the focus on a specific event (i.e., injury), one study (viz. Roy-Davis et al., 2017) reported that participants identified the development of other stressors *as a result* of the original injury experience. Moreover, one study (viz. Wadey et al., 2011) identified one adversity but addressed the perceived benefits resulting from different aspects of the injury experience (i.e., injury onset, rehabilitation, and return to competitive sport).

Six studies specified a range of different negative events and experiences (viz. Howells and Fletcher, 2015, 2016; Galli and Reel, 2012a,b; Sarkar et al., 2015;

Table 3 MMAT Criteria

MMAT—Screening Questions (For All Types)	Qualitative	Quantitative (Randomized Controlled Trials)	Quantitative Nonrandomized	Quantitative Descriptive	Mixed Methods
A. Are there clear qualitative and quantitative research questions (or objectives), or a clear mixed methods question (or objective)?	1.1. Are the sources of qualitative data (archives, documents, informants, observations) relevant to address the research question (objective)?	2.1. Is there a clear description of the randomization (or an appropriate sequence generation)?	3.1. Are participants (organizations) recruited in a way that minimizes selection bias?	4.1. Is the sampling strategy relevant to address the quantitative research question (quantitative aspect of the mixed methods question)?	5.1. Is the mixed methods research design relevant to address the qualitative and quantitative research questions (or objectives), or the qualitative and quantitative aspects of the mixed methods question (or objective)?
B. Do the collected data allow address the research question (objective)? E.g., consider whether the follow-up period is long enough for the outcome to occur (for longitudinal studies or study components)	1.2. Is the process for analyzing qualitative data relevant to address the research question (objective)?	2.2. Is there a clear description of the allocation concealment (or blinding when applicable)?	3.2. Are measurements appropriate (clear origin, or validity known, or standard instrument; and absence of contamination between groups when appropriate) regarding the exposure/intervention and outcomes?	4.2. Is the sample representative of the population under study?	5.2. Is the integration of qualitative and quantitative data (or results) relevant to address the research question (objective)?

—	1.3. Is appropriate consideration given to how findings relate to the context, e.g., the setting, in which the data were collected?	2.3. Are there complete outcome data (80% or above)?	3.3. In the groups being compared (exposed vs nonexposed; with intervention vs without; cases vs controls), are the participants comparable, or do researchers take into account (control for) the difference between these groups?	4.3. Are measurements appropriate (clear origin, or validity known, or standard instrument)?	5.3. Is appropriate consideration given to the limitations associated with this integration, e.g., the divergence of qualitative and quantitative data (or results) in a triangulation design?
—	1.4. Is appropriate consideration given to how findings relate to researchers' influence, e.g., through their interactions with participants?	2.4. Is there low withdrawal/drop-out (below 20%)?	3.4. Are there complete outcome data (80% or above), and, when applicable, an acceptable response rate (60% or above), or an acceptable follow-up rate for cohort studies (depending on the duration of follow-up)?	4.4. Is there an acceptable response rate (60% or above)?	—

Table 4 Studies Included in the Review Scored Against MMAT Criteria

Author(s)	General Screening		Qualitative				Quantitative Descriptive				Mixed Methods			Preliminary Score Out of 4 (cf. Smith et al., 2016)	Quality Percentage Score (cf. Pace et al., 2012)
	A	B	1.1	1.2	1.3	1.4	4.1	4.2	4.3	4.4	5.1	5.2	5.3		
Crawford et al. (2014)	1	1	1	1	1	Cannot tell	—	—	—	—	—	—	—	3	75%
Day (2013)	1	1	1	1	1	1	—	—	—	—	—	—	—	4	100%
Day and Wadey (2016)	1	1	1	1	1	1	—	—	—	—	—	—	—	4	100%
Galli and Reel (2012a)	1	1	1	1	1	1	1	1	0	Cannot tell	1	1	0	2	73%
Galli and Reel (2012b)	1	1	—	—	—	—	0	1	1	Cannot tell	—	—	—	2	50%
Howells and Fletcher (2015)	1	1	1	1	1	1	—	—	—	—	—	—	—	4	100%
Howells and Fletcher (2016)	1	1	1	1	1	1	—	—	—	—	—	—	—	4	100%
McDonough et al. (2011)	1	1	1	1	1	Cannot tell	—	—	—	—	—	—	—	3	75%
Roy-Davis et al. (2017)	1	1	1	1	1	1	—	—	—	—	—	—	—	4	100%

Study															
Salim et al. (2015)	1	1	—	—	—	1	—	Cannot tell	1	Cannot tell	—	—	—	2	50%
Salim et al. (2016)	1	1	1	1	1	—	—	—	—	—	—	—	—	4	100%
Sarkar et al. (2015)	1	1	1	1	0	—	—	—	—	—	—	—	—	3	75%
Tamminen et al. (2013)	1	1	1	1	1	—	—	—	—	—	—	—	—	4	100%
Udry et al. (1997)	1	1	1	1	Cannot tell	—	—	—	—	—	—	—	—	3	75%
Wadey et al. (2011)	1	1	1	1	1	—	—	—	—	—	—	—	—	4	100%
Wadey et al. (2013)	1	1	1	1	Cannot tell	—	—	—	—	—	—	—	—	4	75%
Wadey et al. (2016)	1	1	—	—	—	1	1	1	1	Cannot tell	—	—	—	3	75%

Note: 1 = Yes, 0 = No, Cannot tell = refers to a lack of evidence in the study, — = criteria not relevant to the study.

Table 5 Negative Events and Experiences

Higher-Order Categories	Negative Events and Experiences	Studies
Injury	Nonspecified injury	Galli and Reel (2012a,b), Howells and Fletcher (2016), Sarkar et al. (2015), Tamminen et al. (2013), and Wadey et al. (2013, 2016)
	Lower limb injury	Salim et al. (2015, 2016), Roy-Davis et al. (2017), Udry et al. (1997), and Wadey et al. (2011)
	Skeletal injury and stress fracture	Galli and Reel (2012a), Roy-Davis et al. (2017), and Salim et al. (2015)
	Muscular injury	Salim et al. (2015)
	Facial or head injury	Roy-Davis et al. (2017) and Udry et al. (1997)
	Upper limb and shoulder injury	Galli and Reel (2012a), Howells and Fletcher (2015, 2016), Roy-Davis et al. (2017), Salim et al. (2015, 2016), and Tamminen et al. (2013)
	Back injury	Howells and Fletcher (2015, 2016), Roy-Davis et al. (2017), and Udry et al. (1997)
	Acquired disability	Day and Wadey (2016) and Crawford et al. (2014)
	Acquired spinal cord injury	Day (2013)
	Season-ending injury	Udry et al. (1997)
Physical illness	Breast cancer	McDonough et al. (2011)
	Illness	Howells and Fletcher (2015, 2016)
Developmental experiences	ADHD	Howells and Fletcher (2015)
	Speech impediment	Howells and Fletcher (2015)
	Dyslexia	Howells and Fletcher (2015)
Mental health	Obsessive compulsive disorder	Howells and Fletcher (2015)
	Depression	Galli and Reel (2012a), Howells and Fletcher (2015), and Tamminen et al. (2013)
	Suicidal thoughts	Howells and Fletcher (2015) and Tamminen et al. (2013)
	Self-harm	Howells and Fletcher (2015)
	Substance abuse	Howells and Fletcher (2015)
	Eating disorder	Howells and Fletcher (2015) and Tamminen et al. (2013)

Table 5 Negative Events and Experiences—cont'd

Higher-Order Categories	Negative Events and Experiences	Studies
Interpersonal experiences	Family dysfunction	Howells and Fletcher (2015)
	Sexual abuse (from coach)	Tamminen et al. (2013)
	Relationship difficulties	Howells and Fletcher (2016)
	Adversity to significant other	Howells and Fletcher (2015)
	Interpersonal issues	Galli and Reel (2012b)
	Being bullied	Tamminen et al. (2013)
	Bereavement	Howells and Fletcher (2015) and Sarkar et al. (2015)
Sport-specific experiences	Performance slumps/poor performance	Galli and Reel (2012a), Howells and Fletcher (2016), Sarkar et al. (2015), and Tamminen et al. (2013)
	Repeated nonselection	Sarkar et al. (2015)
	Organizational stressors	Howells and Fletcher (2016)
	Changing sports	Galli and Reel (2012a) and Tamminen et al. (2013)
Performance lifestyle	Living away from home	Howells and Fletcher (2016)
	Financial issues	Galli and Reel (2012a,b)
	Expectations	Galli and Reel (2012a)
	Balancing school with sport	Galli and Reel (2012a)
	Lack of time	Galli and Reel (2012a,b)
	Demands of sport	Galli and Reel (2012a,b)
Other	Personal struggles	Galli and Reel (2012b)
	Political unrest	Sarkar et al. (2015)
	Car accident	Galli and Reel (2012a) and Howells and Fletcher (2016)
	Other	Galli and Reel (2012b)

Tamminen et al., 2013). Of these six studies, two studies explored growth relating to a single and specific adversity (viz. Galli and Reel, 2012a,b) and four studies explored growth relating to multiple adversities (viz. Howells and Fletcher, 2015, 2016; Sarkar et al., 2015; Tamminen et al., 2013). For the purposes of classification (see Table 5), the adversities have been synthesized into higher-order categories comprising injury, physical illness, developmental experiences, mental health, interpersonal experiences, sport-specific experiences, performance lifestyle, and other.

3.3.2 Response to negative events and experiences

Although the focus of the studies was predominantly on growth processes and outcomes, twelve of the studies (viz. Crawford et al., 2014; Day, 2013; Day and Wadey, 2016; Galli and Reel, 2012a; Howells and Fletcher, 2015, 2016; Roy-Davis et al., 2017; Salim et al., 2016; Sarkar et al., 2015; Tamminen et al., 2013; Udry et al., 1997; Wadey et al., 2011) reported the initial response to the negative events and experiences. These have been categorized into cognitive, emotional, physical, and behavioral responses (see Table 6).

3.3.2.1 Cognitive responses

The majority of studies identified an initial response to negative events or experiences that involved a degree of purposeful cognitive processing. This was related to: the impact that the adversity had on the athletes' identities (viz. Crawford et al., 2014; Day and Wadey, 2016; Howells and Fletcher, 2015; Salim et al., 2016; Tamminen et al., 2013), the cause of the experience (viz. Day, 2013, Udry et al., 1997; Wadey et al., 2011), and the implications for the future (viz. Day, 2013; Roy-Davis et al., 2017; Udry et al., 1997). Some participants' thoughts turned to regret and they purposively focused on the extent of their loss (viz. Day, 2013, Roy-Davis et al., 2017, Salim et al., 2016), whereas others were subject to intrusive thoughts and ruminations (Howells and Fletcher, 2015; Salim et al., 2016). Two studies identified cognitive responses that were not interpreted as being overtly negative. To elaborate, Howells and Fletcher (2016) argued that denial was a typical cognitive response to adversity which was used as a short-term palliative coping strategy to protect against the impact of negative emotions and distress, and was adopted by athletes to safeguard identities. As a further illustration, Day and Wadey (2016) identified that a narrative of assimilation meant that one participant rejected his disability as "meaningless" (p. 134).

3.3.2.2 Emotional responses

Participants experienced "emotional disruption" (Tamminen et al., 2013, p. 31) or "emotional upheaval" (Udry et al., 1997, p. 234) that predominantly involved encountering negative emotions such as anger, fury, shock, frustration, guilt, and helplessness (see, e.g., Sarkar et al., 2015). One study (viz. Howells and Fletcher, 2016) found that there was a suppression of negative emotions (i.e., denial) as a short-term coping strategy.

Table 6 Responses to Negative Events and Experiences

Higher-Order Categories	Responses	Studies
Cognitive	Loss/questioning of identity/athletic identity/shattered identity	Crawford et al. (2014), Day and Wadey (2016), Salim et al. (2016), and Tamminen et al. (2013)
	Wake up call	Crawford et al. (2014) and Day (2013)
	Struggled to accept losses	Day (2013)
	Difficulty identifying boundaries of capabilities	Day (2013)
	Focus on/questioning cause of adversity	Day (2013), Udry et al. (1997), and Wadey et al. (2011)
	Attempt to recapture past sense of self	Day and Wadey (2016)
	Rejection of experiences incongruent with pretrauma beliefs	Day and Wadey (2016)
	Upward (accommodation) or downward (assimilation) comparison with others	Day and Wadey (2016)
	Closed door analogy	Howells and Fletcher (2015)
	Development of multiple identities	Howells and Fletcher (2015)
	Denial	Howells and Fletcher (2016)
	Reflection on goals, beliefs, and values	Roy-Davis et al. (2017) and Udry et al. (1997)
	Injury-relevant processing	Udry et al. (1997)
	Thoughts relating to regret	Roy-Davis et al. (2017)
	Intrusive thoughts/ruminations	Salim et al. (2016) and Howells and Fletcher (2015)
	Awareness of pain	Udry et al. (1997)
	Realization of extent of injury	Udry et al. (1997)
	Dwelling on loss	Salim et al. (2016)
	Suicidal thoughts	Crawford et al. (2014)
	Perceptions of isolation	Tamminen et al. (2013) and Udry et al. (1997)
Emotional	Suppression of negative emotions	Howells and Fletcher (2016)
	Feelings of shock	Roy-Davis et al. (2017) and Udry et al. (1997)
	Negative emotions	Salim et al. (2016)
	Frustration	Roy-Davis et al. (2017)

Continued

Table 6 Responses to Negative Events and Experiences—cont'd

Higher-Order Categories	Responses	Studies
	Helplessness	Roy-Davis et al. (2017)
	Gutted	Sarkar et al. (2015)
	Anger	Roy-Davis et al. (2017), Sarkar et al. (2015), and Wadey et al. (2011)
	Guilt	Roy-Davis et al. (2017)
	Emotional disruption (e.g., shattered)	Tamminen et al. (2013)
	Emotional agitation	Udry et al. (1997)
	Self-pity	Udry et al. (1997)
	Loss of confidence	Tamminen et al. (2013)
	Feelings of vulnerability	Day and Wadey (2016)
	Depression	Crawford et al. (2014)
Physical	Stress-related illness	Galli and Reel (2012a)
	Loss of fitness	Salim et al. (2016)
	Tasks now physically impossible or difficult	Day (2013), Day and Wadey (2016), Roy-Davis et al. (2017), and Wadey et al. (2011)
	Focus on mastery	Day and Wadey (2016)
	Success in rehabilitation	Day and Wadey (2016)
Behavioral	Disruption of enjoyable interaction with others	Crawford et al. (2014)
	Difficulty asserting control over personal situation	Day (2013)
	Compete with able bodied athletes	Day and Wadey (2016)
	Maintenance of normality	Howells and Fletcher (2015)
	Avoidance of reminders of stressors	Howells and Fletcher (2016)
	Failure to speak out	Howells and Fletcher (2016) and Salim et al. (2016)
	Withdrawal from others	Tamminen et al. (2013)
	Disclosing thoughts and feelings	Salim et al. (2016)
	Failure to disclose adversities	Howells and Fletcher (2016) and Salim et al. (2016)

3.3.2.3 Physical responses
In part due to the physical nature of sport and the embodied nature of many of the negative events and experiences, several of the studies (viz. Crawford et al., 2014; Day, 2013; Roy-Davis et al., 2017; Wadey et al., 2011) identified a physical response that involved the development of stress-related illnesses (e.g., migraines; viz. Galli and Reel, 2012a), a perception of a loss of fitness (viz. Salim et al., 2016), and the identification that tasks that had previously been mundane or achievable were now difficult or impossible (viz. Day, 2013; Day and Wadey, 2016; Roy-Davis et al., 2017; Wadey et al., 2011).

3.3.2.4 Behavioral responses
The athletes' negative events and experiences resulted in behavioral changes that involved the withdrawal from others (viz. Tamminen et al., 2013), the disruption of enjoyable activities with others (viz. Crawford et al., 2014), and an unwillingness to disclose information about the adversities (viz. Howells and Fletcher, 2016; Salim et al., 2016). One study (viz. Salim et al., 2016) identified both a failure to disclose *and* a willingness to disclose in their participants, which was moderated by hardiness.

3.4 GROWTH-RELATED EXPERIENCES
Growth was generally conceived as a process with the majority of studies focusing on both the mechanisms of growth and indicators of growth. Only one study focused solely on indicators of growth and did not investigate any mechanisms of growth (viz. Galli and Reel, 2012b).

3.4.1 Mechanisms of growth
Following the initial response to negative events and experiences, the majority of the studies identified a secondary response or "transitional process" (Howells and Fletcher, 2015, p. 43) whereby growth was facilitated or instigated by internal or external mechanisms. One study identified that there may be different pathways to growth that either focused on social support or focused on physical activity and competition (viz. McDonough et al., 2011). The majority of studies did not differentiate between the incidence of growth with respect to different adversities. However, there were two exceptions to this. The first was a study that identified athletes with lower limb injuries experienced more growth than those with upper limb injuries and those who participated in team sports experienced more growth than those in individual sports (viz. Salim et al., 2015). The second (viz. Galli and Reel, 2012b) identified that those athletes that reported their most difficult adversity as being related to time demands reported significantly more growth on the appreciation for life subscale of the PTGI than those who reported their most difficult adversity as being related to the mental and physical stress of sport. Collectively, for the purposes of synthesis, the mechanisms of growth identified in the studies have been categorized into internal or external mechanisms (see Table 7).

Table 7 Mechanisms of Growth

Higher-Order Categories	Mechanisms	Studies
Internal	Mental toughness	Crawford et al. (2014)
	Mastery of success	Day (2013) and Day and Wadey (2016)
	Management of risk	Day (2013) and Day and Wadey (2016)
	Taking responsibility/autonomy	Day (2013), Day and Wadey (2016), and Wadey et al. (2016)
	Accommodation	Day and Wadey (2016)
	Assimilation	Day and Wadey (2016)
	Rejection of identity as a disabled athlete (assimilation)	Day and Wadey (2016)
	Development of an athletic identity	Day and Wadey (2016)
	Reflection on what was meaningful in their lives, seeking meaning (significance)	Day (2013) and Howells and Fletcher (2015)
	Maintenance of a positive mentality	Galli and Reel (2012a)
	Being motivated to overcome the stressor/adversity	Day and Wadey (2016), Galli and Reel (2012a), Howells and Fletcher (2015), and Sarkar et al. (2015)
	Questioning the performance narrative	Howells and Fletcher (2015)
	Seeking meaning (comprehension)	Howells and Fletcher (2016) and Wadey et al. (2011)
	Acceptance	Day and Wadey (2016), Howells and Fletcher (2016), and Udry et al. (1997)
	Disclosure of adversity	Howells and Fletcher (2016)
	Enduring distress	Howells and Fletcher (2016)
	Cognitive processing	Howells and Fletcher (2016)
	Metacognition	Roy-Davis et al. (2017)
	Positively appraise their injury	Salim et al. (2015) and Roy-Davis et al. (2017)
	Personality	Roy-Davis et al. (2017)
	Coping styles	Roy-Davis et al. (2017) and Udry et al. (1997)
	Knowledge and prior experiences	Roy-Davis et al. (2017)
	High in hardiness	Salim et al. (2015)
	Positive reframing	Salim et al. (2016)
	Optimism	Udry et al. (1997)
	Reflecting back on the adversity	Wadey et al. (2011)

Table 7 Mechanisms of Growth—cont'd

Higher-Order Categories	Mechanisms	Studies
External	Role of sport: positive e.g., reestablish identity, Crawford et al. (2014); empowerment, McDonough et al. (2011)	Crawford et al. (2014), McDonough et al. (2011), and Tamminen et al. (2013)
	Sanctuary	Howells and Fletcher (2015)
	Support of others (e.g., family, friends, teammates, and coaches) and relatedness	Day and Wadey (2016), Galli and Reel (2012a), Howells and Fletcher (2015, 2016), McDonough et al. (2011), Roy-Davis et al. (2017), Salim et al. (2015), Tamminen et al. (2013), and Wadey et al. (2011, 2016)
	Temporal aspect	Howells and Fletcher (2016)
	Cultural scripts/quest narrative	Howells and Fletcher (2015, 2016) and Roy-Davis et al. (2017)
	Physical resources (e.g., TV, internet, medical care)	Roy-Davis et al. (2017)

3.4.1.1 Internal mechanisms

There were a number of internal processes that facilitated growth. Cognition and cognitive processing (viz. Howells and Fletcher, 2016) was fundamental, with one study (viz. Roy-Davis et al., 2017) focusing on metacognitions to explain how the athletes' thinking was brought under control and another study (viz. Day and Wadey, 2016) exploring the contrasting narratives of positive accommodation and assimilation to distinguish between differing cognitions and subsequent paths of growth. The thinking that accompanied an acceptance of the adversity (viz. Howells and Fletcher, 2016; Udry et al., 1997) involved questioning the performance narrative (viz. Howells and Fletcher, 2015), seeking meaning (comprehension) in the adversity (viz. Howells and Fletcher, 2016; Wadey et al., 2011), positively reframing the experience (viz. Salim et al., 2015; Roy-Davis et al., 2017), reestablishing a sense of autonomy (viz. Day, 2013; Wadey et al., 2016), and engaging in a process of reflection that tended to focus on the significance of the experience in the athletes' lives (viz. Day, 2013; Howells and Fletcher, 2015; Sarkar et al., 2015; Wadey et al., 2011). Personality traits were viewed as important in increasing the likelihood of growth (see, e.g., Roy-Davis et al., 2017), especially hardiness (viz. Salim et al., 2015), optimism (viz. Udry et al., 1997), and mental toughness (viz. Crawford et al., 2014).

3.4.1.2 External mechanisms

There were three key external mechanisms of growth; namely, sport or physical activity, social support, and informational resources. Sport had a significant role in reestablishing identity, empowering the participants, and/or providing a safe place or sanctuary to escape the impact of the adversity (viz. Crawford et al., 2014; Day and Wadey, 2016; Howells and Fletcher, 2015; McDonough et al., 2011; Tamminen et al., 2013). The support of others and feelings of relatedness was evident in the majority of studies (viz. Day and Wadey, 2016; Galli and Reel, 2012a; Howells and Fletcher, 2015, 2016; McDonough et al., 2011; Roy-Davis et al., 2017; Salim et al., 2015; Tamminen et al., 2013; Wadey et al., 2011, 2016) and involved families, friends, teammates, coaches, and sport psychologists. Finally, informational resources were important in facilitating growth and involved exposure to cultural scripts or narratives about how adversity could be negotiated (viz. Howells and Fletcher, 2015, 2016; Roy-Davis et al., 2017). The source of these messages was often in the form of physical resources such as television, medical resources, autobiographies, and the internet (see, e.g., Roy-Davis et al., 2017).

3.4.2 Indicators of growth

Although growth is conceptualized as a process, all of the studies identified indicators of growth that were perceived to have occurred. The indicators of growth were derived from measurement tools (viz. PTGI and SRGS), subjective self-reported identifiers of growth from participants, and interpretations of growth from researchers. These are categorized into three themes: intrapersonal indicators, interpersonal indicators, and physical indicators (see Table 8).

3.4.2.1 Intrapersonal indicators

The majority of indicators of growth that were identified can be classified as being intrapersonal; that is, they involved cognitions, emotions, motivational processes, or an awareness of the self in the context of the wider environment. The role of cognition involving thoughts, perceptions, and judgments, featured heavily in the indicators of growth. Participants engaged in a positive reframing of the negative experiences or derogation of adversity-related experiences to allow them to perceive their adversities in a different light (viz. Day and Wadey, 2016; Galli and Reel, 2012a; Howells and Fletcher, 2016; Salim et al., 2016), partook in active coping (viz. McDonough et al., 2011), and developed a positive bias toward the future (Howells and Fletcher, 2016; McDonough et al., 2011).

Emotions were perceived to be indicators of growth with one study (viz. Wadey et al., 2016) finding a moderate positive correlation between SRG and positive affect (PA), and another (viz. Salim et al., 2015) identifying that emotion-focused coping strategies (i.e., positive reframing and emotional support) mediated the relationship between hardiness and perceived SRG. "Emotional rebound" (Galli and Reel, 2012a, p. 305) characterized the indicators of growth with five studies reporting an increase in positive emotions (viz. Galli and Reel, 2012a; McDonough et al., 2011; Roy-Davis et al., 2017; Salim et al., 2016; Wadey et al., 2011). Motivational processes were important in respect of identifying growth with four studies identifying increased motivation as an indicator of growth (Howells and Fletcher, 2015; McDonough et al., 2011; Sarkar et al., 2015; Wadey et al., 2013).

Table 8 Indicators of Growth

Higher-Order Categories	Growth Identifiers	Studies
Intrapersonal	Deeper understanding of selves	Crawford et al. (2014)
	New life philosophy	Day and Wadey (2016), Galli and Reel (2012a), and Howells and Fletcher (2016)
	Increased personal strength (includes mental toughness and resilience)	Galli and Reel (2012a), McDonough et al. (2011), Roy-Davis et al. (2017), Tamminen et al. (2013), Udry et al. (1997), and Wadey et al. (2011, 2013)
	Increased spirituality	Galli and Reel (2012a), Howells and Fletcher (2015), and McDonough et al. (2011)
	Enhanced life meaning	Day (2013) and Howells and Fletcher (2016)
	Greater appreciation for life	Crawford et al. (2014), Day and Wadey (2016), McDonough et al. (2011), Roy-Davis et al. (2017), and Salim et al. (2016)
	Change in perspective	Crawford et al. (2014) and Galli and Reel (2012a)
	Realization of opportunities	Day (2013)
	New possibilities and opportunities	McDonough et al. (2011)
	Gained perspective	Udry et al. (1997)
	New outlook on life	Day and Wadey (2016) and Wadey et al. (2011, 2013)
	Improved problem solving	Crawford et al. (2014)
	Emotional rebound, e.g., happier	Galli and Reel (2012a)
	Positive emotions and facilitative responses (confidence, hope, optimism, grateful, appreciative, uplifted, interested, excited, and curious)	Roy-Davis et al. (2017)
	Positive reflections on the stressor/derogation of adversity-related experiences/reframing	Day and Wadey (2016), Galli and Reel (2012a), Howells and Fletcher (2016), and Salim et al. (2016)
	Cognitive manipulations (illusory growth)	Howells and Fletcher (2016)
	Motivated illusions	Howells and Fletcher (2016)
	Sport-related intelligence/awareness	Roy-Davis et al. (2017) and Wadey et al. (2011, 2013)
	Self-acceptance	Roy-Davis et al. (2017)
	Heightened positive affect	Salim et al. (2016)
	New appreciation and outlook on sport	Salim et al. (2016)
	Personality development	Udry et al. (1997)

Continued

Table 8 Indicators of Growth—cont'd

Higher-Order Categories	Growth Identifiers	Studies
	Happiness	McDonough et al. (2011)
	Better emotional regulation	Wadey et al. (2011)
	Greater enjoyment in sport	Wadey et al. (2013)
	Belief that can overcome injury	Wadey et al. (2013)
	Increased motivation	Howells and Fletcher (2015), McDonough et al. (2011), Sarkar et al. (2015), and Wadey et al. (2013)
	Positive bias, positive outlook	Howells and Fletcher (2016) and McDonough et al. (2011)
	Active coping	McDonough et al. (2011)
	Improved self-perceptions	McDonough et al. (2011)
	Resilience (assimilation)	Day and Wadey (2016)
Interpersonal	Greater appreciation of friends and family	Crawford et al. (2014), Galli and Reel (2012a), and Tamminen et al. (2013)
	Increased altruism/prosocial behavior	Galli and Reel (2012a), Howells and Fletcher (2015, 2016), Roy-Davis et al. (2017), and Tamminen et al. (2013)
	Enhanced relationships	Day and Wadey (2016), Howells and Fletcher (2015), Roy-Davis et al. (2017), Salim et al. (2016), and Wadey et al. (2011, 2013)
	Greater empathy	Salim et al. (2016)
	Less judgmental	Crawford et al. (2014)
	Prioritizing self over others	McDonough et al. (2011)
	Being able to speak out/disclose	Howells and Fletcher (2016), Salim et al. (2016), and Wadey et al. (2011)
Physical	Better sport functioning/superior performance/physically stronger	Day and Wadey (2016), Galli and Reel (2012a), Howells and Fletcher (2015, 2016), Roy-Davis et al. (2017), Salim et al. (2016), Sarkar et al. (2015), Tamminen et al. (2013), Udry et al. (1997), and Wadey et al. (2011, 2013)
	Increase health and fitness	Crawford et al. (2014) and McDonough et al. (2011)
	Improved health behaviors and knowledge	Roy-Davis et al. (2017) and Salim et al. (2016)
	Gaining sense of mastery	Salim et al. (2016)
	Body self-relationship/body awareness	Roy-Davis et al. (2017) and Wadey et al. (2013)

An increased awareness of the self in the context of sport and the wider world was evident in perceptions about whether growth had occurred. Indicators such as the identification of new possibilities (viz. Day, 2013; Galli and Reel, 2012a; Howells and Fletcher, 2016), a greater appreciation of life (viz. Crawford et al., 2014; Day, 2013; Day and Wadey, 2016; Howells and Fletcher, 2016; McDonough et al., 2011; Roy-Davis et al., 2017; Salim et al., 2016), improved personal strength (viz. Galli and Reel, 2012a; McDonough et al., 2011; Roy-Davis et al., 2017; Tamminen et al., 2013; Udry et al., 1997; Wadey et al., 2011, 2013), an improved level of knowledge related to either the athletes' sports or health-related behaviors (viz. Roy-Davis et al., 2017; Salim et al., 2016; Wadey et al., 2011, 2013), and spiritual change (viz. Galli and Reel, 2012a; Howells and Fletcher, 2015; McDonough et al., 2011) were reported. On this latter point, females reported more spiritual change than males (Galli and Reel, 2012b).

3.4.2.2 Interpersonal indicators

Other indicators were interpersonal in that they involved a change in the interactions or approach toward others. Five studies reported enhanced relationships as being prevalent (viz. Howells and Fletcher, 2015; Roy-Davis et al., 2017; Salim et al., 2016; Wadey et al., 2011, 2013), three studies identified relating to others or positive changes in social relationships as evident (viz. Crawford et al., 2014; Day and Wadey, 2016; Galli and Reel, 2012a; Tamminen et al., 2013), and three studies identified a greater appreciation of friends and family (viz. Crawford et al., 2014; Galli and Reel, 2012a; Tamminen et al., 2013). In one study (viz. Galli and Reel, 2012b), females reported a stronger ability to relate to others than males. There was also evidence in a change in how participants who reported growth approached other people, with five studies reporting increased altruism or prosocial behavior (viz. Galli and Reel, 2012a; Howells and Fletcher, 2015; Howells and Fletcher, 2016; Roy-Davis et al., 2017; Tamminen et al., 2013), one study reporting higher levels of empathy (viz. Salim et al., 2016), and another identifying participants as being less judgmental (viz. Crawford et al., 2014). Engagement with others was an important indicator as three studies identified a move toward disclosure and the ability to speak out to others (viz. Howells and Fletcher, 2016; Salim et al., 2016; Wadey et al., 2011). One study identified growth as occurring when participants acknowledged that as a consequence of their adversity they were now prioritizing themselves over others (viz. McDonough et al., 2011).

3.4.2.3 Physical indicators

Indicators of growth were also physical in nature and tended to relate to the engagement in sport. These indicators were common across all of the 14 studies that addressed indicators of growth. Physical indicators of growth comprised of subcategories involving: improved physical outcomes resulting in better athletic functioning, superior performance, being physically stronger, and/or increased fitness (viz. Crawford et al., 2014; Day, 2013; Day and Wadey, 2016; Galli and Reel, 2012a; Howells and Fletcher, 2015, 2016; McDonough et al., 2011; Roy-Davis et al., 2017; Tamminen et al., 2013; Salim et al., 2016; Sarkar et al., 2015; Udry et al., 1997; Wadey et al., 2011, 2013).

4 DISCUSSION

Building on the increasing attention from sport psychology scholars on the topic of growth following adversity in competitive sport (cf. Tamminen and Neely, 2016), in this paper, we reviewed and synthesized the findings of empirical research conducted in this area. Overall, the findings support the notion that athletes have the potential to benefit from difficulty (Fletcher and Sarkar, 2012; Hardy et al., 2017) and indicate that the concept of growth following adversity provides an appropriate empirical lens to better understand why some athletes succumb to adversity whereas others are able to grow from the experience (cf. Gucciardi, 2017; Howle and Eklund, 2017; Sarkar and Fletcher, 2017). It is worth emphasizing that, although the findings support the potential for growth following adversity, this should in no way undermine the traumatic nature of adversity or imply that individuals should be expected to benefit from adverse events. Rather, the findings offer a narrative of hope that suggests negative responses to adversity are not necessarily terminal and that individuals may benefit in some way from the difficulties they experience. In this section, we address the definitions and theories that have informed the body of research, and we discuss the associated findings related to the adversity- and growth-related experiences of competitive athletes.

4.1 DEFINITIONS AND THEORY

In a similar manner to the terminology used in the wider growth literature, there was disparity in the reviewed studies regarding the terms used to study the adversity- and growth-related experiences of competitive athletes. This variation existed both in the conceptualization of the negative events and experiences, and in the conceptualization of the growth process. Regarding negative events and experiences, although there may be valid reasons for researchers to employ one term in preference to another, these were not always clearly articulated. Across the 17 studies, three terms were used to refer to athletes' negative events and experiences: *stressor*, *adversity*, and *trauma*. These terms have subtly different meanings. Stressors are defined as "the environmental demands (i.e., stimuli) encountered by an individual" (Fletcher et al., 2006, p. 359) encompassing various events, situations, and circumstances, and ranging in severity. At the more severe end of the stressor continuum, adversity typically encompasses "negative life circumstances that are known to be statistically associated with adjustment difficulties" (Luthar and Cicchetti, 2000, p. 858) or, more broadly, as "the state of hardship or suffering associated with misfortune, trauma, distress, difficultly, or a tragic event" (Jackson et al., 2007, p. 3). This conceptual drift from a predominately external "circumstance" to incorporating internal cognitions and affect shifts the emphasis to the relational "state" between an individual and his or her environment. In extending this conceptual trajectory, trauma is an individual's distressed response to adversity or the accumulation of stressors. Although scholars assert that it is possible to identify trauma independent of its cause by focusing on core symptoms associated with impairment (cf. Brewin et al., 2009), others argue that a causal stressor (viz. "the stressor criterion" or "traumatic event") needs to be identified and that this stressor maybe exposure to an actual or threatened event and maybe directly or indirectly experienced (American Psychiatric Association, 2013). Researchers studying growth in competitive sport should, therefore, carefully reflect on the specific focus, findings, and context of their research and adopt terminology accordingly. For example, using the term "stressor" for relatively mild environmental demands and events, and using "adversity" for more severe circumstances typically associated with adjustment difficulties. If the focus is more on an individual's experience or state then, although "adversity" may still be satisfactory, the term "trauma" may more accurately convey an individual's distressed response.

In addition to the variation in the conceptualization of negative events and experiences, disparity was also evident in the nomenclature of growth, although it was apparent that researchers are aware of these issues, as Wadey et al. (2011) articulated: "One issue within this body of research ... is that researchers have referred to this positive concept with a variety of terms, which has perpetuated conceptual ambiguity in this field of study" (p. 143). Notwithstanding the recognition of this issue by sport psychology researchers in 2011, this review identified a number of different terms that were used to refer to the broad construct of growth. These included *perceived benefits, stress-related growth (SRG), adversarial growth,* and *posttraumatic growth (PTG)*. Although Cho and Park (2013) have argued that there are differences in some of the growth-related terminology, which tend to revolve around the level of stress exposure and the severity of the adversity, Joseph et al. (2004) have argued that "the various measures of positive change all appear to be assessing the same broad construct" (p. 94). In a similar way that we do not suggest the utilization of one term to encompass all negative events or experiences, neither do we propose the use of one growth term over another. However, we argue that once a researcher has identified a term to conceptualize a negative event or experience (i.e., stressor, adversity, or trauma), the growth that occurs following that event or experience should be coherently conceptualized using the corresponding growth terminology and, if applicable given the study design, the appropriate measurement tool (e.g., SRGS for SRG; PTGI for PTG). Specifically, when using and exploring the terms stressor, adversity, and trauma, the growth that occurs from that (negative) event or experience should be conceptualized accordingly as SRG, adversarial growth, and PTG, respectively. Furthermore, in addition to definitional and terminology issues, it was apparent that some studies identified in this review were not informed by theories or models of growth. Given that scholars have proposed several theoretical explanations of growth with different characteristics (see Table 1), researchers should pay careful attention to this issue in future growth inquiry. Indeed, transparency and coherence in definitions and theory will be increasingly important to consider for researchers investigating growth following adversity in competitive sport especially with recent research (viz. Roy-Davis et al., 2017) introducing a further definition and theory (Sport Injury-Related Growth; SIRG) to the sport psychology literature.

4.2 ADVERSITY-RELATED EXPERIENCES

A range of adversity-related experiences were detailed in the 17 studies identified in this review, with some researchers focusing on a single adversity and others focusing on multiple adversities. For those who explored growth following a single adversity, the majority concentrated on one (predetermined) type of adversity (e.g., injury) while others focused on a participant-selected adversity (i.e., his or her most negative event or experience). While some scholars (e.g., Roy-Davis et al., 2017) have argued for investigating one specific adversity type, based on biological and evolutionary research, Christopher (2004) argued that it is more useful to focus on the chronic and acute nature of the event or experience as opposed to its specific content. For example, cancer constitutes a progressive and ongoing threat, with an internal source, and may be classified as chronic, whereas injuries have a well-established time frame for onset and termination, are often generated from external sources, and are largely acute in nature. These adversity-related experiences are likely to be appraised in fundamentally different ways. Notwithstanding these issues, focusing on an individual event or experience is perhaps providing a rather simplistic perspective. As Howells and Fletcher (2015) remarked, one adversity inevitably leads to others and, thus, it is injudicious to assume that they can be considered separate from the wider human experience. Accordingly, it may be more ecologically valid to focus on multiple events in relation to an individual's allostatic overload (i.e., the chronic, cumulative effect of [multiple] adversities that exceed an individual's coping skills) because high levels of allostatic load tend to hinder the growth process (see Ruini et al., 2015).

For the majority of athletes in the 17 studies synthesized in this review, the negative events and experiences involved deleterious cognitive and emotional responses that are analogous to the shattering of schemas as described by Janoff-Bulman (1989). The focus on cognition as a response to adversity appears to be fundamental in the growth process since an athlete's perception of a negative event may contribute more to their further development than the event itself (cf. Toering, 2017). The cognitive and emotional responses were accompanied by both physical and behavioral responses that tended to relate to the nature of the adversity but also involved the development of stress-related illnesses. Importantly, although the responses to negative events and experiences have been presented in the results in a linear sequential fashion in line with the zeitgeist of the primacy of cognition and its impact on emotional responses (cf. Lazarus, 1982), it would be informative to explore nonlinear dual processing theories (e.g., parallel distributive processing theory; McClelland and Rogers, 2003) to investigate cognitive appraisal and emotion regulation in response to negative events and experiences to further inform the growth process. Indeed, through the lens of dual-processing theories, it may be possible to further understand the automatic and controlled nature of growth.

4.3 GROWTH-RELATED EXPERIENCES

In respect of the identification of both mechanisms and indicators of growth, the findings of this review are consistent with reviews in the wider growth literature (e.g., Hefferon et al., 2009; Koutrouli et al., 2012). An overarching observation is that growth manifests itself within, and is perceived by, individuals differently. When growth does occur, it appears to do so in different ways and at different rates both within and across individuals, particularly given the everchanging and multilayered nature of stressful encounters. In identifying the mechanisms of growth, mechanisms were characterized as being internal or external. The majority of mechanisms were internal (e.g., cognition-related) and were broadly consistent with those identified in the wider growth literature (e.g., Garnefski et al., 2008). Specifically, growth is facilitated when an individual has the internal resources, such as the cognitive schema, sufficient to transform the negative event or experience into learning (e.g., Sarkar et al., 2015), to attribute meaning (particularly significance) to the adversity (e.g., Day, 2013), and to translate this meaning into adaptive behavior (e.g., Roy-Davis et al., 2017). In terms of external mechanisms, unsurprisingly, social support that was complex, responsive, and flexible enough to mitigate any pathological response was a key external mechanism to promote growth (e.g., McDonough et al., 2011). Interestingly, however, this review also identified an important sport-specific nuance in athletes' growth-related experiences. To illustrate, the role of sport provided a novel insight into external mechanisms of growth. Specifically, sport had a significant role in reestablishing athletes' identities (e.g., Day and Wadey, 2016), empowering the participants (e.g., McDonough et al., 2011), and providing a safe place or sanctuary to escape the impact of the adversity (e.g., Howells and Fletcher, 2015).

Although growth is usually conceptualized as a process, the majority of studies relied on indicators of growth to determine whether growth had occurred. These indicators were categorized as intrapersonal (e.g., greater appreciation for life), interpersonal (e.g., enhanced relationships), or physical (e.g., better sport functioning). Many of the interpersonal and intrapersonal indicators were consistent with those "traditional" indicators of growth (e.g., increased appreciation for life, more meaningful relationships, increased sense of personal strength, a change in priorities, richer existential and spiritual awareness) that are often identifiable through psychometric measures (i.e., PTGI, SRGS). However, there were other indicators of growth (e.g., derogation of adversity-related experiences) that were in line with the theoretical explanation of motivated illusions (cf. Taylor, 1983) and inferred an "alternative" representation of growth that may be characterized as illusory (cf. Sumalla et al., 2009; Maercker and Zoellner, 2004; Zoellner and Maercker, 2006a) or resulting from a process of assimilation (cf. Joseph and Linley, 2005). To date, Howells and Fletcher (2016) are the only researchers to critically consider illusory growth in competitive sport, thereby emphasizing the importance of not only examining indicators of actual growth but also understanding individuals' *perceptions* of growth. Furthermore, of particular interest was the identification of physical indicators of growth (cf. Hefferon et al., 2010). Without the parameters of measures that mainly assess interpersonal and intrapersonal indicators of growth, the qualitative studies identified that (irrespective of the adversity) athletes engaged in competitive sport experienced positive physical indicators of growth emerging from the adversity. Although the specific physical indicators differed across levels, for example, elite athletes identified superior performance (e.g., Sarkar et al., 2015), and club-level participants identified increased fitness (e.g., McDonough et al., 2011), the findings appear unequivocal that growth in competitive sport encompasses physical indicators. Importantly, these physical indicators have not been reported in the wider growth literature and therefore may be specific to growth in sport performers.

4.4 LIMITATIONS AND FUTURE RESEARCH

This systematic review (and the reporting of some of the studies identified as part of this review) is limited by the thematic reporting of the findings using discrete and linear categorization. This may mask the complexities that are evident in the adversity- and growth-related experiences of competitive athletes. To elucidate, in this review, the responses to negative events and experiences were categorized into cognitive, emotional, behavioral, and physical categories but this reporting format can obscure the interactional, cyclical, and multilayered complexities that characterize the responses to adversity. This reinforces the need for studies to be informed by models of growth that focus on these complexities to provide a theoretical framework for in-depth understanding. Moreover, due to the continually evolving nature of growth, researchers should use longitudinal methods by engaging with participants over time, and by collecting data on multiple occasions. An important finding of this review was the identification (via qualitative research) of indicators of growth that had physical characteristics and other indicators that were illusory in nature. Although recent research (viz. Boals and Schuler, in press) has advocated that a revised version of the Stress-Related Growth Scale (SRGS-R) is less prone to reports of illusory growth, continued improved measurement of growth is vital to our understanding of how athletes grow following adversity. These improvements and adaptations should also include aspects that recognize the physical indicators of growth that are characteristic in competitive sport. Lastly, given the growing body of research that exists in this area, intervention studies that can apply these findings in competitive sport are warranted. Although some interventions exist in the wider growth literature (e.g., Tedeschi and McNally, 2011), to the best of our knowledge, no interventions exist in sport. It is hoped that this review and synthesis will facilitate understanding and inform practice in this area. The study of growth following adversity in competitive sport is gaining momentum and the emerging findings have important theoretical and applied implications for the field of sport psychology.

REFERENCES

* References preceded by an asterisk were included in the systematic review.

Affleck, G., Tennen, H., Croog, S., Levine, S., 1987. Causal attribution, perceived benefits, and morbidity after a heart attack: an 8-year study. J. Consult. Clin. Psychol. 55 (1), 29–35. http://dx.doi.org/10.1037//0022-006X.55.1.29.

Almeida, P.L., Luciano, R., Lameiras, J., Buceta, J.M., 2014. Beneficios percibidos de las lesiones deportivas: Estudio cualitativo en futbolistas profesionales y semiprofesionales. Revista de Psicología del Deporte 23 (2), 457–464.

American Psychiatric Association, 2013. Diagnostic and Statistical Manual of Mental Disorders: DSM-5. American Psychiatric Association, Washington, DC.

Bartone, P.T., Ursano, R.J., Wright, K.M., Ingraham, L.H., 1989. The impact of a military air disaster on the health of assistance workers: a prospective study. J. Nerv. Ment. Dis. 177 (6), 317–328.

Boals, A., Schuler, K.L., (in press). Reducing reports of illusory posttraumatic growth: a revised version of the Stress-Related Growth Scale (SRGS-R), Psychol. Trauma. http://dx.doi.org/10.1037/tra0000267. [Epub ahead of print].

Brewin, C.R., Lanius, R.A., Novac, A., Schnyder, U., Galea, S., 2009. Reformulating PTSD for *DSM-V*: life after criterion. J. Trauma. Stress 22 (5), 366–373. http://dx.doi.org/10.1002/jts.20443.

Calhoun, L.G., Tedeschi, R.G., 1998. Posttraumatic growth: future directions. In: Tedeschi, R.G., Park, C.L., Calhoun, L.G. (Eds.), Posttraumatic Growth: Theory and Research on Change in the Aftermath of Crisis. Lawrence Erlbaum, Mahwah, NJ, pp. 215–238.

Calhoun, L.G., Cann, A., Tedeschi, R.G., 2010. The posttraumatic growth model: sociocultural considerations. In: Weiss, T., Berger, R. (Eds.), Posttraumatic Growth and Culturally Competent Practice: Lessons Learned From Around the Globe. Wiley Online Library, Hoboken, NJ, pp. 1–14. http://dx.doi.org/10.1002/9781118270028.ch1.

Carver, C.S., 1997. You want to measure coping but your protocol' too long: consider the brief cope. Int. J. Behav. Med. 4 (1), 92–100. http://dx.doi.org/10.1207/s15327558ijbm0401_6.

Centre for Reviews and Dissemination, 2009. Systematic Reviews: CRD's Guidance for Undertaking Reviews in Health Care. University of York, York.

Chickering, A.W., Reisser, L., 1993. Education and Identity, second ed. Jossey-Bass, Hoboken, NJ.

Cho, D., Park, C.L., 2013. Growth following trauma: overview and current status. Terapia Psicologica 31 (1), 69–79. http://dx.doi.org/10.4067/S0718-48082013000100007.

Christopher, M., 2004. A broader view of trauma: a biopsychosocial-evolutionary view of the role of the traumatic stress response in the emergence of pathology and/or growth. Clin. Psychol. Rev. 24 (1), 75–98. http://dx.doi.org/10.1016/j.cpr.2003.12.003.

Collins, J.A., Fauser, B.C.J.M., 2005. Balancing the strengths of systematic and narrative reviews. Hum. Reprod. Update 11, 103–104. http://dx.doi.org/10.1093/humupd/dmh058.

Collins, D., MacNamara, Á., 2012. The rocky road to the top. Sports Med. 42, 907–914. http://dx.doi.org/10.1007/BF03262302.

Cook, D.J., Mulrow, C.D., Haynes, R.B., 1997. Systematic reviews: synthesis of best evidence for clinical decisions. Ann. Intern. Med. 126, 376–380.

*Crawford, J.J., Gayman, A.M., Tracey, J., 2014. An examination of post-traumatic growth in Canadian and American ParaSport athletes with acquired spinal cord injury. Psychol. Sport Exerc. 15 (4), 399–406. http://dx.doi.org/10.1016/j.psychsport.2014.03.008.

*Day, M.C., 2013. The role of initial physical activity experiences in promoting posttraumatic growth in Paralympic athletes with an acquired disability. Disabil. Rehabil. 35 (24), 2064–2072. http://dx.doi.org/10.3109/09638288.2013.805822.

*Day, M.C., Wadey, R., 2016. Narratives of trauma, recovery, and growth: the complex role of sport following permanent acquired disability. Psychol. Sport Exerc. 22, 131–138. http://dx.doi.org/10.1016/j.psychsport.2015.07.004.

Deci, E., Ryan, R., 1985. Intrinsic Motivation and Self-Regulation in Human Behaviour. Plenum Press, New York, NY.

Fletcher, D., Sarkar, M., 2012. A grounded theory of psychological resilience in Olympic champions. Psychol. Sport Exerc. 13 (5), 669–678. http://dx.doi.org/10.1016/j.psychsport.2012.04.007.

Fletcher, D., Hanton, S., Mellalieu, S.D., 2006. An organizational stress review: conceptual and theoretical issues in competitive sport. In: Hanton, S., Mellalieu, S.D. (Eds.), Literature Reviews in Sport Psychology. Nova Science, Hauppauge, NY, pp. 321–374.

Forsdyke, D., Smith, A., Jones, M., Gledhill, A., 2016. Psychosocial factors associated with outcomes of sports injury rehabilitation in competitive athletes: a mixed studies systematic review. Br. J. Sports Med. 50, 537–544. http://dx.doi.org/10.1136/bjsports-2015-094850.

*Galli, N., Reel, J.J., 2012a. 'It was hard, but it was good': a qualitative exploration of stress-related growth in division I intercollegiate athletes. Qual. Res. Sport Exerc. Health 4 (3), 297–319. http://dx.doi.org/10.1080/2159676X.2012.693524.

*Galli, N., Reel, J.J., 2012b. Can good come from bad? An examination of adversarial growth in Division I NCAA athletes. J. Intercollegiate Sport 5, 199–212.

Galli, N., Vealey, R.S., 2008. Bouncing back from adversity: athletes' experiences of resilience. Sport Psychol. 22 (3), 316–335.

Galli, N., Podlog, L., Wadey, R., Mellalieu, S.D., Hall, M.S., 2013. In: Need satisfaction, affect, and stress-related growth following sport injury: a mediation analysis. Paper Presented at North American Society for the Psychology of Sport and Physical Activity (NASPSPA), New Orleans, LA.

Garnefski, N., Kraaij, V., Schroevers, M.J., Somsen, G.A., 2008. Post-traumatic growth after a myocardial infarction: a matter of personality, psychological health, or cognitive coping? J. Clin. Psychol. Med. Settings 15 (4), 270–277. http://dx.doi.org/10.1007/s10880-008-9136-5.

Gledhill, A., Harwood, C., Forsdyke, D., 2017. Psychosocial factors associated with talent development in football: a systematic review. Psychol. Sport Exerc. 31, 93–112. http://dx.doi.org/10.1016/j.psychsport.2017.04.002.

Gucciardi, D.F., 2017. The psychosocial development of world-class athletes: additional considerations for understanding the whole person and salience of adversity. Prog. Brain Res. 232, 127–132. http://dx.doi.org/10.1016/bs.pbr.2016.11.006.

Hardy, L., Barlow, M., Evans, L., Rees, T., Woodman, T., Warr, C., 2017. Great British medallists: psychosocial biographies of super-elite and elite athletes from Olympic sports. Prog. Brain Res. 232, 1–119

Harris, J.D., Quatman, C.E., Manring, M.M., Siston, R.A., Flanigan, D.C., 2014. How to write a systematic review. Am. J. Sports Med. 42 (11), 2761–2768. http://dx.doi.org/10.1177/0363546513497567.

Hefferon, K., Grealy, M., Mutrie, N., 2009. Post-traumatic growth and life threatening physical illness: a systematic review of the qualitative literature. Br. J. Health Psychol. 14 (2), 343–378. http://dx.doi.org/10.1348/135910708X332936.

Hefferon, K., Grealy, M., Mutrie, N., 2010. Transforming from cocoon to butterfly: the potential role of the body in the process of posttraumatic growth. J. Humanist. Psychol. 50 (2), 224–247. http://dx.doi.org/10.1177/0022167809341996.

Heil, J., 1993. Psychology of Sport Injury. Human Kinetics, Champaign, IL.

Hong, Q.N., Pluye, P., Bujold, M., Wassef, M., 2017. Convergent and sequential synthesis designs: implications for conducting and reporting systematic reviews of qualitative and quantitative evidence. Syst. Rev. 6, 61–75. http://dx.doi.org/10.1186/s13643-017-0454-2.

*Howells, K., Fletcher, D., 2015. Sink or swim: adversity- and growth-related experiences in Olympic swimming champions. Psychol. Sport Exerc. 16, 37–48. http://dx.doi.org/10.1016/j.psychsport.2014.08.004.

*Howells, K., Fletcher, D., 2016. Adversarial growth in Olympic swimmers: constructive reality or illusory self-deception? J. Sport Exerc. Psychol. 38, 173–186. http://dx.doi.org/10.1123/jsep.2015-0159.

Howle, T.C., Eklund, R.C., 2017. On elite and super-elite Great British athletes: some theoretical implications from Hardy et al.'s (2017) findings. Prog. Brain Res. 232, 121–125.

Jackson, D., Firtko, A., Edenborough, M., 2007. Personal resilience as a strategy for surviving and thriving in the face of workplace adversity: a literature review. J. Adv. Nurs. 60 (1), 1–9. http://dx.doi.org/10.1111/j.1365–2648.2007.04412.x.

Janoff-Bulman, R., 1989. Assumptive worlds and the stress of traumatic events: applications of the schema construct. Soc. Cogn. 7 (2), 113–136. http://dx.doi.org/10.1521/soco.1989.7.2.113.

Joseph, S., Linley, A., 2005. Positive adjustment to threatening events: an organismic valuing theory of growth through adversity. Rev. Gen. Psychol. 9 (3), 262–280. http://dx.doi.org/10.1037/1089-2680.9.3.262.

Joseph, S., Linley, P.A., 2006. Growth following adversity: theoretical perspectives and implications for clinical practice. Clin. Psychol. Rev. 26 (8), 1041–1053. http://dx.doi.org/10.1016/j.cpr.2005.12.006.

Joseph, S., Williams, R., Yule, W., 1993. Changes in outlook following disaster: the preliminary development of a measure to assess positive and negative responses. J. Trauma. Stress 6 (2), 271–279. http://dx.doi.org/10.1007/BF00974121.

Joseph, S., Linley, A.P., Harris, G.J., 2004. Understanding positive change following trauma and adversity: structural clarification. J. Loss Trauma 10 (1), 83–96. http://dx.doi.org/10.1080/15325020490890741.

Joseph, S., Murphy, D., Regel, S., 2012. An affective-cognitive processing model of posttraumatic growth. Clin. Psychol. Psychother. 19 (4), 316–324. http://dx.doi.org/10.1002/cpp.1798.

Koutrouli, N., Anagnostopoulos, F., Potamianos, G., 2012. Posttraumatic stress disorder and posttraumatic growth in breast cancer patients: a systematic review. Women Health 52 (5), 503–516. http://dx.doi.org/10.1080/03630242.2012.679337.

Kubler-Ross, E., 1969. On Death and Dying. Macmillan, New York, NY.

Lazarus, R.S., 1982. Thoughts on the relations between emotion and cognition. Am. Psychol. 37 (9), 1019–1024. http://dx.doi.org/10.1037/0003-066X.37.9.1019.

Linley, P.A., Joseph, S., 2004. Positive change following trauma and adversity: a review. J. Trauma. Stress 17 (1), 11–21. http://dx.doi.org/10.1023/B:JOTS.0000014671.27856.7e.

Luthar, S.S., Cicchetti, D., 2000. The construct of resilience: implications for interventions and social policies. Dev. Psychopathol. 12 (4), 857–885. http://dx.doi.org/10.1017/S0954579400004156.

Maercker, A., Zoellner, T., 2004. The Janus face of self-perceived growth: toward a two-component model of posttraumatic growth. Psychol. Inq. 15 (1), 41–48.

McClelland, J.L., Rogers, T.T., 2003. The parallel distributed processing approach to semantic cognition. Nat. Rev. Neurosci. 4, 310–322. http://dx.doi.org/10.1038/nrn1076.

*McDonough, M.H., Sabiston, C.M., Ullrich-French, S., 2011. The development of social relationships, social support, and posttraumatic growth in a dragon boating team for breast cancer survivors. J. Sport Exerc. Psychol. 33 (5), 627–648.

Pace, R., Pluye, P., Bartlett, G., Macaulay, A.C., Salsberg, J., Jagosh, J., Seller, R., 2012. Testing the reliability and efficiency of the pilot Mixed Methods Appraisal Tool (MMAT) for systematic mixed studies review. Int. J. Nurs. Stud. 49 (1), 47–53. http://dx.doi.org/10.1016/j.ijnurstu.2011.07.002.

Park, C., Cohen, L., Murch, R., 1996. Assessment and prediction of stress-related growth. J. Pers. 64 (1), 71–105. http://dx.doi.org/10.1111/j.1467-6494.1996.tb00815.x.

Pluye, P., Hong, Q.N., 2014. Combining the power of stories and the power of numbers: mixed methods research and mixed studies reviews. Annu. Rev. Public Health 35, 29–45. Retrieved on [31 January 2017] from, http://arjournals.annualreviews.org/eprint/qFxpDWrNzjzwjfkgtd4V/full/10.1146/annurev-publhealth-032013-182440.

Pluye, P., Robert, E., Cargo, M., Bartlett, G., O'Cathain, A., Griffiths, F., Boardman, F., Gagnon, M.P., Rousseau, M.C., 2011. Proposal: a mixed methods appraisal tool for systematic mixed studies reviews. Retrieved on [31 January 2017] from, http://mixedmethodsappraisaltoolpublic.pbworks.com.

Podlog, L., Lochbaum, M., Stevens, T., 2010. Need satisfaction, well-being, and perceived return-to-sport outcomes among injured athletes. J. Appl. Sport Psychol. 22 (2), 167–182. http://dx.doi.org/10.1080/10413201003664665.

Pope, C., Mays, N., Popay, J., 2007. Synthesising Qualitative and Quantitative Health Research: A Guide to Methods. Open University Press, Maidenhead, UK.

*Roy-Davis, K., Wadey, R., Evans, L., 2017. A grounded theory of sport injury-related growth. Sport Exerc. Perform. Psychol. 6 (1), 35–52. http://dx.doi.org/10.1037/spy0000080.

Ruini, C., Offidani, E., Vescovelli, F., 2015. Life stressors, allostatic overload, and their impact on posttraumatic growth. J. Loss Trauma 20 (2), 109–122. http://dx.doi.org/10.1080/15325024.2013.830530.

Sabiston, C.M., McDonough, M.H., Crocker, P.R., 2007. Psychosocial experiences of breast cancer survivors involved in a dragon boat program: exploring links to positive psychological growth. J. Sport Exerc. Psychol. 29 (4), 419–438.

*Salim, J., Wadey, R., Diss, C., 2015. Examining the relationship between hardiness and perceived stress-related growth in a sport injury context. Psychol. Sport Exerc. 19, 10–17. http://dx.doi.org/10.1016/j.psychsport.2014.12.004.

*Salim, J., Wadey, R., Diss, C., 2016. Examining hardiness, coping and stress-related growth following sport injury. J. Appl. Sport Psychol. 28 (2), 154–169. http://dx.doi.org/10.1080/10413200.2015.1086448.

Sarkar, M., Fletcher, D., 2017. Adversity-related experiences *are* essential for Olympic success: additional evidence and considerations. Prog. Brain Res. 232, 159–165.

*Sarkar, M., Fletcher, D., Brown, D.J., 2015. What doesn't kill me…: adversity-related experiences are vital in the development of superior Olympic performance. J. Sci. Med. Sport 18 (4), 475–479. http://dx.doi.org/10.1016/j.jsams.2014.06.010.

Smith, S.G., Sestak, I., Forster, A., Partridge, A., Side, L., et al., 2016. Factors affecting uptake and adherence to breast cancer chemoprevention: a systematic review and meta-analysis. Ann. Oncol. 27 (4), 575–590. http://dx.doi.org/10.1093/annonc/mdv590.

Smith, N., Kinnafick, F.E., Cooley, S.J., Sandal, G.M., (in press). Reported growth following mountaineering expeditions: the role of personality and perceived stress, Environ. Behav. http://dx.doi.org/10.1177/0013916516670447. [Epub ahead of print].

Souto, R.Q., Khanassov, V., Hong, Q.N., Bush, P.L., Vedel, I., Pluye, P., 2015. Systematic mixed studies reviews: updating results on the reliability and efficiency of the mixed methods appraisal tool. Int. J. Nurs. Stud. 52 (1), 500–501. http://dx.doi.org/10.1016/j.ijnurstu.2014.08.010.

Sumalla, E.C., Ochoa, C., Blanco, I., 2009. Posttraumatic growth in cancer: reality or illusion? Clin. Psychol. Rev. 29 (1), 24–33. http://dx.doi.org/10.1016/j.cpr.2008.09.006.

Swann, C., Keegan, R.J., Piggott, D., Crust, L., 2012. A systematic review of the experience, occurrence, and controllability of flow states in elite sport. Psychol. Sport Exerc. 13, 807–819. http://dx.doi.org/10.1016/j.psychsport.2012.05.006.

Tamminen, K., Neely, K.C., 2016. Positive growth in sport. In: Holt, N.L. (Ed.), Positive Youth Development Through Sport, second ed. Routledge, New York, NY, pp. 193–204.

*Tamminen, K.A., Holt, N.L., Neely, K.C., 2013. Exploring adversity and the potential for growth among elite female athletes. Psychol. Sport Exerc. 14 (1), 28–36. http://dx.doi.org/10.1016/j.psychsport.2012.07.002.

Taylor, S.E., 1983. Adjustment to threatening events: a theory of cognitive adaptation. Am. Psychol. 38 (11), 1161–1173. http://dx.doi.org/10.1037/0003-066X.38.11.1161.

Tedeschi, R.G., Calhoun, L.G., 1995. Trauma & Transformation: Growing in the Aftermath of Suffering. Sage, Thousand Oaks, CA.

Tedeschi, R.G., Calhoun, L.G., 1996. The posttraumatic growth inventory: measuring the positive legacy of trauma. J. Trauma. Stress 9 (3), 455–471. http://dx.doi.org/10.10007/BF02103658.

Tedeschi, R.G., Calhoun, L.G., 2004. Posttraumatic growth: conceptual foundations and empirical evidence. Psychol. Inq. 15 (1), 1–18. http://dx.doi.org/10.1207/s15327965pli1501_01.

Tedeschi, R.G., McNally, R.J., 2011. Can we facilitate posttraumatic growth in combat veterans? Am. Psychol. 66 (1), 19–24. http://dx.doi.org/10.1037/a0021896.

Toering, T., 2017. Eventual sport performance level: what about the role of type of sport, perception of critical life events, and practice quality? Prog. Brain Res. 232, 181–185.

*Udry, E., Gould, D., Bridges, D., Beck, L., 1997. Down but not out: athlete responses to season-ending injuries. J. Sport Exerc. Psychol. 19 (3), 229–248.

Van Tulder, M., Furlan, A., Bombardier, C., Bouter, L., Editorial Board of the Cochrane Collaboration Back Review Group, 2003. Updated method guidelines for systematic reviews in the Cochrane Collaboration Back Review Group. Spine 28 (12), 1290–1299.

*Wadey, R., Evans, L., Evans, K., Mitchell, I., 2011. Perceived benefits following sport injury: a qualitative examination of their antecedents and underlying mechanisms. J. Appl. Sport Psychol. 23 (2), 142–158. http://dx.doi.org/10.1080/10413200.2010.543119.

*Wadey, R., Clark, S., Podlog, L., McCullough, D., 2013. Coaches' perceptions of athletes' stress-related growth following sport injury. Psychol. Sport Exerc. 14 (2), 125–135. http://dx.doi.org/10.1016/j.psychsport.2012.08.004.

*Wadey, R., Podlog, L., Galli, N., Mellalieu, S.D., 2016. Stress-related growth following sport injury: examining the applicability of the organismic valuing theory. Scand. J. Med. Sci. Sports 26, 1132–1139. http://dx.doi.org/10.1111/sms.12579.

Watson, D., Clark, L.A., Tellegen, A., 1988. Development and validation of brief measures of positive and negative affect: the PANAS scales. J. Pers. Soc. Psychol. 54 (6), 1063–1070. http://dx.doi.org/10.1037/0022-3514.54.6.1063.

Wiese-Bjornstal, D.M., Smith, A.M., Shaffer, S.M., Morrey, M.A., 1998. An integrated model of response to sport injury: psychological and sociological dynamics. J. Appl. Sport Psychol. 10, 46–69. http://dx.doi.org/10.1080/10413209808406377.

Zoellner, T., Maercker, A., 2006a. Posttraumatic growth in clinical psychology—a critical review and introduction of a two component model. Clin. Psychol. Rev. 26 (5), 626–653. http://dx.doi.org/10.1016/j.cpr.2006.01.008.

CHAPTER

Effects of acute high-intensity exercise on cognitive performance in trained individuals: A systematic review

Sarah E. Browne*,†,1, Mark J. Flynn†, Barry V. O'Neill†, Glyn Howatson*,‡, Phillip G. Bell*,†, Crystal F. Haskell-Ramsay*

*Northumbria University, Newcastle upon Tyne, Tyne and Wear, United Kingdom
†GSK Human Performance Lab, Brentford, United Kingdom
‡Water Research Group, North West University, Potchefstroom, South Africa
1Corresponding author: Tel.: 191-2274875, e-mail address: browne.sarah@northumbria.ac.uk

Abstract

Background: High-intensity exercise is generally considered to have detrimental effects on cognition. However, high fitness levels are suggested to alleviate this effect.

Objectives: The specific objective of this review was to evaluate the literature on the effect of acute high-intensity exercise on cognitive performance in trained individuals.

Methods: Studies were sourced through electronic databases, reference lists of retrieved articles, and manual searches of relevant reviews. Included studies examined trained participants, included a high-intensity exercise bout, used a control or comparison group/condition, and assessed cognitive performance via general laboratory tasks during or ≤10min following exercise cessation.

Results: Ten articles met the inclusion criteria. Results indicated that the effect of acute high-intensity exercise on cognitive performance in trained individuals is dependent on the specific cognitive domain being assessed. Generally, simple tasks were not affected, while the results on complex tasks remain ambiguous. Accuracy showed little tendency to be influenced by high-intensity exercise compared to measures of speed.

Conclusion: Multiple factors influence the acute exercise–cognition relationship and thus future research should be highly specific when outlining criteria such as fitness levels, exercise intensity, and exercise mode. Furthermore, greater research is needed assessing more cognitive domains, greater exercise durations/types, and trained populations at high intensities.

CHAPTER 9 High-intensity exercise and cognition in trained individuals: A review

Keywords

High intensity, Acute exercise, Cognition, Trained, Athlete, Fitness

1 INTRODUCTION

The health benefits provided through participation in regular exercise are well established, with an abundance of evidence demonstrating increases in physical well-being and improvements to many metabolic parameters (Penedo and Dahn, 2005). What's more, additional benefits are provided for those engaged in exercise for the purpose of training, where exercise is undertaken at higher intensities and/or greater frequencies/durations (Pollock et al., 1998). A growing body of research has demonstrated that these benefits are not just limited to physical health but extend to improvements to brain function and cognition (Hillman et al., 2008). Numerous intervention studies have found superior cognitive performance in trained subjects compared to untrained (Colcombe and Kramer, 2003; Tomporowski and Ellis, 1986), signifying a positive influence of physical fitness on cognition. Furthermore, converging evidence from a number of neuroimaging and neurophysiological techniques provide support for these functional relationships, showing exercise and aerobic fitness level to be associated with profound changes in sensory, motor, and autonomic regions of the brain (Kramer and Erickson, 2007), alongside larger regional brain volumes (Tseng et al., 2013), reinforced neural networks (Nakata et al., 2010), and increased neuroplasticity (Knaepen et al., 2010).

There are a number of narrative, systematic, and meta-analytic reviews assessing the relationship between acute exercise and cognition (Brisswalter et al., 2002; Chang et al., 2012; Etnier et al., 1997, 2006; Kashihara et al., 2009; Lambourne and Tomporowski, 2010; Tomporowski, 2003; Tomporowski and Ellis, 1986). However, despite advances in our understanding, a consensus in this area is lacking. For example, studies have found positive effects (Hogervorst et al., 1996; McMorris and Graydon, 1997), negative effects (Chmura et al., 1997; Fery et al., 1997), and no effects (Bard and Fleury, 1978; McMorris and Graydon, 2000) of exercise on cognition. Literature reviews have proposed that these conflicting results are due to a number of moderating variables that influence the exercise–cognition relationship; consequently conclusions on the overall effect of exercise on cognition can only be drawn when moderators are controlled. Indeed, this highlights the complex relationship between exercise and cognitive performance and thus, alongside broad all-encompassing reviews, there is a need for more detailed systematic reviews to help articulate the effects of exercise on cognitive performance within explicit parameters.

Two of the most recent meta-analyses indicate that the main moderators to be considered when examining the relationship between exercise and cognition include exercise intensity, exercise duration, exercise mode, cognitive task type, participant fitness, and study design (Chang et al., 2012; Lambourne and Tomporowski, 2010).

Surprisingly the number of intervention trials that have examined the relationship between high-intensity exercise and cognition is relatively small, especially when compared to the wealth of research investigating the effect of moderate intensity exercise on cognition (Chang et al., 2012; Kashihara et al., 2009). Generally, it is accepted that moderate exercise promotes positive changes in cognitive performance (Chang et al., 2012; Tomporowski, 2003); however, a consensus on the effect of high-intensity exercise has not been reached.

Compared to low- and moderate-intensity exercise, high intensities initiate significant metabolic, mechanical, and biochemical disturbances both peripherally and centrally. These physiological changes include; a significant disruption to intramuscular homeostasis (Jones et al., 2008), a disproportionate increase in the rate of peripheral fatigue development (Burnley et al., 2012), and an increase in the release of catecholamines including adrenaline and noradrenaline (McMorris et al., 2015). The large increase in adrenaline and thus arousal has led to much research investigating the inverted-U hypothesis, which postulates a reduction in cognitive performance at higher exercise intensities (Gutin, 1973). Despite this theory setting a general notion that high-intensity exercise has detrimental effects on cognition, empirical studies have consistently failed to detect a clear relationship (Tomporowski, 2003). Although there is evidence supporting the inverted-U theory (Aks, 1998; Brisswalter et al., 1995; Chmura et al., 1994; Gutin, 1973; McMorris and Graydon, 2000; Reilly and Smith, 1986), this effect is not always observed in athletic populations and individuals with greater fitness levels (Hüttermann and Memmert, 2014; Pesce, 2009). Consequently, one of the main methodological problems often proposed to explain the diversity of experimental results is the failure to control for physical fitness (Brisswalter et al., 2002; Tomporowski, 2003). In support of this, neuroimaging studies have shown greater metabolic workloads require increased brain activation of the motor cortices which come at the expense of other brain regions due to limited resource capacity (Dietrich and Sparling, 2004). This observation indicates that the influence of physical exercise on the efficacy of cognitive processes may be mediated by the level of activation induced by physical exertion. Within this framework, it is suggested that individuals with greater fitness levels who are accustomed to greater metabolic workloads create less physiological task constraints, which enables greater resource allocation for cognitive tasks (Brisswalter et al., 2002). Thus, it is hypothesized that the cognitive performance of individuals with greater fitness levels will not be negatively affected by high-intensity exercise as postulated by the inverted-U theory.

Trained individuals from various backgrounds such as sports, the military, and emergency services regularly engage in situations where they must respond quickly and make critical decisions during and following exposure to high physical workloads. The ability to maintain cognitive performance on such occasions can have large consequences, and thus prior to providing any recommendations or advice, an increased understanding within this area is required to establish the effect of high-intensity exercise on a range of cognitive domains. The limited number of studies investigating high-intensity exercise and cognition is surprising considering theories such as the inverted-U postulating detrimental effects. Furthermore, the need for more reviews to control moderators is emphasized when the impact of these moderators on the acute exercise–cognition relationship is considered. Adding more focus to these reviews via this method will enable the provision of more detailed conclusions regarding the effect of different exercise intensities on cognitive domains in specific populations. Thus, it is the purpose of this systematic review to critically assess the effect of acute high-intensity exercise on cognitive performance in trained individuals.

2 METHOD

This systematic review was performed following Cochrane Collaboration recommendations and criteria (Higgins and Green, 2011), which are in line with guidelines from the Preferred Reporting Items for Systematic Reviews and Meta-Analyses (PRISMA) statement (Moher et al., 2009).

2.1 ELIGIBILITY CRITERIA

PICO (population, intervention, comparison, and outcome) criteria were used to determine eligibility for this review. Accordingly, the following inclusion criteria were applied: studies included trained/highly fit participants; high-intensity exercise was the independent variable; a control and/or comparison group was used; performance on at least one general laboratory cognitive task was the dependent variable; and cognitive tests were administered either during or ≤10 min following exercise cessation. Included studies were designed to test the effect of high-intensity exercise (intervention) on cognitive task performance (outcome); consequently studies were not included if exercise was not the main intervention (i.e., assessment of pharmacological or nutritional interventions). In addition, only full-text original studies written in English were included.

2.1.1 Trained/highly fit participants: Definition
When aerobic fitness was provided, a population was deemed trained if they could be classified as having "excellent" or "superior" fitness according to the ACSM guidelines (American College of Sports Medicine, 2014). This provided an age- and gender-adjusted criteria for inclusion. When aerobic fitness was not provided, inclusion was based upon the description of the sample provided by the author(s). Studies were included if they examined at least one trained group; if studies examined low/moderately fit participants but compared them to a trained group they were included.

2.1.2 High-intensity exercise: Definition
In line with previous reviews an "acute" exercise period was defined as "exercise performed within a single day" (Chang et al., 2012). Consistent with the definitions used by McMorris et al. (2015), high-intensity exercise was defined as exercise

≥80% maximum power output (W_{max}). If W_{max} values were not reported but maximum oxygen uptake ($V\cdot O_{2max}$) or percentage maximum heart rate (%HR_{max}) were, the conversion formula provided by Arts and Kuipers (1994) was used to determine eligibility. This procedure has previously been applied by both McMorris and Hale (2012) and Schapschroer et al. (2016). If other indicators of intensity were provided, e.g., RPE or percentage heart rate reserve (%HRR), the exercise physiology literature was examined to ascertain whether or not the intensity was sufficient to qualify for inclusion. Exercise was deemed high intensity when exercise went to voluntary exhaustion or when maximal effort was required. Where exercise was intermittent, duration and time working at high intensities were used to determine eligibility.

2.2 INFORMATION SOURCES AND SEARCH STRATEGY

The search strategy included several steps to ensure that all possible relevant articles were obtained. First, an online search of electronic databases was conducted. To build the search criteria for database searches a PICO search strategy was employed (Higgins and Green, 2011); an example of the strategy can be seen in Table 1.

Table 1 Example of PubMed Search Strategy

Concept Search Strategy	Line	Entry
Trained individuals	1	Trained
	2	Athlete*
	3	Skill*
	4	Expert
	5	Recreational athlete
	6	1 or 2 or 3 or 4 or 5
Exercise intensity	7	Strenuous exercise
	8	High-intensity exercise
	9	Physical exertion
	10	Physical load
	11	Fatiguing exercise
	12	7 or 8 or 9 or 10 or 11
Cognitive performance	13	Cogniti*
	14	Executive function
	15	Memory
	16	Psychomotor
	17	Reaction time
	18	Attention
	19	Decision making
	20	13 or 14 or 15 or 16 or 17 or 18 or 19
	21	6 and 12 and 20

*truncation of word

To avoid database bias, searches were conducted on seven different electronic databases: Academic Search Complete, PsycARTICLES, PsycINFO, PubMed, Scopus, SPORTDiscus, and Web of Science. In addition, reference lists within retrieved articles were manually reviewed as well as reference lists from previous reviews relevant to the exercise and cognition literature (Chang et al., 2012; Etnier et al., 1997; Lambourne and Tomporowski, 2010; McMorris and Hale, 2012; Schapschroer et al., 2016; Tomporowski, 2003; Tomporowski and Ellis, 1986). Electronic database searches were done on February 4, 2017 and studies published anytime until the day of searching were considered.

2.3 STUDY SELECTION AND DATA COLLECTION PROCESS

Two study team members (SB and MF) screened for initial exclusion via titles and abstracts. If it was unclear from the title or abstract whether a study met the inclusion criteria, a secondary exclusion was conducted based on a review of full-text articles. If the full text was not available, first authors were contacted to obtain the manuscript. Both study team members independently scanned full-text articles to determine whether they met the inclusion criteria. Any disagreements were resolved by discussion and if an agreement could not be attained, inclusion was decided by a third study team member (CHR). The data collection process is presented in Fig. 1.

2.4 QUALITY ASSESSMENT

All studies included in the review were subject to quality assessment as suggested by the Cochrane guidelines (Higgins and Green, 2011). The quality of the studies was assessed by two members of the study team (Sarah E. Browne and Mark J. Flynn) who graded them with respect to their methodological strength using the quantitative assessment tool "QualSyst" (Kmet et al., 2004). To assess scientific rigor, QualSyst assesses 14 items that are scored depending on the degree to which the specific criteria were met (yes=2, partial=1, no=0) (Table 2). Items not applicable to a particular study design were marked "NA" and excluded from the calculation of the summery score. The total sum of all relevant items was divided by the total possible score to give each study a final summary score. Quality assessment was completed independently by two members of the study team (SB and MF), and disagreements were solved by consensus or by a third study team member (Crystal F. Haskell-Ramsay). A final summary score of $\geq 75\%$ indicated strong quality, a score of 55%–74% indicated moderate quality, and a score $\leq 54\%$ indicated weak quality.

2.5 ANALYSIS

Cognitive tasks were identified based upon the particular test that was administered and were subsequently classified into a general cognitive task category (dependent variable). Since only a few tests measure a single cognitive construct (Lezak, 2004), two of the most well-established compendiums for neuropsychological assessment were used to identify and categorize tasks (Lezak, 2004; Strauss et al., 2006). Both

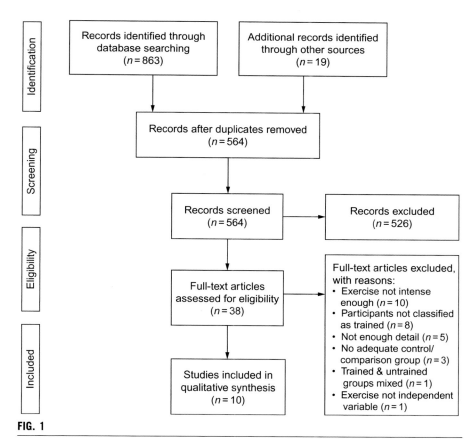

FIG. 1

PRISMA flowchart illustrating the literature search and selection process at each stage.

From Moher, D., Liberati, A., Tetzlaff, J., Altman, D.G., PRISMA Group, 2009. Preferred reporting items for systematic reviews and meta-analyses: the PRISMA statement. PLoS Med. 6, e1000097.

of these resources have been used for similar purposes in previous reviews (Chang et al., 2012; Roig et al., 2013) and provide precise definition and categorization of cognitive tests into different domains. Each cognitive task and the time it was done was classified as an outcome measure and the number of outcome measures were tallied. Consistent with similar reviews, the direction of each outcome measure was coded as positive (+), negative (−), and no effect (o).

3 RESULTS
3.1 STUDY SELECTION

Each step of the systematic search with the number of studies reviewed at each stage and main reasons for exclusion are shown in Fig. 1. A total of 863 articles were located through the systematic search; of these 318 were duplicates and therefore

Table 2 Quality Assessment

Study	Question Described	Appropriate Study Design	Appropriate Study Selection	Characteristics Described	Random Allocation	Researchers Blinded	Subjects Blinded	Outcome Measures Well Defined and Robust to Bias	Appropriate Sample Size	Analytic Methods Well Described	Estimate of Variance Reported	Controlled for Confounding	Results Reported in Detail	Conclusion Supported by Results	Rating
Brisswalter	2	2	2	2	0	N/A	N/A	2	1	2	2	0	2	2	Strong
Draper	2	2	1	1	0	N/A	N/A	2	1	1	2	0	2	1	Moderate
Guizani	2	2	2	2	N/A	N/A	N/A	2	1	2	2	0	2	2	Strong
Labelle	2	2	2	2	2	N/A	N/A	2	2	2	2	0	2	2	Strong
Llorens	2	2	2	2	1	N/A	N/A	2	1	2	2	1	2	2	Strong
Lo Bue-Estes	2	2	2	2	N/A	N/A	N/A	2	0	2	2	2	1	2	Strong
Reilly	2	2	2	1	0	N/A	N/A	2	1	2	2	0	1	1	Moderate
Smith	2	2	2	1	0	N/A	N/A	2	2	2	2	2	2	2	Strong
Tomporowski	2	2	1	2	0	N/A	N/A	2	1	0	1	0	1	2	Moderate
Whyte	2	2	2	2	2	N/A	N/A	2	2	2	2	2	2	2	Strong

2 indicates yes, 1 indicates partial, 0 indicates no, N/A, not applicable. Quality scores: ≥75% strong, 55%–74% moderate, ≤54% weak.

removed. An additional 19 articles were identified from additional records including relevant reviews and handsearching through the reference lists of the articles found through the database search. A total of 564 articles were screened by title and abstract, leading to the exclusion of 526 articles which did not meet the inclusion criteria. The main reasons for exclusion included: "cognitive performance not being the dependent variable," "exercise not being the independent variable," "cognitive tasks not being general laboratory based tasks," and "studies examining injuries or diseases." The remaining 38 articles were assessed for inclusion by reading the full text; this resulted in a further 28 articles being excluded. The primary reason for exclusion was the exercise intervention not meeting the required intensity (Del Percio et al., 2009; Elsworthy et al., 2016; Hancock and McNaughton, 1986; Hogervorst et al., 1996; Hüttermann and Memmert, 2014; Lemmink and Visscher, 2005; Pesce and Audiffren, 2011; Pesce et al., 2007, 2011; SjÖBerg, 1980). Several studies were excluded because participants did not meet the "trained" criteria (Aks, 1998; Fery et al., 1997; Fleury and Bard, 1987; Fleury et al., 1981; Levitt and Gutin, 1971; McMorris and Keen, 1994; Thomas et al., 2016; Wang et al., 2013). Five studies did not provide enough detail to enable inclusion with regard to either trained status (Malomsoki and Szmodis, 1970; Nibbeling et al., 2014; Strauss and Carlock, 1966) or exercise intensity (Guizani et al., 2006b; Reddy et al., 2014). Three studies did not provide an adequate control or comparison group (Chmura et al., 1994; Luft et al., 2009; Thomson et al., 2009). One study mixed trained and untrained participants in the intervention groups (Tsorbatzoudis et al., 1998), and exercise in one study was not the main independent variable (Coco et al., 2009). A total of 10 studies remained and were included in the review.

3.2 QUALITY ASSESSMENT

The quality assessment scores for the final 10 articles determined that 7 were of strong quality and 3 were of moderate quality (see Table 2). The strategy of blinding researchers and participants was not a realistic expectation of the current literature due to the obvious intervention; these two criteria were deemed "not appropriate" for all studies. There were two common methodological weaknesses among the studies. First, only three studies reported random allocation which referred to studies randomly allocating participants to each exercise intensity. Two studies were not able to be assessed against these criteria due to neither counterbalancing exercise intensities. The second common weakness was in the controlling of confounding variables. To fulfill these criteria, studies were reviewed to determine if they controlled for: participants taking medications and/or nutritional supplementation that may interfere with cognitive processes; high-intensity exercise and caffeine consumption prior to testing and recent concussion or injury. Five studies failed to report any type of control to minimize external conditions on cognitive function.

3.3 DESCRIPTIVE CHARACTERISTICS OF INCLUDED STUDIES

The characteristics of the studies regarding participants, exercise interventions, and cognitive tests are shown in Table 3. Overall, the data from 130 participants (89 males, 29 females) who met the "trained" criteria were included in this review. Mean age ranged from 19 to 31 years (mean 23.3 ± 3.8 years), and from the seven studies that provided specific information regarding the fitness level of the participants, mean $V \cdot O_{2max/peak}$ values was $57.5 \, mL \, kg^{-1} \, min^{-1}$ (mean range 50.6–$66.0 \, mL \, kg^{-1} \, min^{-1}$). Nine studies used participants involved in sports including; running, triathlon, soccer, hockey, athletics, fencing, rugby, Gaelic football, and hurling. Only two modes of exercise were used in study protocols with cycling being the most common ($n=6$) followed by running ($n=4$); the average exercise time at high intensity was $5.6 \pm 3.0 \, min$.

In total, four studies used a mixed between-subjects and within-subjects design, three studies used a within-subjects crossover design, and two used an independent between-subjects only design. Seven studies compared high-intensity exercise to moderate and low intensities (Brisswalter et al., 1997; Draper et al., 2010; Guizani et al., 2006a; Labelle et al., 2013; Lo Bue-Estes et al., 2008; Reilly and Smith, 1986; Smith et al., 2016): of these five studies included a rest condition (Brisswalter et al., 1997; Draper et al., 2010; Guizani et al., 2006a; Lo Bue-Estes et al., 2008; Smith et al., 2016) with two failing to counterbalance the order of the rest or exercise intensities (Guizani et al., 2006a; Lo Bue-Estes et al., 2008). The three remaining studies did not compare high-intensity exercise to other intensities but instead investigated one high-intensity exercise session and compared this to a rest condition (Llorens et al., 2015) or nonexercising group implementing counterbalanced (Whyte et al., 2015) and noncounterbalanced (Tomporowski et al., 1987) orders.

3.4 EFFECT OF ACUTE HIGH-INTENSITY EXERCISE OF COGNITIVE FUNCTION

Results for the effect of high-intensity exercise on cognitive performance in trained groups are presented in Table 4. In total, 4 cognitive domains were assessed across 10 studies with some studies assessing more than 1 domain. Reaction time and information processing were combined under the category "information processing" and were regarded as simple cognitive tests. Executive function, attention, and memory were considered complex.

3.4.1 Information processing

Five studies assessed information processing: of these four used RT tasks and one used a pursuit rotor task. Reaction time was assessed via SRT and CRT in four studies (Brisswalter et al., 1997; Draper et al., 2010; Guizani et al., 2006a; Lo Bue-Estes et al., 2008). In all four studies, high-intensity exercise was found to have no effect on speed or accuracy of SRT performance. CRT was assessed in two of these four

Table 3 Summary of Studies Examining the Effect of High-Intensity Exercise on Cognitive Performance

Author	Design	Participants	Exercise			Cognitive Performance Outcome				Comments
			Type and Intensity	Time at HI		Task	General Category	Time of Testing	Main Results	
Brisswalter et al. (1997)	Mixed	Trained: n=10 A=23.3±1.5 M/F=10/0 $\dot{V}O_2$=64.1±2.3 Untrained: n=10 A=23.7±1.8 M/F=10/0 $\dot{V}O_2$=42.2±3.0	Cycle ergometer 20%, 40%, 60%, 80% W_{max}	10 min		SRT	Information processing	Pre, post, and during	No effect on speed or accuracy during or post-HI exercise in trained group compared to lower intensity exercise	Rest condition Counterbalanced order
Draper et al. (2010)	Crossover	Trained: n=12 A=31.5±5 M/F=12/0 $\dot{V}O_2$=NR	Cycle ergometer 80% VT, 25% ΔVT, 75% ΔVT	3 min + time of task		SRT, CRT	Information processing	During	Speed: Faster CRT during HI exercise (75% ΔVT) compared to rest. No effect on movement time or total response time for SRT or CRT Accuracy: No effect of HI exercise on SRT or CRT compared to rest	Rest condition Counterbalanced order Speed assessed as RT, movement time, and total response time; considered as six outcome measures

Continued

Table 3 Summary of Studies Examining the Effect of High-Intensity Exercise on Cognitive Performance—cont'd

			Exercise		Cognitive Performance Outcome				
Author	Design	Participants	Type and Intensity	Time at HI	Task	General Category	Time of Testing	Main Results	Comments
Guizani et al. (2006a)	Mixed	*Trained: n=12* A=19.0±2.9 M/F=NR V̇O$_2$=50.7±5.6 *Untrained: n=12* A=20.8±3.9 M/F=NR V̇O$_2$=36.9±4.6	*Cycle ergometer* 20%, 40%, 60%, 80% W_{max}	6 min	SRT, CRT	Information processing	During	Speed: Faster CRT at 80% W_{max} in trained group compared to rest. No effect on SRT Accuracy: No effect of HI exercise compared to rest	Rest condition No counterbalanced order Simple effects analyses were conducted despite no interaction
Labelle et al. (2013)	Mixed	*Trained: n=16* A=24.6±2.5 M/F=9/7 V̇O$_2$=50.6±7.9 *Untrained: n=21* A=23.2±2.6 M/F=10/11 V̇O$_2$=38.3±5.2	*Cycle ergometer* 40%, 60%, 80% W_{max}	6.5 min	Stroop task	Executive function	During	No effect of HI exercise in trained group for speed or accuracy compared to lower intensities or untrained group	No rest condition Counterbalanced order
Llorens et al. (2015)	Mixed	*Trained: n=14* A=19–28 M/F=14/0 V̇O$_2$=58.4±3.0 *Untrained: n=13* A=19–28 M/F=13/0 V̇O$_2$=41.3±6.3	*Cycle ergometer* Incremental test to exhaustion	NR	Spatial attention task	Attention	Post	Speed: Faster RT after exercise in trained group compared to rest condition Accuracy: No effect	Rest condition Counterbalanced order

Study	Design	Participants	Exercise	Duration	Task	Cognitive domain	Timing	Results	Notes
Lo Bue-Estes et al. (2008)	Mixed	Trained: n = 9 A = 20.8 ± 0.9 M/F = 0/9 $\dot{V}O_2$ = 55.3 ± 7.9 Control: n = 8 A = 21.1 ± 2.2 M/F = 0/8 $\dot{V}O_2$ = NR	Treadmill 25%, 50%, 75%, 100% $\dot{V}O_2$	1 min	SRT, CPT, CS, WM, VSM, CSD	Information processing, attention, memory	Pre, post, and during	No change postexercise in SRT, CPT, CS, VSM, or CSD. Negative effect of HI on working memory at 100% $\dot{V}O_2$ and postexercise compared to preexercise	Rest group (control) No counterbalanced order Experimental group mean scores adjusted using control group mean scores on complex cognitive tasks. WM only test done "during" exercise. All but SRT were measured as throughput from the ANAM; a measure of correct hits (accuracy) in a set period of time
Reilly and Smith (1986)	Crossover	Trained: n = 10 A = 20.0 ± 0.8 M/F = 10/0 $\dot{V}O_2$ = 57.6 ± 7.7	Cycle ergometer 10%, 25%, 40%, 55%, 70%, 85% $\dot{V}O_2$	NR	Pursuit rotor task	Information processing	During	Reduced performance at 85% $\dot{V}O_2$ compared to control (10% $\dot{V}O_2$)	No rest condition (10% $\dot{V}O_2$ used as control) Counterbalanced order
Smith et al. (2016)	Crossover	Trained: n = 15 A = 28.0 ± 5.0 M/F = 6/9 $\dot{V}O_2$ = NR	Treadmill 70% HRR (until RPE 15–17) and 90% HRR (until RPE 18–19)	~4 min	Go/NoGo	Executive function	During	Speed: Slower RT during high-intensity exercise compared to rest and moderate intensity Accuracy: Higher omission and decision errors during high-intensity exercise compared to rest	Rest condition Counterbalanced order Accuracy provided two outcome measures split into omission and decision error rate

Continued

Table 3 Summary of Studies Examining the Effect of High-Intensity Exercise on Cognitive Performance—cont'd

Author	Design	Participants	Exercise Type and Intensity	Time at HI	Cognitive Performance Outcome Task	General Category	Time of Testing	Main Results	Comments
Tomporowski et al. (1987)	Between-subjects	Trained: $n=12$ A = 19–23 M/F = 8/4 $\dot{V}O_2 = 66.0 \pm$ NR Untrained: $n=12$ A = 17–29 M/F = 8/4 $\dot{V}O_2 = 41.1 \pm$ NR	Treadmill Incremental up to 80% $\dot{V}O_2$	NR	Free-recall memory	Memory	Post	and moderate intensity No effect of exercise on free-recall memory compared to untrained group	No rest condition No counterbalanced order Intensity of exercise was based on HR corresponding to % $\dot{V}O_2$
Whyte et al. (2015)	Between-subjects	Trained: $n=20$ A = 21.1 ± 1.3 M/F = 20/0 $\dot{V}O_2$ = NR Control: $n=20$ A = 21.2 ± 1.3 M/F = 20/0 $\dot{V}O_2$ = NR	Intermittent running and jumping Until RPE ≥18	~6 min	SDMT, stroop	Attention, executive function	Pre and post	Reduced stroop performance postexercise compared to the control group. No change in SDMT performance postexercise in either group	Rest group (control) Counterbalanced order average % of HR_{max} during exercise = 94.6%

~, approximately; A, age; ANAM, automated neuropsychological assessment metrics; between-subjects, independent between subjected design; CP, continual processing task; crossover, within-subjects crossover design; CRT, choice reaction time; CS, code substitution; CSD, code substitution delay; Δ, the difference between VT and $\dot{V}O_{2peak}$; HI, high intensity; HR_{max}, heart rate maximum; HRR, heart rate reserve; M/F, male/female; mixed, mixed between- and within-subjects design; n, number of subjects; NR, not reported; RPE, rate of perceived exertion; RT, reaction time; SDMT, symbol digits modality test; SRT, simple reaction time; $\dot{V}O_2$, maximal/peak oxygen consumption; VSM, visual spatial memory; VT, ventilator threshold; W_{max}, maximum power output; WM, working memory. Age is provided as mean ± standard deviation or range. $\dot{V}O_2$ value are normalized to body weight (mL·kg^{-1}·min^{-1}). Counterbalanced order refers to exercise intensity. Only participants whose data for the outcomes of interest were analyzed.

Table 4 Effects of High-Intensity Exercise on Each Cognitive Task Category

Cognitive Task Category	Studies	Total Outcome Measures	Outcome Measures					
			Post			During		
			+	o	−	+	o	−
Information processing								
Speed	n=4	n=11	−	n=2	−	n=2	n=7	−
Accuracy	n=4	n=7	−	n=1	−	−	n=5	n=1
Executive function								
Speed	n=2	n=2	−	−	−	−	n=1	n=1
Accuracy	n=3	n=4	−	−	n=1	−	n=1	n=2
Memory								
Speed	n=0	n=0	−	−	−	−	−	−
Accuracy	n=2	n=6	−	n=4	n=1	−	−	n=1
Attention								
Speed	n=1	n=1	n=1	−	−	−	−	−
Accuracy	n=3	n=3	−	n=3	−	−	−	−

studies (Draper et al., 2010; Guizani et al., 2006a), and in both, an improvement in speed of CRT was observed following high-intensity exercise. In contrast, the one remaining study assessed information processing using a pursuit rotor task (Reilly and Smith, 1986) and found a negative effect of exercise. In total, there were 18 outcome measures: 11 measuring speed and 7 measuring accuracy. Overall, no effect of high-intensity exercise on speed or accuracy was observed in the majority of outcome measures.

3.4.2 Executive function
Of the 10 studies included, 3 assessed executive function. Labelle et al. (2013) measured both speed and accuracy on a stroop task during exercise providing 2 outcome measures which showed no effect of high-intensity exercise. Similarly, Whyte et al. (2015) also used a stroop task but administered the test pre- and postexercise and measured correct responses (regarded as accuracy) only. These results demonstrated a reduced performance following high-intensity exercise. Differently, Smith et al. (2016) used a Go/NoGo task to measure speed and two types of accuracy (omission and decision errors) providing three outcome measures which all showed a deterioration in cognitive performance during high-intensity exercise. A total of six outcome measures were provided for executive function.

3.4.3 Memory
Two studies assessed memory. Lo Bue-Estes et al. (2008) assessed memory using an automated neuropsychological assessment metrics (ANAM) cognitive testing system. Results from the ANAM (throughput) provide a corrected response rate

measuring the number of correct responses in a set period of time; due to the nature of the task and to align these results with that of others, this review has regarded throughput as accuracy. Four domains of memory were assessed (short term, longer term, visual spatial, working) postexercise. In addition, working memory was assessed during as well as postexercise, providing a total of five outcome measures. All tasks other than working memory showed no effect of high-intensity exercise. Working memory was negatively affected during exercise at 100% $\dot{V}O_{2max}$ and immediately following high-intensity exercise when compared to the preexercise-rested state. Tomporowski et al. (1987) on the other hand assessed free-recall memory postexercise at 80% $\dot{V}O_{2max}$ and found no effect of exercise on memory compared to an untrained group. Overall, four outcome measures for memory found no effect of high-intensity exercise, while two found negative effects.

3.4.4 Attention

Three studies assessed attention following high-intensity exercise. Lo Bue-Estes et al. (2008) used a continual processing task assessing accuracy (as throughput from the ANAM) and found no effect. Llorens et al. (2015) used a spatial attention task assessing both speed and accuracy and similarly found no effect of high-intensity exercise on accuracy performance; contrarily a positive effect of RT on the memory task was observed. Whyte et al. (2015) on the other hand used a symbol digit modality test (SDMT) to assess correct responses (regarded as accuracy) following a high-intensity intermittent running and jumping protocol. The results of this study indicated no effect of high-intensity exercise on SDMT performance for both speed and accuracy. In total, four outcome measures were provided. No effect was found for accuracy, while one positive effect and one negligible effect were observed for speed.

3.4.5 Time of testing

The time at which cognitive tasks were administered presents one of the main methodological differences between studies. When assessing the effect of acute exercise on cognitive function, tasks can be administered during and/or after the cessation of exercise. Within the 10 studies included, 5 studies assessed cognitive function during exercise (Draper et al., 2010; Guizani et al., 2006a; Labelle et al., 2013; Reilly and Smith, 1986; Smith et al., 2016), 2 studies performed tasks both during and pre–postexercise (Brisswalter et al., 1997; Lo Bue-Estes et al., 2008), 1 study performed tasks before and immediately following exercise (Whyte et al., 2015), and the remaining 2 studies performed cognitive tasks postexercise only (Llorens et al., 2015; Tomporowski et al., 1987). There was no obvious pattern of a time-related effect on any cognitive domain; however, as there were fewer outcome measures for postexercise compared to during, this review cannot establish any clear conclusions regarding the time of testing.

4 DISCUSSION

This focused review set out with a specific aim to systematically review the existing literature surrounding the effect of acute high-intensity exercise on cognitive performance in trained individuals. Overall the majority of studies included suggested that both speed and accuracy in tasks requiring simple cognitive processing are not influenced during or following high-intensity exercise in trained individuals. Conversely, the results regarding more complex cognitive functions are more ambiguous. These results support others that emphasize the importance of considering the specific cognitive task type that is being assessed when synthesizing results, as the influence of exercise on cognition has been shown to be dependent on the specific cognitive domain that is being assessed (Lambourne and Tomporowski, 2010; McMorris et al., 2016).

Of the cognitive domains assessed, information processing requires the least resource due to the simple nature of the tasks. Although these tasks are primarily controlled by the prefrontal cortex, they require substantially less neural activity than complex tasks (McMorris and Hale, 2012). Within the information processing domain, reaction time was most commonly assessed with 4 studies measuring a total of 17 outcome measures for both speed and accuracy. Reaction time is a popular measure in the literature on acute exercise and cognition, particularly in studies designed to assess the inverted-U hypothesis (Lambourne and Tomporowski, 2010). Moreover, reaction time on simple tasks has previously been concluded to be sensitive to the effects of acute exercise (Tomporowski, 2003). In the most recent comprehensive meta-analysis on acute exercise and cognition, Chang et al. (2012) concluded there to be no significant effect of exercise on reaction time; this effect however was averaged over a range of exercise intensities, leading the authors to suggest that there may have been an influence of exercise intensity on reaction time that went unnoticed. The current review however found no effect on reaction time speed in 9 of the 11 observed outcome measures. Interestingly, the two remaining outcome measures were from the only two studies included in the review that assessed CRT with both these studies observing a positive effect of high-intensity exercise on CRT performance (Draper et al., 2010; Guizani et al., 2006a). This supports the conclusions of Draper et al. (2010) whose results indicate that the SRT and CRT are affected differently by exercise and should ideally be considered individually. Due to the limited number of studies in this review, all RT measures were considered as a whole as organized by Lezak (2004). To establish more evidence for differential effects on these two tasks more research is required.

Collectively, the lack of effect observed for information processing tasks disagrees with the results of Lambourne and Tomporowski (2010) and fails to conform to the heavily investigated inverted-U theory (Gutin, 1973). A potential reason for this, and an important criterion within this review, that differentiates it from others is the high fitness level of participants. The current results may therefore provide the support for the cardiovascular fitness hypothesis which states that individuals with greater fitness levels are affected less cognitively than those with lower levels of fitness (Etnier et al., 1997; Tomporowski and Ellis, 1986). While there is additional support for this hypothesis (Brisswalter et al., 1997; Hüttermann and Memmert, 2014; Levitt and Gutin, 1971; McMorris and Keen, 1994), Weingarten (1973) suggests beneficial effects of fitness on cognition are only observed during more complex tasks and thus do not impact simple cognitive processes. If one were to follow this direction, the results may alternatively support that of others who suggest reaction time may not be a particularly reliable or sensitive marker of cognitive performance relative to acute exercise (Chang et al., 2012).

Largely, the results for accuracy on information processing tasks indicate no effect of high-intensity exercise, supporting previous findings from both McMorris and Hale (2012) and Schapschroer et al. (2016). This result is surprising considering evidence, particularly in sporting conditions, demonstrating reductions in whole-body psychomotor skill accuracy with fatiguing physical exertion (Russell and Kingsley, 2011). One suggestion has indicated the likeliness of a ceiling effect for accuracy in healthy individuals when assessed in tasks such as reaction time (McMorris et al., 2016). Alternatively, it has been suggested that consistent failures to observe accuracy effects are due to the nature of the cognitive tasks. Many cognitive tasks for both simple and complex cognitive functions have been designed to measure performance through speed of processing, with accuracy measures merely there to safeguard against the speed–accuracy trade-off (McMorris and Hale, 2012). If accuracy is to be validly assessed, it is argued that tests must be used that control participants focus on solving the tasks with a reduced emphasis on speed. Interestingly, when accuracy is assessed via whole-body psychomotor skills, such as those performed in sporting situations, heavy exercise has been found to have a large effect (McMorris et al., 2015). This may potentially indicate that the diminishing accuracy observed with increasing levels of fatigue during sporting situations may be reasoned more toward reductions in neurophysiological mechanisms rather than cognitive components. Deciphering the extent to which precise areas contribute to reductions in whole-body performance provides a challenging yet important task for future research.

Complex cognitive functions are described as that of central executive processes involving several functions including planning, scheduling, working memory, multitasking, cognitive flexibility, and abstract thinking (Hillman et al., 2008; McMorris and Hale, 2012). These functions are heavily dependent on the activation of the prefrontal cortex which constitutes the highest level of cortical hierarchy (Fuster, 2001). Compared to the simple cognitive tasks assessed in this review, the impact of high-intensity exercise on higher cognitive processes is less clear. As a whole, 6 studies assessed executive function, attention, and memory providing a total of 16 outcome measures for both speed and accuracy (Fig. 2). Attention was assessed by three studies; while little can be concluded on speed of attention as it was only assessed by one study, all accuracy outcome measures indicated no difference with high-intensity exercise. Similarly the results for memory demonstrated a greater tendency to show no effect of high-intensity exercise; however, as five of the six outcome measures

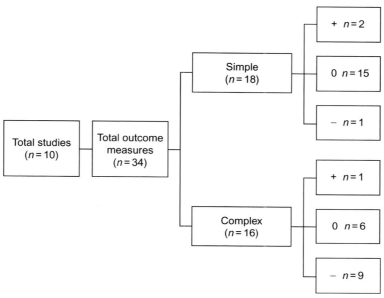

FIGURE 2

Overview of the effect of high-intensity exercise on simple and complex cognitive tasks. "+", positive effect on cognitive performance; "o", no change; "−", deterioration.

were obtained from one study, this review cannot establish any clear conclusions on this cognitive domain. Observations on measures of memory following acute exercise have failed to yield reliable results with some authors, suggesting that memory may not be a particularly sensitive construct to the effects of acute exercise (Chang et al., 2012). As demonstrated by Lo Bue-Estes et al. (2008) however, exercise often has diverse effects on different types of memory, and thus when assessing this construct, the specific type of memory the task is assessing is an important consideration. The lack of studies assessing higher cognitive functions such as attention and memory with high-intensity exercise is surprising considering the complex environments that trained individuals, such as athletes and military personal, are regularly confronted with when under high physical loads. The ability to assess unknown situations and make appropriate decisions relies heavily on central executive processes, and thus if strategies to improve decision making in these situations are to be established, a greater body of research on the effect of high physical loads on cognitive function is required.

Interestingly, measures of executive function have been shown to be particularly sensitive to exercise in general, with effects being significantly larger than other areas of cognition (Chang et al., 2012). This is interesting considering that substantial age-related deteriorations of executive function are positively influenced by physical training and fitness (Hillman et al., 2008), potentially indicating that executive function may be particularly sensitive to acute exercise and fitness. Of the complex tasks assessed in this review, executive function displayed the most sensitive results with regard to acute high-intensity exercise, with speed displaying both negative and negligible results, while accuracy demonstrated a greater tendency to be negatively affected by high-intensity exercise. While caution must be taken with any concluding remarks due to the limited sample of studies, these results do align with the previous research, suggesting that complex tasks are more likely to be affected by exercise than simple tasks during high-intensity exercise (Dietrich, 2006; McMorris and Graydon, 2000). These views have largely been derived from the transient hypofrontality theory, which proposes that high-intensity exercise causes a change in physiological state, momentarily disrupting brain homeostasis causing a modification in the brains resource allocation (Dietrich, 2003). Based upon this theory, the neural resources used for conducting exercise compete with the same resources necessary to perform cognitive processing. As the maintenance of high-intensity exercise requires large increases in neural resource demands to sustain activation within the motor and sensory regions, essential metabolic resources such as oxygen and glucose are reduced in other brain regions (Labelle et al., 2013). This reallocation of neural resources results in the temporary inhibition of brain regions not essential to performing exercise, such as areas of the prefrontal cortex involved in higher cognitive functions (Dietrich, 2006; Dietrich and Sparling, 2004).

During the review process of full-text articles, there were a number of studies that marginally failed to meet the classification for high-intensity exercise imposed by this review. The issue surrounding classification of exercise intensity raises an important matter within the exercise literature. Currently, there is a lack of consensus on the definition of "high-intensity" exercise. This has led to the use of a variety of exercise intensities being classified as "high intensity" (Hancock and McNaughton, 1986; Hogervorst et al., 1996; Hüttermann and Memmert, 2014). In trying to elucidate the effect of high-exercise intensities on cognitive performance, it is essential that consistency is maintained across studies. To maintain consistency and enable comparison of results, the current review classified high-intensity exercise in line with previous systematic and meta-analytic reviews as $\geq 80\%$ maximum power output or equivalent (McMorris and Hale, 2012; McMorris et al., 2015; Schapschroer et al., 2016). Exercise intensities this high have been shown to induce substantial increases in brain concentrations of neurotransmitters dopamine and noradrenaline beyond optimal thresholds (McMorris et al., 2016), and thus distinguish high- from moderate- and low-exercise intensities. Going forward, studies should aim to use consistent terminology when assessing the effect of acute exercise to provide more consensus within the literature.

Interestingly, the current results are dissimilar to those of a recent review which supported favorable effects of high-intensity exercise on both general and sport-specific cognitive tasks (Schapschroer et al., 2016). However, within the review by Schapschroer et al. (2016), only two studies investigated the effect of high-intensity exercise on the performance of general cognitive tasks, of which both were included in the current review (Guizani et al., 2006a; Llorens et al., 2015).

In agreement, the two respective studies did find positive effects of high-intensity exercise on cognitive performance. Notably however, these effects were found for the speed of performance on CRT tasks which, as previously stated, requires more evidence to establish differential effects, if any, between SRT and CRT during and/or following exercise. The difference in overall conclusions between both reviews is likely due to methodological differences. First, the general conclusions drawn from Schapschroer et al. (2016) encompass both general and sport-specific cognitive tests. Second, the review by Schapschroer et al. (2016) was solely based on athletic populations. These methodological differences are important considering research surrounding the "expert performance approach" which has evidenced consistently better performances by athletes on sport-specific cognitive tasks. Consequently, it is argued that measures of fundamental cognitive ability in simulated sport environments are confounded by athletes' superior declarative and procedural knowledge (Voss et al., 2010). As the current review only specified a fitness criterion for inclusion, the type of cognitive task was controlled for to reduce potential confounding.

4.1 CONSIDERATIONS FOR FUTURE RESEARCH

An important consideration for further research that has been highlighted in previous reviews is the exercise modality used when performing high-intensity exercise (Lambourne and Tomporowski, 2010; Schapschroer et al., 2016). This review highlighted a preference of the use of cycling compared to other modes of exercise. The amount of studies assessing trained individuals on exercise modes that they are not accustomed to is surprising as it reduces the ecological validity of the findings (Guizani et al., 2006a; Llorens et al., 2015; McMorris and Graydon, 1996, 1997). Lambourne and Tomporowski (2010) assessed the effect of cycling vs treadmill exercise and found cycling to elicit larger and more positive effects on cognitive performance compared to treadmill running which found negative effects. As running requires greater metabolic energy, it is plausible that sensory afferents influence the integration of cortical activation and lower the signal to noise to a greater extent than cycling exercise, thus resulting in less efficient cognitive processing (Lambourne and Tomporowski, 2010). This highlights an important consideration when designing studies. For example, assessing the effect of exercise on cognition using cycling protocols with soccer players may not provide ecologically valid results. To increase validity, where possible studies should consider the accustomed or usual exercise mode of the target population and use this when designing protocols. Notably, the systematic search of the literature found no study which assessed cognitive function following maximal strength exercises. This is surprising considering its relevance to many sports and thus more work in this area is needed.

A potential flaw of the literature as a whole is the use of short duration exercise only. The average exercise duration at high intensities in the current review was 5.6 min. While it is appreciated that exercise at high-intensities cannot be sustained for long periods of time, many sports require intermittent bursts at high intensities over prolonged periods or efforts that require the maintenance of a high percentage of $V\cdot O_{2max}$ for sustained durations. Research using transcranial magnetic stimulation has shown that shorter high-intensity exercise lasting around 6 min results in greater peripheral fatigue, whereas longer durations (>30 min) cause greater disturbances in central fatigue (Thomas et al., 2015). The large disturbances in central fatigue caused by prolonged high-intensity exercise could have a pronounced effect on cognitive function, thus presenting a further area of investigation.

Within this review, accounting for methodological differences between studies was challenging. For example, three studies did not used a rest group or condition (Brisswalter et al., 1997; Labelle et al., 2013; Reilly and Smith, 1986) but instead compared high-intensity exercise to low and moderate intensities. While these studies enable a comparison of high-intensity exercise with that of others, they fail to enable any conclusions to be drawn about high-intensity exercise compared to a rested state. Furthermore, two studies did not employ counterbalancing of exercise intensities (Guizani et al., 2006a; Lo Bue-Estes et al., 2008). These methodological differences therefore make it difficult to assess if changes in cognitive performance were due to the impact of high-intensity exercise or because of factors such as learning or time effects. In addition, attention should be given to the control of confounding variables. Within the quality criteria check, only four studies were deemed to have adequately controlled for confounding. One study met the partial control criteria, while five studies did not report any measures of controlling for confounding. Cognitive performance in healthy individuals is readily influenced by small changes in day-to-day living as well as things like prior exercise and nutrition; as such care should be taken to control and report confounding variables.

4.2 LIMITATIONS OF THE CURRENT REVIEW

Due to the limited number of studies, specific cognitive tasks could not be assessed separately and instead were assigned to general cognitive task categories. Although we believe this was a reasonable approach to examining the research question, this does reduce specificity of results which may have led to differences within each category being overlooked.

Another important limitation of this review is the methodology applied regarding exercise intensity. This review followed that of others and classified the physical intensity of exercise based on W_{max}, $V\cdot O_{2max}$, or HR_{max}. Studies that used other values of intensity such as HRR or RPE were thus compared to the studies that provided data on W_{max}, $V\cdot O_{2max}$, or HR_{max} and eligibility was accordingly determined. Although this limitation was relevant to only three studies which we were confident met the eligibility criteria, this strategy undoubtedly contains subjectivity as previously acknowledged (McMorris and Hale, 2012; Schapschroer et al., 2016).

The strategy to classify populations within studies as "trained" was also partially reliant on subjectivity. The first criterion was an objective measure of participants $V\cdot O_{2max}$, which was required to meet "excellent" or "superior" classification via the ACSM guidelines (American College of Sports Medicine, 2014). When $V\cdot O_{2max}$ was not provided however, inclusion was based upon the description of the sample provided by the author(s) which inevitably invites subjectivity. However, limiting study inclusion to VO_{2max} provision may have excluded studies which clearly included trained participants as demonstrated by their level of sport or hours of training per week. Thus it was felt that criteria applied provided the best opportunity of including all relevant articles. While everything was done to ensure fitness levels met the required criteria, confidence cannot be guaranteed.

Finally, only articles written in English were included in this systematic review. This is a criterion applied within many reviews and has been deemed an acceptable method due to English being the most commonly understood and published language. Nonetheless, articles important to this review may be written in other languages.

5 CONCLUSION

Previously, there has been no clear consensus on the effects of high-intensity exercise on cognitive function. It is probable that this is due to the surprisingly limited pool of studies that have assessed cognitive performance during and following high-intensity exercise. Sporting and military situations provide just a few examples of when trained individuals are required to simultaneously handle high levels of physical exertion and cognitive loads; thus, it is important that the impact of high-intensity exercise, particularly on complex cognitive functions, is sufficiently researched. The results of the current review indicate there to be no significant effects of high-intensity exercise on measures of simple cognitive processing in individuals with high fitness levels; the effects on complex functions however remain ambiguous. Since multiple factors influence the acute exercise–cognition relationship, future research should control for potential moderators and thus be highly specific when determining and defining exercise intensities, participant fitness, and cognitive domains. The small number of studies included in this review highlights an area in need of more research, particularly surrounding prolonged high-intensity exercise and resistance exercise with a greater array of cognitive domains being assessed.

REFERENCES

American College of Sports Medicine, 2014. *ACSM's Guidelines for Exercise Testing and Prescription*, ninth ed. Lippincott Williams & Wilkins, Baltimore, MD.

Aks, D.J., 1998. Influence of exercise on visual search: implications for mediating cognitive mechanisms. Percept. Mot. Skills 87, 771–783.

Arts, F., Kuipers, H., 1994. The relation between power output, oxygen uptake and heart rate in male athletes. Int. J. Sports Med. 15, 228–231.

Bard, C., Fleury, M., 1978. Influence of imposed metabolic fatigue on visual capacity components. Percept. Mot. Skills 47, 1283–1287.

Brisswalter, J., Durand, M., Delignieres, D., Legros, P., 1995. Optimal and non-optimal demand in a dual task of pedalling and simple reaction time: effects on energy expenditure and cognitive performance. J. Hum. Mov. Stud. 29, 15–34.

Brisswalter, J., Arcelin, R., Audiffren, M., Delignieres, D., 1997. Influence of physical exercise on simple reaction time: effect of physical fitness. Percept. Mot. Skills 85, 1019–1027.

Brisswalter, J., Collardeau, M., René, A., 2002. Effects of acute physical exercise characteristics on cognitive performance. Sports Med. 32, 555–566.

Burnley, M., Vanhatalo, A., Jones, A.M., 2012. Distinct profiles of neuromuscular fatigue during muscle contractions below and above the critical torque in humans. J. Appl. Physiol. 113, 215–223.

Chang, Y.-K., Labban, J., Gapin, J., Etnier, J.L., 2012. The effects of acute exercise on cognitive performance: a meta-analysis. Brain Res. 1453, 87–101.

Chmura, J., Nazar, K., Kaciuba-Uścilko, H., 1994. Choice reaction time during graded exercise in relation to blood lactate and plasma catecholamine thresholds. Int. J. Sports Med. 15, 172–176.

Chmura, J., Krysztofiak, H., Ziemba, A.W., Nazar, K., Kaciuba-Uścilko, H., 1997. Psychomotor performance during prolonged exercise above and below the blood lactate threshold. Eur. J. Appl. Physiol. Occup. Physiol. 77, 77–80.

Coco, M., Di Corrado, D., Calogero, R.A., Perciavalle, V., Maci, T., Perciavalle, V., 2009. Attentional processes and blood lactate levels. Brain Res. 1302, 205–211.

Colcombe, S., Kramer, A.F., 2003. Fitness effects on the cognitive function of older adults: a meta-analytic study. Psychol. Sci. 14, 125–130.

Del Percio, C., Babiloni, C., Infarinato, F., Marzano, N., Iacoboni, M., Lizio, R., Aschieri, P., Ce, E., Rampichini, S., Fano, G., Veicsteinas, A., Eusebi, F., 2009. Effects of tiredness on visuo-spatial attention processes in elite karate athletes and non-athletes. Arch. Ital. Biol. 147, 1–10.

Dietrich, A., 2003. Functional neuroanatomy of altered states of consciousness: the transient hypofrontality hypothesis. Conscious. Cogn. 12, 231–256.

Dietrich, A., 2006. Transient hypofrontality as a mechanism for the psychological effects of exercise. Psychiatry Res. 145, 79–83.

Dietrich, A., Sparling, P.B., 2004. Endurance exercise selectively impairs prefrontal-dependent cognition. Brain Cogn. 55, 516–524.

Draper, S., Mcmorris, T., Parker, J.K., 2010. Effect of acute exercise of differing intensities on simple and choice reaction and movement times. Psychol. Sport Exerc. 11, 536–541.

Elsworthy, N., Burke, D., Dascombe, B.J., 2016. Physical and psychomotor performance of Australian football and rugby league officials during a match simulation. J. Sports Sci. 34, 420–428.

Etnier, J.L., Salazar, W., Landers, D.M., Petruzzello, S.J., Han, M., Nowell, P., 1997. The influence of physical fitness and exercise upon cognitive functioning: a meta-analysis. J. Sport Exerc. Psychol. 19, 249–277.

Etnier, J.L., Nowell, P.M., Landers, D.M., Sibley, B.A., 2006. A meta-regression to examine the relationship between aerobic fitness and cognitive performance. Brain Res. Rev. 52, 119–130.

Fery, Y.A., Ferry, A., Vom Hofe, A., Rieu, M., 1997. Effect of physical exhaustion on cognitive functioning. Percept. Mot. Skills 84, 291–298.

Fleury, M., Bard, C., 1987. Effects of different types of physical activity on the performance of perceptual tasks in peripheral and central vision and coincident timing. Ergonomics 30, 945–958.

Fleury, M., Bard, C., Carriere, L., 1981. The effects of physical or perceptual work loads on a coincidence/anticipation task. Percept. Mot. Skills 53, 843–850.

Fuster, J.N.M., 2001. The prefrontal cortex—an update: time is of the essence. Neuron 30, 319–333.

Guizani, S.M., Bouzaouach, I., Tenenbaum, G., Ben Kheder, A., Feki, Y., Bouaziz, M., 2006a. Simple and choice reaction times under varying levels of physical load in high skilled fencers. J. Sports Med. Phys. Fitness 46, 344–351.

Guizani, S.M., Tenenbaum, G., Bouzaouach, I., Ben Kheder, A., Feki, Y., Bouaziz, M., 2006b. Information-processing under incremental levels of physical loads: comparing racquet to combat sports. J. Sports Med. Phys. Fitness 46, 335–343.

Gutin, B., 1973. Exercise-induced activation and human performance: a review. Res. Q. Am. Assoc. Health Phys. Educ. Recreat. 44, 256–268.

Hancock, S., Mcnaughton, L., 1986. Effects of fatigue on ability to process visual information by experienced orienteers. Percept. Mot. Skills 62, 491–498.

Higgins, J.P.T., Green, S. (Eds.), 2011. *Cochrane Handbook for Systematic Reviews of Interventions,* Version 5.1.0 [updated March 2011]. The Cochrane Collaboration. Available from www.handbook.cochrane.org.

Hillman, C.H., Erickson, K.I., Kramer, A.F., 2008. Be smart, exercise your heart: exercise effects on brain and cognition. Nat. Rev. Neurosci. 9, 58–65.

Hogervorst, E., Riedel, W., Jeukendrup, A., Jolles, J., 1996. Cognitive performance after strenuous physical exercise. Percept. Mot. Skills 83, 479–488.

Hüttermann, S., Memmert, D., 2014. Does the inverted-u function disappear in expert athletes? An analysis of the attentional behavior under physical exercise of athletes and non-athletes. Physiol. Behav. 131, 87–92.

Jones, A.M., Wilkerson, D.P., Dimenna, F., Fulford, J., Poole, D.C., 2008. Muscle metabolic responses to exercise above and below the "critical power" assessed using 31P-MRS. Am. J. Physiol. Regul. Integr. Comp. Physiol. 294, R585–R593.

Kashihara, K., Maruyama, T., Murota, M., Nakahara, Y., 2009. Positive effects of acute and moderate physical exercise on cognitive function. J. Physiol. Anthropol. 28, 155–164.

Kmet, L.M., Lee, R.C., Cook, L.S., 2004. Standard Quality Assessment Criteria for Evaluating Primary Research Papers From a Variety of Fields. Alberta Heritage Foundation for Medical Research Edmonton, Alberta, Canada.

Knaepen, K., Goekint, M., Heyman, E.M., Meeusen, R., 2010. Neuroplasticity—exercise-induced response of peripheral brain-derived neurotrophic factor. Sports Med. 40, 765–801.

Kramer, A.F., Erickson, K.I., 2007. Effects of physical activity on cognition, well-being, and brain: human interventions. Alzheimers Dement. 3, S45–S51.

Labelle, V., Bosquet, L., Mekary, S., Bherer, L., 2013. Decline in executive control during acute bouts of exercise as a function of exercise intensity and fitness level. Brain Cogn. 81, 10–17.

Lambourne, K., Tomporowski, P., 2010. The effect of exercise-induced arousal on cognitive task performance: a meta-regression analysis. Brain Res. 1341, 12–24.

Lemmink, K.A., Visscher, C., 2005. Effect of intermittent exercise on multiple-choice reaction times of soccer players. Percept. Mot. Skills 100, 85–95.

Levitt, S., Gutin, B., 1971. Multiple choice reaction time and movement time during physical exertion. Res. Q. 42, 405–410.

Lezak, M.D., 2004. Neuropsychological Assessment. Oxford University Press, USA.

Llorens, F., Sanabria, D., Huertas, F., 2015. The influence of acute intense exercise on exogenous spatial attention depends on physical fitness level. Exp. Psychol. 62, 20–29.

Lo Bue-Estes, C., Willer, B., Burton, H., Leddy, J.J., Wilding, G.E., Horvath, P.J., 2008. Short-term exercise to exhaustion and its effects on cognitive function in young women. Percept. Mot. Skills 107, 933–945.

Luft, C.D., Takase, E., Darby, D., 2009. Heart rate variability and cognitive function: effects of physical effort. Biol. Psychol. 82, 164–168.

Malomsoki, J., Szmodis, I., 1970. Visual response time changes in athletes during physical effort. Int. Z. Angew. Physiol. 29, 65–72.

Mcmorris, T., Graydon, J., 1996. The effect of exercise on the decision-making performance of experienced and inexperienced soccer players. Res. Q. Exerc. Sport 67, 109–114.

Mcmorris, T., Graydon, J., 1997. The effect of exercise on cognitive performance in soccer-specific tests. J. Sports Sci. 15, 459–468.

Mcmorris, T., Graydon, J., 2000. The effect of incremental exercise on cognitive performance. Int. J. Sports Physiol. 31, 66–81.

McMorris, T., Hale, B.J., 2012. Differential effects of differing intensities of acute exercise on speed and accuracy of cognition: a meta-analytical investigation. Brain Cogn. 80, 338–351.

McMorris, T., Keen, P., 1994. Effect of exercise on simple reaction times of recreational athletes. Percept. Mot. Skills 78, 123–130.

McMorris, T., Hale, B.J., Corbett, J., Robertson, K., Hodgson, C.I., 2015. Does acute exercise affect the performance of whole-body, psychomotor skills in an inverted-U fashion? A meta-analytic investigation. Physiol. Behav. 141, 180–189.

McMorris, T., Turner, A., Hale, B.J., Sproule, J., 2016. Beyond the catecholamines hypothesis for an acute exercise-cognition interaction: a neurochemical perspective. In: McMorris, T. (Ed.), Exercise-Cognition Interaction: Neuroscience Perspectives. Academic, New York, pp. 65–104.

Moher, D., Liberati, A., Tetzlaff, J., Altman, D.G., The PRISMA Group, 2009. Preferred reporting items for systematic reviews and meta-analyses: the PRISMA statement. PLoS Med. 6 (7), e1000097.

Nakata, H., Yoshie, M., Miura, A., Kudo, K., 2010. Characteristics of the athletes' brain: evidence from neurophysiology and neuroimaging. Brain Res. Rev. 62, 197–211.

Nibbeling, N., Oudejans, R.R., Ubink, E.M., Daanen, H.A., 2014. The effects of anxiety and exercise-induced fatigue on shooting accuracy and cognitive performance in infantry soldiers. Ergonomics 57, 1366–1379.

Penedo, F.J., Dahn, J.R., 2005. Exercise and well-being: a review of mental and physical health benefits associated with physical activity. Curr. Opin. Psychiatry 18, 189–193.

Pesce, C., 2009. An integrated approach to the effect of acute and chronic exercise on cognition: the linked role of individual and task constraints. In: McMorris, T., Tomporowski, P.D., Audiffren, M. (Eds.), Exercise and Cognitive Function (pp. 213–226). Wiley and Sons, West Sussex.

Pesce, C., Audiffren, M., 2011. Does acute exercise switch off switch costs? A study with younger and older athletes. J. Sport Exerc. Psychol. 33, 609–626.

Pesce, C., Tessitore, A., Casella, R., Pirritano, M., Capranica, L., 2007. Focusing of visual attention at rest and during physical exercise in soccer players. J. Sports Sci. 25, 1259–1270.

Pesce, C., Cereatti, L., Forte, R., Crova, C., Casella, R., 2011. Acute and chronic exercise effects on attentional control in older road cyclists. Gerontology 57, 121–128.

Pollock, M.L., Gaesser, G.A., Butcher, J.D., Després, J.-P., Dishman, R.K., Franklin, B.A., Garber, C.E., 1998. ACSM position stand: the recommended quantity and quality of exercise for developing and maintaining cardiorespiratory and muscular fitness, and flexibility in healthy adults. Med. Sci. Sports Exerc. 30, 975–991.

Reddy, S., Eckner, J.T., Kutcher, J.S., 2014. Effect of acute exercise on clinically measured reaction time in collegiate athletes. Med. Sci. Sports Exerc. 46, 429–434.

Reilly, T., Smith, D., 1986. Effect of work intensity on performance in a psychomotor task during exercise. Ergonomics 29, 601–606.

Roig, M., Nordbrandt, S., Geertsen, S.S., Nielsen, J.B., 2013. The effects of cardiovascular exercise on human memory: a review with meta-analysis. Neurosci. Biobehav. Rev. 37, 1645–1666.

Russell, M., Kingsley, M., 2011. Influence of exercise on skill proficiency in soccer. Sports Med. 41, 523–539.

Schapschroer, M., Lemez, S., Baker, J., Schorer, J., 2016. Physical load affects perceptual-cognitive performance of skilled athletes: a systematic review. Sports Med. Open 2, 37.

Sjöberg, H., 1980. Physical fitness and mental performance during and after work. Ergonomics 23, 977–985.

Smith, M., Tallis, J., Miller, A., Clarke, N.D., Guimaraes-Ferreira, L., Duncan, M.J., 2016. The effect of exercise intensity on cognitive performance during short duration treadmill running. J. Hum. Kinet. 50, 27–35.

Strauss, P.S., Carlock, J., 1966. Effects of load-carrying on psychomotor performance. Percept. Mot. Skills 23, 315–320.

Strauss, E., Sherman, E., Spreen, O.A., 2006. A Compendium of Neuropsychological Tests: Administration, Norms, and Commentary. Oxford University Press, New York, NY.

Thomas, K., Goodall, S., Stone, M., Howatson, G., Gibson, A.S.C., Ansley, L., 2015. Central and peripheral fatigue in male cyclists after 4-, 20-, and 40-km time trials. Med. Sci. Sports Exerc. 47, 537–546.

Thomas, R., Johnsen, L.K., Geertsen, S.S., Christiansen, L., Ritz, C., Roig, M., Lundbye-Jensen, J., 2016. Acute exercise and motor memory consolidation: the role of exercise intensity. PLoS One 11, e0159589.

Thomson, K., Watt, A., Liukkonen, J., 2009. Differences in ball sports athletes speed discrimination skills before and after exercise induced fatigue. J. Sports Sci. Med. 8, 259–264.

Tomporowski, P.D., 2003. Effects of acute bouts of exercise on cognition. Acta Psychol. (Amst.) 112, 297–324.

Tomporowski, P.D., Ellis, N.R., 1986. Effects of exercise on cognitive processes: a review. Psychol. Bull. 99, 338.

Tomporowski, P.D., Ellis, N.R., Stephens, R., 1987. The immediate effects of strenuous exercise on free-recall memory. Ergonomics 30, 121–129.

Tseng, B.Y., Uh, J., Rossetti, H.C., Cullum, C.M., Diaz-Arrastia, R.F., Levine, B.D., Lu, H., Zhang, R., 2013. Masters athletes exhibit larger regional brain volume and better cognitive performance than sedentary older adults. J. Magn. Reson. Imaging 38, 1169–1176.

Tsorbatzoudis, H., Barkoukis, V., Danis, A., GROUIOS, G., 1998. Physical exertion in simple reaction time and continuous attention of sport participants. Percept. Mot. Skills 86, 571–576.

Voss, M.W., Kramer, A.F., Basak, C., Prakash, R.S., Roberts, B., 2010. Are expert athletes 'expert' in the cognitive laboratory? A meta-analytic review of cognition and sport expertise. Appl. Cogn. Psychol. 24, 812–826.

Wang, C.-C., Chu, C.-H., Chu, I.H., Chan, K.-H., Chang, Y.-K., 2013. Executive function during acute exercise: the role of exercise intensity. J. Sport Exerc. Psychol. 35, 358–367.

Weingarten, G., 1973. Mental performance during physical exertion: the benefit of being physically fit. Int. J. Sport Psychol. 4, 16–26

Whyte, E.F., Gibbons, N., Kerr, G., Moran, K.A., 2015. Effect of a high-intensity, intermittent-exercise protocol on neurocognitive function in healthy adults: implications for return-to-play management after sport-related concussion. J. Sport Rehabil. 16, 2014-0201.

CHAPTER 10

Changes in brain activity during action observation and motor imagery: Their relationship with motor learning

Nobuaki Mizuguchi*,†,‡, Kazuyuki Kanosue*,1

*Faculty of Sport Sciences, Waseda University, Tokorozawa, Saitama, Japan
†Faculty of Science and Technology, Keio University, Yokohama, Kanagawa, Japan
‡The Japan Society for the Promotion of Science, Chiyoda-ku, Tokyo, Japan
1Corresponding author: Tel.: +81-4-2947-6826; Fax: +81-4-2947-6826,
e-mail address: kanosue@waseda.jp

Abstract

Many studies have demonstrated that training utilizing action observation and/or motor imagery improves motor performance. These two techniques are widely used in sports and in the rehabilitation of movement-related disorders. Motor imagery has also been used for brain–machine/computer interfaces (BMI/BCI). During both action observation and motor imagery, motor-related regions such as the premotor cortex and inferior parietal lobule are activated. This is common to actual execution and are involved with the underlying mechanisms of motor learning without execution. Since it is easier to record brain activity during action observation and motor imagery than that during actual sport movements, action observation, and motor imagery of sports skills or complex whole body movements have been utilized to investigate how neural mechanisms differ across the performance spectrum ranging from beginner to expert. However, brain activity during action observation and motor imagery is influenced by task complexity (i.e., simple vs complex movements). Furthermore, temporal changes in brain activity during actual execution along the long time course of motor learning are likely nonlinear and would be different from that during action observation or motor imagery. Activity in motor-related regions during action observation and motor imagery is typically greater in experts than in nonexperts, while the activity during actual execution is often smaller in experts than in nonexperts.

Keywords

Brain activity, Motor learning, Plasticity, Expertise, Performance, Athlete, Expert, Whole body movement, Sport

1 BASIC ASPECTS OF ACTION OBSERVATION

Beginners often observe the actions of experts in order to improve their motor performance. It is well established that such observational learning enhances motor performance (e.g., Mattar and Gribble, 2005). For example, action observation without physical training can improve novel complex whole body movements such as dance sequences (Cross et al., 2009). Rohbanfard and Proteau (2011) found that the learning effect of observing a sequential reaching movement of an expert (who had practiced 3000 trials over 15 days) was greater than the learning effect achieved by observing the movement of a novice. However, the observation of both expert and novice actions (i.e., the combination) was more effective than observation of the expert action alone. The authors concluded that the observers profited not only by observing a well-practiced motor schema, but also by observing likely movement errors (Rohbanfard and Proteau, 2011). In contrast, the performance of dart experts was decremented after observing novices with poor dart throwing form (Ikegami and Ganesh, 2014). These findings suggest that the effect of observing motor actions is complex and that care should be taken when selecting an action to be utilized for action observation learning.

Brain areas activated during action observation include the ventral and dorsal premotor cortex (PMv, PMd), inferior parietal lobule (IPL), superior parietal lobule (SPL), superior temporal sulcus (STS), and dorsolateral prefrontal cortex (DLPFC) (Calvo-Merino et al., 2005; Caspers et al., 2010; Mizuguchi et al., 2016a). These regions, in combination, are called the action observation network (AON). A subset of the above regions, the PMv, IPL, and STS are identified as the mirror neuron system (Rizzolatti and Craighero, 2004). Activity in the frontoparietal AON can be evaluated as a μ-suppression of electroencephalogram (EEG) over the sensorimotor cortex (Arnstein et al., 2011). In addition, corticospinal excitability as assessed by evaluating the amplitude of the motor evoked potential (MEP) is enhanced during action observation (e.g., Fadiga et al., 1995).

A neuroimaging study found that the right IPL and inferior frontal cortex were associated with action outcome rather than kinematic parameters (Hamilton and Grafton, 2008). In an earlier study, these authors noted that the anterior intraparietal sulcus represents the goal of an observed action (Hamilton and Grafton, 2006). Recent electrophysiological studies indicate that neuronal activity during action observation involves a bottom-up stimulus-driven process in the front-parietal AON and a top-down rule-based process in the DLPFC (Amoruso et al., 2016; Ubaldi et al., 2015). It is worth noting that brain activity during action observation is altered by instruction or by context. For example, the frontoparietal AON shows a higher degree of activation when the participants are asked to imitate an observed movement after its observation (Buccino et al., 2004).

2 BASIC ASPECTS OF MOTOR IMAGERY

It is well documented that motor imagery training (or mental practice) improves motor skills (Lotze and Halsband, 2006; Mizuguchi et al., 2012a; Pascual-Leone et al., 1995). In one example, Pascual-Leone et al. (1995) reported that motor imagery training of sequential finger tapping reduces temporal tapping errors. Motor imagery training also facilitates muscle strength, but without muscle hypertrophy (Lebon et al., 2010; Ranganathan et al., 2004; Yue and Cole, 1992). The improvement of motor performance with motor imagery training has also been reported in athletes (Battaglia et al., 2014). Indeed, professional players perform motor imagery training more often than amateurs (Cumming and Hall, 2002). Combining motor imagery training with physical practice has generally proved to be more effective than physical practice alone (e.g., Allami et al., 2008). However, in a case where participants were asked to image a novel movement that they had not previously done, doing motor imagery training alone did not improve their performance (Mulder et al., 2004).

The ability to perform motor imagery varies across people, and this capability would have a major effect on the effectiveness of motor imagery training (Isaac, 1992). The ability to perform motor imagery can be assessed with various methodologies, which include questionnaires, mental chronometry, mental rotation tasks, and peripheral autonomic responses as evaluated by skin resistance (Guillot et al., 2008; Guillot and Collet, 2005; McAvinue and Robertson, 2008; Vingerhoets et al., 2002). However, the results produced by different methods do not necessarily coincide with each other. For example, the vividness of motor imagery as assessed by a questionnaire showed a poor correlation with the score of a mental chronometry task (Williams et al., 2015). Thus, different psychological tests likely reflect the different aspects of motor imagery ability. Guillot et al. (2008) evaluated the ability for motor imagery using a combination of methods which included questionnaires, mental chronometry, and the measurement of peripheral autonomic responses. Then, they demonstrated that the ability to perform motor imagery was associated with activity in the front-parietal motor regions (Guillot et al., 2008). In addition, corticospinal excitability during motor imagery was increased to a greater degree in participants with a high motor imagery ability as compared to those with a lower ability (Lebon et al., 2012; Williams et al., 2012).

Other studies have found that the quality of motor imagery is improved by appropriate sensory signals such as proprioceptive (postural) signals, tactile, or visual inputs (Mizuguchi et al., 2009, 2011, 2012b, 2015; Sakamoto et al., 2009; Vargas et al., 2004). For example, touching the imaged object produces an increase in corticospinal excitability and activity in the front-parietal regions during motor imagery as well as scores of psychological tests for motor imagery of action with an object (Mizuguchi et al., 2009, 2011, 2012b, 2015).

During motor imagery, the supplementary motor area (SMA), PM, primary motor cortex, IPL, SPL, basal ganglia, and cerebellum are activated (Decety et al., 1994; Guillot et al., 2009; Hétu et al., 2013; Lotze et al., 1999; Mizuguchi et al., 2013a, 2014a,b; Naito et al., 2002), which are similar to the activation during actual motor execution (Hanakawa et al., 2003, 2008; Zabicki et al., 2017). Enhancement of corticospinal excitability during motor imagery is likely muscle specific (Fadiga et al., 1999). Furthermore, this enhancement is associated with the imagined movement phase and force level (Hashimoto and Rothwell, 1999; Mizuguchi et al., 2013b). These findings indicate that the neural processing that is involved with motor imagery emulates the processing of an actual execution.

However, the function of some regions will differ between motor imagery and actual execution. The primary motor cortex does not show strong activation during motor imagery as it does during an actual motor execution (see review by Hétu et al., 2013). Kasess et al. (2008), utilizing a dynamic causal modeling analysis, suggest that this weak activation occurs because the SMA suppresses activity in the primary motor cortex. When using transcranial magnetic stimulation (TMS), Kuhtz-Buschbeck et al. (2003) found a significant enhancement of corticospinal excitability during motor imagery. However, when utilizing functional magnetic resonance imaging (fMRI), they did not find a significant activation in the primary motor cortex. This suggests that the sensitivity for detecting activity in the primary motor cortex is higher for TMS than for fMRI. A difference in the opposite direction occurred for the SPL, which was more activated during motor imagery than during execution (Hanakawa et al., 2003). This higher activity likely came about because more attention was paid to the imagined action than the actual execution.

Other studies have demonstrated that activated regions differ between types of motor imagery (i.e., kinesthetic and visual motor imagery). Activity in the parietal regions was similar between kinesthetic and visual motor imagery, but differed between motor and visual regions (Guillot et al., 2009; Mizuguchi et al., 2017). Therefore, the instruction to participants involved with motor imagery experiments must be explicit and the type of motor imagery desired clearly characterized.

3 BRAIN ACTIVITY RELATING TO TASK COMPLEXITY

In this section, we first briefly summarize the influence of task complexity on brain activity during actual motor execution. We will then evaluate the effect of task complexity on brain activity during action observation and motor imagery.

3.1 ACTUAL EXECUTION

With the use of handgrips, precision and phasic grip generate stronger activity in the SMA and primary motor cortex than do power and static grip (King et al., 2014). Likewise, the MEP amplitude is greater during precision grip than during power grip, even when the level of background muscle activity is the same (Tazoe and Perez, 2017). The SMA and PM are more activated during complex hand movements (e.g., sequential finger tapping) than simple hand movements (Meister et al., 2005; Winstein et al., 1997). In addition, the SMA, PM, SPL, thalamus, and cerebellum are more activated during complex spatiotemporal interlimb (i.e., antiphase) coordination than they are during simple interlimb coordination (i.e., in-phase) (Debaere et al., 2004). These findings suggest that activity in the motor-related regions increases as the demands of the task increase in complexity and accuracy.

3.2 ACTION OBSERVATION

The anterior intraparietal cortex shows a higher level of activation during action observation of complex object-manipulation movements (e.g., grasping a key, putting it into a lock and turning it) than during simple movements (e.g., grasping of an object) (Biagi et al., 2010). Since, activity in the anterior intraparietal cortex is important for understanding the nature of a particular action (Hamilton and Grafton, 2006), enhancement of activity in this area is likely associated with the development of an understanding of complex actions. Observation of a balance task that involved a perturbation elicited a greater activation of the visual cortex and superior temporal gyrus that extended to the IPL than did that of a simple balance task (Taube et al., 2015). However, there was no change in activity in the SMA, PM, or cerebellum (Taube et al., 2015). Since activity in the visual cortex and IPL is modulated by attention (Chong et al., 2008; Kanwisher and Wojciulik, 2000), the higher visual activity during observation of the complex balance task likely reflected the greater level of attention paid to the unstable movement.

Although neuroimaging studies rarely found activity in the primary motor cortex during action observation, MEP amplitude induced by TMS to the primary motor cortex did increase during action observation of complex sequential finger tapping as compared with that seen with simple sequential finger tapping (Roosink and Zijdewind, 2010). It may be that detection sensitivity of activity changes in the primary motor cortex assessed with imaging technique is lower than that assessed with TMS (Kuhtz-Buschbeck et al., 2003).

On the other hand, brain activity during action observation of complex whole body movements likely differs depending on skill levels of observers. For example, basketball novices would rely on a higher-order decision-making strategy for the evaluation of the shot quality of others (Abreu et al., 2012), while professional basketball players would utilize their own motor system to detect errors (Abreu et al., 2012; Aglioti et al., 2008). Therefore, the underlying neuronal mechanism for error detection of complex movements would differ before and after acquiring a complex motor skill.

3.3 MOTOR IMAGERY

The PM, posterior parietal cortex, and cerebellum are more activated during motor imagery of a sequential finger-thumb opposition task than they are during motor imagery of a simple finger compressing task (Kuhtz-Buschbeck et al., 2003). Corticospinal excitability also increased more during motor imagery of complex sequential finger tapping than simple sequential finger tapping (Roosink and Zijdewind, 2010). Therefore, the effect of task complexity on brain activity is similar for motor imagery and motor execution in that in both cases activity in motor-related regions increases as task complexity increases.

However, for motor imagery of a very complex whole body movement (essentially an impossible movement for the participants), the primary visual cortex (V1) was activated more than it was for a simple whole body movement (Mizuguchi et al., 2016b). Since V1 activity is likely related to the creation of a visual image (Kosslyn et al., 1999), visual motor imagery was likely recruited unintentionally to compensate for the less vivid kinesthetic motor imagery of difficult whole body movements that participants have no motor representation. Indeed, motor imagery training was less effective than action observation in the early phase of motor learning of a new complex four-limb, hand–foot coordination task (Gonzalez-Rosa et al., 2015).

4 ACTIVITY CHANGE IN ASSOCIATION WITH MOTOR LEARNING
4.1 ACTUAL EXECUTION

The brain of human adults can be reorganized by motor learning (e.g., Draganski et al., 2004). In addition, it is well known that brain activity increases or decreases with motor learning depending on the learning stage (Dayan and Cohen, 2011; Kelly and Garavan, 2005). This is thought to occur because multiple reorganization processes are involved and proceed under different time scales (Wymbs and Grafton, 2015). Dayan and Cohen (2011) demonstrated that learning stage can be divided into "fast" and "slow" phases. The fast stage is defined as the initial learning phase during which motor performance is improved rapidly. The duration of both learning stages is highly task dependent. For example, the fast stage of a simple finger tapping task could last for minutes, whereas the fast stage of a complex musical piece might last for months (Dayan and Cohen, 2011).

In the fast stage of motor learning, brain activity in the prefrontal, parietal associative regions, and primary motor cortex decreased (Dayan and Cohen, 2011; Debarnot et al., 2014). These decreases are likely to be associated with a decreasing cognitive load and a reduction of unnecessary muscle contractions as the novel movement performed becomes more efficiently.

In the slow stage of learning, activity in the primary sensorimotor regions, the SMA, and dorsolateral striatum is increased with the steady advances in motor learning (Dayan and Cohen, 2011). Enhancement of activity in the sensorimotor regions likely reflects an expansion of the motor and somatosensory maps that represent specific body parts (Karni et al., 1995; Pleger et al., 2003). Indeed, representation of the fingers in the primary motor and somatosensory cortices is expanded in musical experts (Elbert et al., 1995; Rosenkranz et al., 2007). This expansion of representation, of course, has a definite limitation, which is termed a ceiling effect.

During the period of the slow learning state, changes in neural efficiency occur simultaneously with expansion of representation. However, neural efficiency would progress more slowly than does the expansion of representation. Thus, after the expansion of representation has plateaued, additional motor trainings (over years) continue to decrease activity in the motor-related regions (Fig. 1). As a consequence of this, activity in the motor-related regions of experts is found to be lower than that of nonexperts and novices (Nakata et al., 2010). Indeed, a top level footballer recruited limited motor regions during an ankle movement (Naito and Hirose, 2014).

4.2 ACTION OBSERVATION

Although activity in the AON is greater during observation of a novel movement in novices, this activity decreases during the initial phase of learning (Liew et al., 2013). However, continued training increased activity in the AON during action observation (Kirsch and Cross, 2015). This trend seems to be similar to changes that occur during actual execution. That is, activity in the AON decreases during the fast stage of learning, whereas the activity increases during the slow stage of learning.

In sports, players often need to acquire a new motor skill. However, as far as we know, no study has directly examined changes in AON activity that occur during the acquisition of a novel, complex whole body movement (essentially an impossible movement for the participants). We performed a longitudinal pilot study designed to examine changes in activity in the AON that occur during the acquisition of a gymnastic movement termed a "kip." The completely naïve participants trained 2-month and acquired the movement. Before the training sessions (pre-fMRI), brain activation related to observation of a kip was identified only in the bilateral V5. After training and acquisition of the movement, however, the bilateral V5, the right PMv, and the SMA were activated during observation (Fig. 2). Thus, our results confirmed our suspicion that the acquisition of a novel, difficult whole body movement would lead to increased activity in the AON.

In contrast to what occurs during an actual execution, brain activity in motor-related regions such as the frontoparietal AON during action observation remains at a higher level in experts. Calvo-Merino et al. (2006) reported that the AON in male dancers showed greater activation during observation of male-specific ballet moves as compared to those of female-specific ballet moves, and vice versa. An EEG study also confirmed that event-related desynchronization induced by observing dance movements was greater in dancers than in nondancers (Orgs et al., 2008). Since the frontoparietal AON is associated with action understanding (Hamilton and

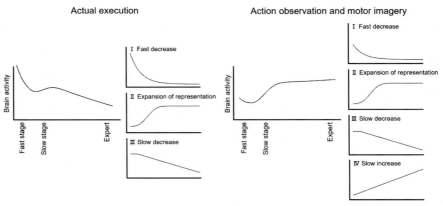

FIG. 1

Schema of changes in brain activity associated with motor learning. The *left panel* indicates activity in the motor-related regions during the actual execution. The *right panel* indicates activity in the action observation network during action observation, or activity in the motor-related regions during motor imagery. In the "fast stage" of motor learning, decreasing brain activity is dominant during actual execution, action observation, and motor imagery. This is likely related to the improvement of synaptic efficacy and/or a decreasing attentional demand (I). In the "slow stage" of motor learning, expansion of representation (II) and neural efficiency (III) progress simultaneously. However, the effect of expansion of representation dominates until a certain point. Thus, total activity would increase during this stage. During long-term training (becoming an expert) brain activity shows a slow, steady decrease during actual execution. On the other hand, brain activity during action observation remains at a higher level in experts. This high level of activity is likely related to deep understanding that experts possess regarding the observed action (IV). Since this high level of action understanding likely increases activity in the action observation network, the total activity increases rather than decreases. Brain activity during motor imagery also remains at a higher level in experts. This activity is likely associated with the fidelity of motor imagery because activity in the motor-related regions would be greater during faithful motor imagery (IV). Note that the time scales in each factor are task specific.

Grafton, 2006, 2008), activity in the AON was likely greater during action observation of well-learned familiar movements than it was during action observation of unfamiliar movements. Note that neural efficiency would have continued to progress for motor learning. Therefore, the effect of action understanding (increased activity) might have been greater than that of neural efficiency (decreased activity). In addition, complexity of the observed action might also be related to the differences seen in brain activity. That is, since studies which compare brain activity between experts and nonexperts often use difficult complex actions, smaller activity in the AON of nonexperts would reflect a low ability for action understanding (Yang, 2015).

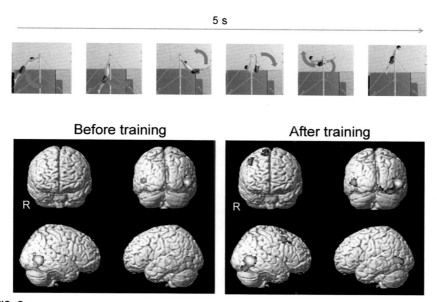

FIG. 2

Activity changes in the supplementary motor area (SMA) and premotor cortex (PM) following motor learning of a novel complex whole body movement (kip) (unpublished data). The participants were initially completely naïve regarding the performance of a kip ($n=8$). To avoid the effect of observing a novel movement, the participants made several separate observations before the pre-fMRI scan. Before the training sessions, brain activation related to the kip observation was identified only in the bilateral V5. After 2-month of training and acquisition of the movement, however, the bilateral V5, the right PM, and the SMA were activated during kip observation.

In sports games, players follow not only the action of other players but also that of an object (such as a ball). Cricket experts can accurately predict the trajectory of a thrown ball from the movement of a cricket bowler (Land and McLeod, 2000), and elite basketball players can detect the potential for error of a shot earlier and more accurately than can visual experts such as coaches or sports journalists (Aglioti et al., 2008). Interestingly, the players likely utilized their own motor system because motor regions were more activated when they detected a shot that was likely to miss. On the other hand, for situations involving an action prediction or judgment (i.e., scoring movement quality), activity in the AON becomes smaller as learning progresses (Babiloni et al., 2010; Cross et al., 2013). To sum is up, the type of instruction and task as well as motor expertise all have a major effect on the brain activity that occurs action observation.

4.3 MOTOR IMAGERY

Activity in the PM, IPL, and temporal lobe during motor imagery of a novel sequential finger tapping decreases after practice (Lacourse et al., 2005). The trend of brain activity decreases during the fast stage of motor learning is likely similar to that which occurs during actual execution or action observation. However, in contradiction to an actual execution, most studies report that brain activity in the motor-related regions during motor imagery increases in association with motor learning over the years (experts vs novices).

Although professional violinists show less activation in the prefrontal cortex, parietal regions, and cerebellar hemispheres during motor imagery of Mozart's violin concerto than do amateur violinists, the PMv and SPL are more activated (Lotze et al., 2003). In addition, Olsson and colleagues report that during motor imagery of a high jump, the PM and cerebellum are activated in high jumpers, while only visual areas were activated in novices (Olsson et al., 2008). Furthermore, corticospinal excitability during motor imagery of a tennis stroke increases in expert tennis players but not in novices (Fourkas et al., 2008). Thus, brain activity in motor-related regions during motor imagery is greater in experts than in nonexperts. Although activity in motor regions during an actual execution decreases as long-term motor learning progresses, brain activity during motor imagery likely remains high or increases still further. This might be explained by characteristic of motor imagery. That is, the accuracy of motor imagery is likely increased by motor learning or the acquisition of a motor repertoire. Since the ability or quality of motor imagery is likely improved as motor learning progresses, activity in motor-related regions would be increased, since the activity is associated with motor imagery ability (Guillot et al., 2008). Therefore, brain activity in motor-related regions during motor imagery is likely to be increased rather than decreased during long-term training.

5 CONCLUSION

Brain activity during action observation and motor imagery changes as motor learning progresses. Activity changes in motor-related regions reflect multiple reorganization processes. This include increases in both neural efficiency and in an expansion of representation with different time scales. During the fast stage of motor learning of difficult movements, brain activity in wider regions decreases all in actual execution, action observation, and motor imagery. With additional learning (i.e., the slow stage), activity in the motor-related regions increases, which might be associated with the increased motor representation. These activity intensifications occur with actual execution, action observation, and motor imagery. However, through learning motor skills over a very long time (i.e., experts), activity in motor-related regions decreases slightly during an actual execution, while brain activity remains higher during action observation and motor imagery. In addition, the temporal changes in brain activity is task dependent. In summary, changes in the neuronal dynamics of action observation and motor imagery that occur during motor learning are not the same as those which occur during actual execution, especially in experts.

ACKNOWLEDGMENTS

The authors are grateful to Dr. Jun Tsuchiya, Dr. Hiroki Nakata, and Dr. Takatoshi Higuchi for conducting the unpublished experiment. The authors also thank Dr. Larry Crawshaw for English editing. This work was supported by JSPS KAKENHI Grant Numbers JP26750242 and JP16J01324 to N.M. and JP26242065 to K.K. All authors declare no conflicts of interest.

REFERENCES

Abreu, A.M., Macaluso, E., Azevedo, R.T., Cesari, P., Urgesi, C., Aglioti, S.M., 2012. Action anticipation beyond the action observation network: a functional magnetic resonance imaging study in expert basketball players. Eur. J. Neurosci. 35 (10), 1646–1654.

Aglioti, S.M., Cesari, P., Romani, M., Urgesi, C., 2008. Action anticipation and motor resonance in elite basketball players. Nat. Neurosci. 11 (9), 1109–1116.

Allami, N., Paulignan, Y., Brovelli, A., Boussaoud, D., 2008. Visuo-motor learning with combination of different rates of motor imagery and physical practice. Exp. Brain Res. 184 (1), 105–113.

Amoruso, L., Finisguerra, A., Urgesi, C., 2016. Tracking the time course of top-down contextual effects on motor responses during action comprehension. J. Neurosci. 36 (46), 11590–11600.

Arnstein, D., Cui, F., Keysers, C., Maurits, N.M., Gazzola, V., 2011. μ-suppression during action observation and execution correlates with BOLD in dorsal premotor, inferior parietal, and SI cortices. J. Neurosci. 31 (40), 14243–14249.

Babiloni, C., Marzano, N., Infarinato, F., Iacoboni, M., Rizza, G., Aschieri, P., Cibelli, G., Soricelli, A., Eusebi, F., Del Percio, C., 2010. "Neural efficiency" of experts' brain during judgment of actions: a high-resolution EEG study in elite and amateur karate athletes. Behav. Brain Res. 207 (2), 466–475.

Battaglia, C., D'Artibale, E., Fiorilli, G., Piazza, M., Tsopani, D., Giombini, A., Calcagno, G., di Cagno, A., 2014. Use of video observation and motor imagery on jumping performance in national rhythmic gymnastics athletes. Hum. Mov. Sci. 38, 225–234.

Biagi, L., Cioni, G., Fogassi, L., Guzzetta, A., Tosetti, M., 2010. Anterior intraparietal cortex codes complexity of observed hand movements. Brain Res. Bull. 81 (4–5), 434–440.

Buccino, G., Vogt, S., Ritzl, A., Fink, G.R., Zilles, K., Freund, H.J., Rizzolatti, G., 2004. Neural circuits underlying imitation learning of hand actions: an event-related fMRI study. Neuron 42 (2), 323–334.

Calvo-Merino, B., Glaser, D.E., Grèzes, J., Passingham, R.E., Haggard, P., 2005. Action observation and acquired motor skills: an fMRI study with expert dancers. Cereb. Cortex 15 (8), 1243–1249.

Calvo-Merino, B., Grèzes, J., Glaser, D.E., Passingham, R.E., Haggard, P., 2006. Seeing or doing? Influence of visual and motor familiarity in action observation. Curr. Biol. 16 (19), 1905–1910.

Caspers, S., Zilles, K., Laird, A.R., Eickhoff, S.B., 2010. ALE meta-analysis of action observation and imitation in the human brain. Neuroimage 50 (3), 1148–1167.

Chong, T.T., Williams, M.A., Cunnington, R., Mattingley, J.B., 2008. Selective attention modulates inferior frontal gyrus activity during action observation. Neuroimage 40 (1), 298–307.

Cross, E.S., Kraemer, D.J., Hamilton, A.F., Kelley, W.M., Grafton, S.T., 2009. Sensitivity of the action observation network to physical and observational learning. Cereb. Cortex 19 (2), 315–326.

Cross, E.S., Stadler, W., Parkinson, J., Schütz-Bosbach, S., Prinz, W., 2013. The influence of visual training on predicting complex action sequences. Hum. Brain Mapp. 34 (2), 467–486.

Cumming, J., Hall, C., 2002. Deliberate imagery practice: the development of imagery skills in competitive athletes. J. Sports Sci. 20 (2), 137–145.

Dayan, E., Cohen, L.G., 2011. Neuroplasticity subserving motor skill learning. Neuron 72 (3), 443–454.

Debaere, F., Wenderoth, N., Sunaert, S., Van Hecke, P., Swinnen, S.P., 2004. Cerebellar and premotor function in bimanual coordination: parametric neural responses to spatiotemporal complexity and cycling frequency. Neuroimage 21 (4), 1416–1427.

Debarnot, U., Sperduti, M., Di Rienzo, F., Guillot, A., 2014. Experts bodies, experts minds: how physical and mental training shape the brain. Front. Hum. Neurosci. 8, 280.

Decety, J., Perani, D., Jeannerod, M., Bettinardi, V., Tadary, B., Woods, R., Mazziotta, J.C., Fazio, F., 1994. Mapping motor representations with positron emission tomography. Nature 371 (6498), 600–602.

Draganski, B., Gaser, C., Busch, V., Schuierer, G., Bogdahn, U., May, A., 2004. Neuroplasticity: changes in grey matter induced by training. Nature 427 (6972), 311–312.

Elbert, T., Pantev, C., Wienbruch, C., Rockstroh, B., Taub, E., 1995. Increased cortical representation of the fingers of the left hand in string players. Science 270 (5234), 305–307.

Fadiga, L., Fogassi, L., Pavesi, G., Rizzolatti, G., 1995. Motor facilitation during action observation: a magnetic stimulation study. J. Neurophysiol. 73 (6), 2608–2611.

Fadiga, L., Buccino, G., Craighero, L., Fogassi, L., Gallese, V., Pavesi, G., 1999. Corticospinal excitability is specifically modulated by motor imagery: a magnetic stimulation study. Neuropsychologia 37 (2), 147–158.

Fourkas, A.D., Bonavolontà, V., Avenanti, A., Aglioti, S.M., 2008. Kinesthetic imagery and tool-specific modulation of corticospinal representations in expert tennis players. Cereb. Cortex 18 (10), 2382–2390.

Gonzalez-Rosa, J.J., Natali, F., Tettamanti, A., Cursi, M., Velikova, S., Comi, G., Gatti, R., Leocani, L., 2015. Action observation and motor imagery in performance of complex movements: evidence from EEG and kinematics analysis. Behav. Brain Res. 281, 290–300.

Guillot, A., Collet, C., 2005. Duration of mentally simulated movement: a review. J. Mot. Behav. 37 (1), 10–20.

Guillot, A., Collet, C., Nguyen, V.A., Malouin, F., Richards, C., Doyon, J., 2008. Functional neuroanatomical networks associated with expertise in motor imagery. Neuroimage 41 (4), 1471–1483.

Guillot, A., Collet, C., Nguyen, V.A., Malouin, F., Richards, C., Doyon, J., 2009. Brain activity during visual versus kinesthetic imagery: an fMRI study. Hum. Brain Mapp. 30 (7), 2157–2172.

Hamilton, A.F., Grafton, S.T., 2006. Goal representation in human anterior intraparietal sulcus. J. Neurosci. 26 (4), 1133–1137.

Hamilton, A.F., Grafton, S.T., 2008. Action outcomes are represented in human inferior frontoparietal cortex. Cereb. Cortex 18 (5), 1160–1168.

Hanakawa, T., Immisch, I., Toma, K., Dimyan, M.A., Van Gelderen, P., Hallett, M., 2003. Functional properties of brain areas associated with motor execution and imagery. J. Neurophysiol. 89 (2), 989–1002.

Hanakawa, T., Dimyan, M.A., Hallett, M., 2008. Motor planning, imagery, and execution in the distributed motor network: a time-course study with functional MRI. Cereb. Cortex 18 (12), 2775–2788.

Hashimoto, R., Rothwell, J.C., 1999. Dynamic changes in corticospinal excitability during motor imagery. Exp. Brain Res. 125 (1), 75–81.

Hétu, S., Grégoire, M., Saimpont, A., Coll, M.P., Eugène, F., Michon, P.E., Jackson, P.L., 2013. The neural network of motor imagery: an ALE meta-analysis. Neurosci. Biobehav. Rev. 37 (5), 930–949.

Ikegami, T., Ganesh, G., 2014. Watching novice action degrades expert motor performance: causation between action production and outcome prediction of observed actions by humans. Sci. Rep. 4, 6989.

Isaac, A.R., 1992. Mental practice—does it work in the field? Sport Psychol. 6 (2), 192–198.

Kanwisher, N., Wojciulik, E., 2000. Visual attention: insights from brain imaging. Nat. Rev. Neurosci. 1 (2), 91–100.

Karni, A., Meyer, G., Jezzard, P., Adams, M.M., Turner, R., Ungerleider, L.G., 1995. Functional MRI evidence for adult motor cortex plasticity during motor skill learning. Nature 377 (6545), 155–158.

Kasess, C.H., Windischberger, C., Cunnington, R., Lanzenberger, R., Pezawas, L., Moser, E., 2008. The suppressive influence of SMA on M1 in motor imagery revealed by fMRI and dynamic causal modeling. Neuroimage 40 (2), 828–837.

Kelly, A.M., Garavan, H., 2005. Human functional neuroimaging of brain changes associated with practice. Cereb. Cortex 15 (8), 1089–1102.

King, M., Rauch, H.G., Stein, D.J., Brooks, S.J., 2014. The handyman's brain: a neuroimaging meta-analysis describing the similarities and differences between grip type and pattern in humans. Neuroimage 102 (Pt. 2), 923–937.

Kirsch, L.P., Cross, E.S., 2015. Additive routes to action learning: layering experience shapes engagement of the action observation network. Cereb. Cortex 25 (12), 4799–4811.

Kosslyn, S.M., Pascual-Leone, A., Felician, O., Camposano, S., Keenan, J.P., Thompson, W.L., Ganis, G., Sukel, K.E., Alpert, N.M., 1999. The role of area 17 in visual imagery: convergent evidence from PET and rTMS. Science 284 (5411), 167–170.

Kuhtz-Buschbeck, J.P., Mahnkopf, C., Holzknecht, C., Siebner, H., Ulmer, S., Jansen, O., 2003. Effector-independent representations of simple and complex imagined finger movements: a combined fMRI and TMS study. Eur. J. Neurosci. 18 (12), 3375–3387.

Lacourse, M.G., Orr, E.L., Cramer, S.C., Cohen, M.J., 2005. Brain activation during execution and motor imagery of novel and skilled sequential hand movements. Neuroimage 27 (3), 505–519.

Land, M.F., McLeod, P., 2000. From eye movements to actions: how batsmen hit the ball. Nat. Neurosci. 3 (12), 1340–1345.

Lebon, F., Collet, C., Guillot, A., 2010. Benefits of motor imagery training on muscle strength. J. Strength Cond. Res. 24 (6), 1680–1687.

Lebon, F., Byblow, W.D., Collet, C., Guillot, A., Stinear, C.M., 2012. The modulation of motor cortex excitability during motor imagery depends on imagery quality. Eur. J. Neurosci. 35 (2), 323–331.

Liew, S.L., Sheng, T., Margetis, J.L., Aziz-Zadeh, L., 2013. Both novelty and expertise increase action observation network activity. Front. Hum. Neurosci. 7, 541.

Lotze, M., Halsband, U., 2006. Motor imagery. J. Physiol. Paris 99 (4–6), 386–395.

Lotze, M., Montoya, P., Erb, M., Hulsmann, E., Flor, H., Klose, U., Birbaumer, N., Grodd, W., 1999. Activation of cortical and cerebellar motor areas during executed and imagined hand movement: an fMRI study. J. Cogn. Neurosci. 11 (5), 491–501.

Lotze, M., Scheler, G., Tan, H.-R.M., Braun, C., Birbaumer, N., 2003. The musician's brain: functional imaging of amateurs and professionals during performance and imagery. Neuroimage 20 (3), 1817–1829.

Mattar, A.A., Gribble, P.L., 2005. Motor learning by observing. Neuron 46 (1), 153–160.

McAvinue, L.P., Robertson, I.H., 2008. Measuring motor imagery ability: a review. Eur. J. Cogn. Psychol. 20 (2), 232–251.

Meister, I., Krings, T., Foltys, H., Boroojerdi, B., Müller, M., Töpper, R., Thron, A., 2005. Effects of long-term practice and task complexity in musicians and nonmusicians performing simple and complex motor tasks: implications for cortical motor organization. Hum. Brain Mapp. 25 (3), 345–352.

Mizuguchi, N., Sakamoto, M., Muraoka, T., Kanosue, K., 2009. Influence of touching an object on corticospinal excitability during motor imagery. Exp. Brain Res. 196 (4), 529–535.

Mizuguchi, N., Sakamoto, M., Muraoka, T., Nakagawa, K., Kanazawa, S., Nakata, H., Moriyama, N., Kanosue, K., 2011. The modulation of corticospinal excitability during motor imagery of actions with objects. PLoS One 6 (10), e26006.

Mizuguchi, N., Nakata, H., Uchida, Y., Kanosue, K., 2012a. Motor imagery and sport performance. J. Phys. Fitness Sports Med. 1 (1), 103–111.

Mizuguchi, N., Sakamoto, M., Muraoka, T., Moriyama, N., Nakagawa, K., Nakata, H., Kanosue, K., 2012b. Influence of somatosensory input on corticospinal excitability during motor imagery. Neurosci. Lett. 514 (1), 127–130.

Mizuguchi, N., Nakata, H., Hayashi, T., Sakamoto, M., Muraoka, T., Uchida, Y., Kanosue, K., 2013a. Brain activity during motor imagery of an action with an object: a functional magnetic resonance imaging study. Neurosci. Res. 76 (3), 150–155.

Mizuguchi, N., Umehara, I., Nakata, H., Kanosue, K., 2013b. Modulation of corticospinal excitability dependent upon imagined force level. Exp. Brain Res. 230 (2), 243–249.

Mizuguchi, N., Nakata, H., Kanosue, K., 2014a. Activity of right premotor-parietal regions dependent upon imagined force level: an fMRI study. Front. Hum. Neurosci. 8, 810.

Mizuguchi, N., Nakata, H., Kanosue, K., 2014b. Effector-independent brain activity during motor imagery of the upper and lower limbs: an fMRI study. Neurosci. Lett. 581, 69–74.

Mizuguchi, N., Yamagishi, T., Nakata, H., Kanosue, K., 2015. The effect of somatosensory input on motor imagery depends upon motor imagery capability. Front. Psychol. 6, 104.

Mizuguchi, N., Nakata, H., Kanosue, K., 2016a. The right temporoparietal junction encodes efforts of others during action observation. Sci. Rep. 6, 30274.

Mizuguchi, N., Nakata, H., Kanosue, K., 2016b. Motor imagery beyond the motor repertoire: activity in the primary visual cortex during kinesthetic motor imagery of difficult whole body movements. Neuroscience 315, 104–113.

Mizuguchi, N., Nakamura, M., Kanosue, K., 2017. Task-dependent engagements of the primary visual cortex during kinesthetic and visual motor imagery. Neurosci. Lett. 636, 108–112.

Mulder, T., Zijistra, S., Zijlstra, W., Hochstenbach, J., 2004. The role of motor imagery in learning a totally novel movement. Exp. Brain Res. 154 (2), 211–217.

Naito, E., Hirose, S., 2014. Efficient foot motor control by Neymar's brain. Front. Hum. Neurosci. 8, 594.

Naito, E., Kochiyama, T., Kitada, R., Nakamura, S., Matsumura, M., Yonekura, Y., Sadato, N., 2002. Internally simulated movement sensations during motor imagery activate cortical motor areas and the cerebellum. J. Neurosci. 22 (9), 3683–3691.

Nakata, H., Yoshie, M., Miura, A., Kudo, K., 2010. Characteristics of the athletes' brain: evidence from neurophysiology and neuroimaging. Brain Res. Rev. 62 (2), 197–211.

Olsson, C.J., Jonsson, B., Larsson, A., Nyberg, L., 2008. Motor representations and practice affect brain systems underlying imagery: an fMRI study of internal imagery in novices and active high jumpers. Open Neuroimag. J. 2, 5–13.

Orgs, G., Dombrowski, J.H., Heil, M., Jansen-Osmann, P., 2008. Expertise in dance modulates alpha/beta event-related desynchronization during action observation. Eur. J. Neurosci. 27 (12), 3380–3384.

Pascual-Leone, A., Nquyet, D., Cohen, L.G., Brasil-Neto, J.P., Cammarota, A., Hallett, M., 1995. Modulation of muscle responses evoked by transcranial magnetic stimulation during the acquisition of new fine motor skills. J. Neurophysiol. 74 (3), 1037–1045.

Pleger, B., Foerster, A.F., Ragert, P., Dinse, H.R., Schwenkreis, P., Malin, J.P., Nicolas, V., Tegenthoff, M., 2003. Functional imaging of perceptual learning in human primary and secondary somatosensory cortex. Neuron 40 (3), 643–653.

Ranganathan, V.K., Siemionow, V., Liu, J.Z., Sahgal, V., Yue, G.H., 2004. From mental power to muscle power—gaining strength by using the mind. Neuropsychologia 42 (7), 944–956.

Rizzolatti, G., Craighero, L., 2004. The mirror-neuron system. Annu. Rev. Neurosci. 27, 169–192.

Rohbanfard, H., Proteau, L., 2011. Learning through observation: a combination of expert and novice models favors learning. Exp. Brain Res. 215 (3–4), 183–197.

Roosink, M., Zijdewind, I., 2010. Corticospinal excitability during observation and imagery of simple and complex hand tasks: implications for motor rehabilitation. Behav. Brain Res. 213 (1), 35–41.

Rosenkranz, K., Williamon, A., Rothwell, J.C., 2007. Motorcortical excitability and synaptic plasticity is enhanced in professional musicians. J. Neurosci. 27 (19), 5200–5206.

Sakamoto, M., Muraoka, T., Mizuguchi, N., Kanosue, K., 2009. Combining observation and imagery of an action enhances human corticospinal excitability. Neurosci. Res. 65 (1), 23–27.

Taube, W., Mouthon, M., Leukel, C., Hoogewoud, H.M., Annoni, J.M., Keller, M., 2015. Brain activity during observation and motor imagery of different balance tasks: an fMRI study. Cortex 64, 102–114.

Tazoe, T., Perez, M.A., 2017. Cortical and reticular contributions to human precision and power grip. J. Physiol. 595 (8), 2715–2730.

Ubaldi, S., Barchiesi, G., Cattaneo, L., 2015. Bottom-up and top-down visuomotor responses to action observation. Cereb. Cortex 25 (4), 1032–1041.

Vargas, C.D., Olivier, E., Craighero, L., Fadiga, L., Duhamel, J.R., Sirigu, A., 2004. The influence of hand posture on corticospinal excitability during motor imagery: a transcranial magnetic stimulation study. Cereb. Cortex 14 (11), 1200–1206.

Vingerhoets, G., de Lange, F.P., Vandemaele, P., Deblaere, K., Achten, E., 2002. Motor imagery in mental rotation: an fMRI study. Neuroimage 17 (3), 1623–1633.

Williams, J., Pearce, A.J., Loporto, M., Morris, T., Holmes, P., 2012. The relationship between corticospinal excitability during motor imagery and motor imagery ability. Behav. Brain Res. 226 (2), 369–375.

Williams, S.E., Guillot, A., Di Rienzo, F., Cumming, J., 2015. Comparing self-report and mental chronometry measures of motor imagery ability. Eur. J. Sport Sci. 15 (8), 703–711.

Winstein, C.J., Grafton, S.T., Pohl, P.S., 1997. Motor task difficulty and brain activity: investigation of goal-directed reciprocal aiming using positron emission tomography. J. Neurophysiol. 77 (3), 1581–1594.

Wymbs, N.F., Grafton, S.T., 2015. The human motor system supports sequence-specific representations over multiple training-dependent timescales. Cereb. Cortex 25 (11), 4213–4225.

Yang, J., 2015. The influence of motor expertise on the brain activity of motor task performance: a meta-analysis of functional magnetic resonance imaging studies. Cogn. Affect. Behav. Neurosci. 15 (2), 381–394.

Yue, G., Cole, K.J., 1992. Strength increases from the motor program: comparison of training with maximal voluntary and imagined muscle contractions. J. Neurophysiol. 67 (5), 1114–1123.

Zabicki, A., de Haas, B., Zentgraf, K., Stark, R., Munzert, J., Krüger, B., 2017. Imagined and executed actions in the human motor system: testing neural similarity between execution and imagery of actions with a multivariate approach. Cereb. Cortex. 27 (9), 4523–4536.

CHAPTER 11

Moving concussion care to the next level: The emergence and role of concussion clinics in the UK

Osman H. Ahmed[*,†,1], Mike Loosemore[‡], Katy Hornby[‡], Bhavesh Kumar[‡],
Richard Sylvester[‡,§], Hegoda Levansri Makalanda[¶], Tim Rogers[||], David Edwards[||,#]
Akbar de Medici[‡]

[*]Faculty of Health and Social Sciences, Bournemouth University, Bournemouth, United Kingdom
[†]The FA Centre for Disability Football Research, St George's Park, Burton-Upon-Trent, United Kingdom
[‡]Institute of Sport and Exercise Health, University College London, London, United Kingdom
[§]National Hospital of Neurology and Neurosurgery, London, United Kingdom
[¶]The Royal London Hospital, London, United Kingdom
[||]Cognacity, London, United Kingdom
[#]University of Zululand, KwaDlangezwa, South Africa
[1]Corresponding author: Tel.: +44-1202-968147; Fax: +44-1202-962736,
e-mail address: osman.hassan.ahmed@gmail.com

Abstract

Concussion is a worldwide issue in sports medicine at present, and in recent years has evolved into a major consideration for sports in the United Kingdom (UK). Governing bodies, sports clinicians, and indeed athletes themselves are dealing with the implications that this injury brings. In parallel with this, innovative means of managing this condition are emerging. The creation of specialized concussion clinics (which mirror those present in the United States and Canada) is one means of enhancing concussion care in the UK. In this chapter, the emergence of concussion clinics in the UK will be discussed. The specific roles of the multidisciplinary teams working in these clinics will be outlined (including the disciplines of sports medicine, radiology, neurology, physiotherapy, and psychology/psychiatry), and the approaches used in the management of concussion in this setting will be explored. Future recommendations for the growth and development of clinic-based concussion care in the UK will also be discussed.

Keywords

Concussion, Mild traumatic brain injury, Sports medicine, Neurology, Public health

1 INTRODUCTION

The management of concussion in sport has arguably been the subject of more discussion than any other sports injury in recent years. Associated with this discussion has been an evolution in how concussion has been managed (Williams and Danan, 2016); previous misconceptions such as "not letting someone with concussion fall asleep" have been shown to be outdated (BBC News, 2016), and a more nuanced approach to the management of this injury is slowly being adopted in the United Kingdom (UK). With the large focus in the UK on contact sports which have a high rate of concussion (such as football and rugby), it is unsurprising that the UK has become associated with the "Concussion crisis" which is suggested as being a global issue (Lancet Neurology, 2014).

One enterprise that has facilitated the process of concussion management in other countries (primarily in the United States of America (USA) and Canada) has been the introduction of concussion clinics. These clinics offer a specialized approach to the management of this injury, utilizing a multidisciplinary team (MDT) with medical staff who have extensive experience relating to concussion. In the USA alone, there are dedicated concussion clinics in at least 35 states (The New York Times, 2013) and Canada has also seen a similar emergence of concussion clinics within its borders. At present, there has been only limited research undertaken relating to concussion clinics (Mihalik et al., 2013), with one appraisal of these clinics showing variance in the levels of training of the individuals working there (Ellis et al., 2017). There have also been concerns raised as to nefarious individuals who are seeking to exploit the public's fear over concussions via these clinics, including reports of clinics being operated by staff with no obvious background in concussion or neurology in general (STAT, 2015).

Although the UK does not have the proliferation of concussion clinics to the extent of North America, they are starting to become more widespread and are likely to be a "major player" in the UK's approach to the management of this injury in the coming years. In this chapter, we will discuss the key components of concussion clinics and outline the members of the MDT and treatment/assessment approaches that they incorporate. We conclude with a five-point plan for the development of concussion clinics in the UK, with the suggestion that these clinics can help to greatly improve the quality of care for athletes with concussion.

2 CONCUSSION CLINICS IN THE UK

Historically in the UK, concussive injuries have been managed by the general practitioner (GP) and the emergency department (ED). Management from the ED typically consists of the provision of an advice card of potential symptoms from the National Institute for Health and Care Excellence (NICE) (NICE, 2014) along with general advice prior to the patient being discharged; however, any concerns over the possibility of a more substantial neurotraumatic injury (such as a progressive brain bleed) mean a patient will be admitted for a computerized tomography (CT) scan and subsequent observation. The main priority of the medical team in these circumstances is to exclude a potentially life-threatening event, rather than assess for a concussion or deliver appropriate advice regarding return to play (RTP), work, or study. Concerns over the immediate effects (Alsalaheen et al., 2016) and long-term effects (Finkbeiner et al., 2016) of concussion are not included in the traditional National Health Service (NHS) model of head injury management (NICE, 2014). In parallel with this, the knowledge of those individuals responsible for the immediate management of concussion in the ED has been shown to be inconsistent in the UK, with emergency care clinicians demonstrated to have insufficient knowledge to diagnose and manage the condition appropriately (Phillips et al., 2016).

Despite these limitations in concussion knowledge and management at the primary and secondary levels in the UK, in elite sport there is a greater appreciation of the potential harm associated with concussion (Neurology Central, 2016). Several of the major global sporting federations (including FIFA, the IOC, World Rugby, the IIHF, and the FEI) have been involved in the International Consensus Conference series on concussion in sport, with the best-practice management guidelines established by this group providing a degree of clarity on the diagnosis, management, and RTP following concussion for sport clinicians (McCrory et al., 2013). There has also been consideration given to the pitch-side evaluation of concussion. In Rugby Union, rule changes have been introduced with substitutions and time allowances for head injury assessments (World Rugby, 2015). Several sporting organizations have begun to adopt a proactive role with regards to concussion education. The level of concussion knowledge within the general public in the UK has been shown to be limited (Weber & Edwards, 2012), and thus concussion education campaigns from organizations such as the Rugby Football Union (2016) have been beneficial. As yet, the effect of these interventions on the translation of concussion-related knowledge has not been established (Provvidenza et al., 2013).

In contrast with the UK, across the Atlantic in the USA there is a much more recognized understanding of concussion (Sarmiento et al., 2014). The recent Hollywood movie "Concussion" (Smith and Stewart, 2016) and a multitude of comedy shows (Ahmed et al., 2014) have also helped to raise public awareness of the implications of this injury. Unlike in the UK, litigation has been a major factor in this process with the highly publicized $765 million settled lawsuit against the National Football League (NFL) helping to increase concussion-related research and provide medical help to more than 18,000 former players (NFL, 2013). A consequence of this situation is the rapid increase in the number and popularity of concussion clinics; however, without any governing structure, the quality of some of these clinics has been brought into question (CBC News—Health, 2016).

In the UK all NHS services are regulated by the Care Quality Commission (CQC), the independent regulator which ensures that services meet a fundamental standard of care (Care Quality Commission, 2015). The private health sector in the UK is also regulated by the CQC; however, the procedures involved in establishing a new clinical service are less complicated than the NHS, and thus this sector has been quicker to adapt to the need for dedicated concussion clinics. One of the first concussion clinics in the UK was established by the Institute of Sport Exercise and Health (ISEH) in London (ISEH, 2016), as part of the National Centre for Sports and Exercise Medicine (NCSEM) and as a legacy of the 2012 Olympic Games. This clinic, which opened in 2015, aims to address the needs of sports persons who suffer from prolonged or unusual symptoms of concussion and has the input of multiple medical specialists. These complex cases may require the input of a comprehensive MDT to help them recover and return to normal activities, and dedicated concussion clinics such as these can act as a "one-stop shop" to address the holistic needs of the patient. There are also several other concussion clinics working across the UK (Concussion Clinic, 2016; Neurocog Concussion Clinic, 2016), and given the growing recognition of concussion in sport it is likely that this number will continue to increase.

3 MULTIDISCIPLINARY APPROACH TO CONCUSSION MANAGEMENT

The effective management of concussion by a dedicated concussion clinic requires synergistic work between all members of the MDT. Pabian et al. (2016) have outlined the importance of the MDT in this process, and in a typical concussion clinic the individuals involved in the care of an athlete may extend to: sports physicians; radiologists; neurologists; physiotherapists; and psychologists. The respective input from each of these professionals will now be discussed.

3.1 SPORTS PHYSICIAN

The sports physician plays a central role in the clinic-based management of concussion. Using their widespread experiences from the field of play as well as from primary and secondary care medicine, they have a unique insight into the requirements of the athlete. Acting as a case manager, the sports physician is able to triage referrals to the complex concussion clinic according to clinical, sporting, or occupational urgency of the case. Depending on the capacity of the clinic, it may be necessary for the sports physician to offer an initial consultation to the athlete prior to referral to the clinic in order to expediently assess, advise, and protect the athlete from further injury (Burman et al., 2016). As such, the sports physician must be skilled at clinically excluding serious medical differential diagnoses, and instigating appropriate investigations and management as necessary.

Sports physicians with greater expertise may be able to initially identify other common causes of trauma-induced symptoms compatible with concussion such as cervical, vestibular, ocular, or auditory sources, or an unmasking of a susceptibility to conditions including migraines and mood disorders (Singh et al., 2016). These cases may benefit from direct referral to other MDT members, for example, physiotherapists, neuropsychologists, or neurologists. Working within a combined MDT clinic alongside a specialist neurologist, the sports physician offers expertise in the clinical assessment, management, and communication with athletes and their support team. They can analyze the mechanism of injury and rapidly establish clinical features that may provide more information regarding the injury, and discern subsequent progress and return-to-sport targets. In formulating a bespoke management plan, the sports physician may consider additional biopsychosocial modifiers such as the demands of the sport, the athlete's position and style of play, athlete personality and fears, and the medical support available (Stewart et al., 2012).

Working as an external consultant in a clinic setting should empower the sports physician to be objective and clear in their verbal and written communication regarding their RTP advice. This is especially prudent given the ethical issues associated with the management of concussion in sport (Hudson and Spradley, 2016). Medical support staff working directly for an athlete or sports organization could be subject to pressure to expedite athlete return to sport by those individuals with a vested interest. This pressure may come from the athlete themselves, coaches, owners of the sports club, family members of the athlete, sponsors of the athlete, and even their own medical colleagues (Kroshus et al., 2015a). In this situation, sports physicians should follow their code of ethics (International Federation of Sports Medicine, 2017; McNamee et al., 2016) and educate these external parties as to the best-practice medical management of the athlete. Referral to a specialist concussion clinic can prove a powerful resource to help manage such instances of conflict.

3.2 RADIOLOGIST

Several tools are available at the radiologists' disposal to assist in the diagnosis of symptoms in a concussion clinic. The mainstay of acute imaging in the immediate aftermath of concussion is undertaken using CT, with the decision to obtain this usually arising at the ED and in conjunction with the National Institute for Health and Care Excellence (NICE) guidelines (National Institute for Health and Care Excellent, 2014). CT can identify lesions such as skull fractures, contusions, oedema, and extraaxial hemorrhage that may require acute neurosurgical intervention, hospitalization, or follow up. These CT studies are often normal when patients are referred to concussion clinics. Magnetic resonance imaging (MRI) is invariably the next imaging test performed. The standard brain imaging sequences (T1-weighted and T2-weighted fast spin echo sequences, T2-weighted gradient echo sequence, and fluid-attenuated inversion recovery (FLAIR) sequences) are more sensitive than CT and can demonstrate complications such as posttraumatic encephalomalacia, reactive gliosis, and hemosiderin (degraded blood products) (Parizel et al., 2005). Conventional MRI techniques assess the anatomical structures and can detect axonal injury indirectly via microhemorrhage associated with capillary injury; however, a normal MRI cannot exclude a mild TBI.

Susceptibility-weighted imaging (SWI) is an advanced, high-resolution gradient echo sequence that is far more sensitive to hemorrhage from diffuse axonal injury (DAI) and deoxyhemoglobin in venous blood (Kou et al., 2010). This sequence identifies significantly more small hemorrhages than the conventional gradient echo imaging sequence (Liu et al., 1999). Diffusion-weighted imaging (DWI) measures the motion of protons in water molecules. This is normally random in nature but can be altered in certain situations including after shearing injuries, hemorrhage, or oedema. The DWI signal is then increased, and this can be measured by a reduction in apparent diffusion coefficient. This is most sensitive in the acute phase of concussion (Liu et al., 1999). Modern MRI protocols will encompass SWI and DWI with the conventional anatomical sequences. Advanced MR imaging techniques have also evolved that allow a functional component of imaging to also be performed, and as these become more available, they are gaining more routine use in the clinical setting.

Diffusion tensor imaging is an advanced sequence, derived from DWI, but can also quantify a vector. In theory, the ability to assess directional diffusion characteristics allows this sequence to outline water molecule movement along axons, and thus examine their disruption in TBI. It is able to identify damage not seen on conventional MRI sequences (Wilde et al., 2008). Diffusion characteristics are typically measured in 3, 32, or 64 dimensions and mapped with image reconstructions. Abnormalities are defined by fractional anisotropy (FA), axial diffusivity, or mean diffusivity. These are typically localized to the white matter tracts of the corpus callosum or corticospinal tracts (Rutgers et al., 2008). Acutely, the changes will reflect cytotoxic oedema but may then evolve to represent axonal degeneration over time (Rutgers et al., 2008). FA may be reduced in DAI, but increased in concussion, and may correlate with clinical impairment, outcomes from injury, and prognosis (Mayer et al., 2010; Wilde et al., 2008). It is suggested that abnormalities in concussion can be detected up to 6 months after injury (Mayer et al., 2010). Magnetic resonance spectroscopy imaging (MRSI) analyzes the metabolites in a volume of brain tissue. Neuronal death may occur after concussion, varying the levels of *N*-acetylasparate (NAA) and choline. Elevations in choline are due to myelin and cell membrane injury with reduction in NAA due to axonal injury (Ross et al., 1998). MRSI can detect metabolic changes associated in concussion with normal structural imaging, and may assist in predicting outcomes and neuropsychological function after rehabilitation (Holshouser et al., 2005).

In recent years, innovations in imaging have included newer MRI techniques including diffusion kurtosis tensor imaging (Lancaster et al., 2016), magnetic transfer imaging, magnetoencephalography, and functional MRI (fMRI) (Gonzalez and Walker, 2011). These are currently still under investigation and not in general clinical use; however, as the concussion becomes better understood, physiologic imaging is likely to become more influential.

3.3 NEUROLOGIST

Although there is currently limited evidence that neurology input improves outcome following sports-related concussion, early interventions offered by a specialist service significantly reduce social morbidity and severity of postconcussion symptoms in nonsports-related head injury (Wade et al., 1998). A neurologist with experience in assessing and managing patients with traumatic brain injury can provide critical input to a concussion clinic in several key areas. The diagnosis of concussion, given the current absence of objective biomarkers of brain injury, is difficult (Sharp and Jenkins, 2015). Other conditions including migraine, peripheral vestibular injury, and psychological dysfunction can lead to symptoms that can mimic those that occur acutely following a concussion. Additionally, a minority of those with suspected concussion will have sustained a more severe brain injury but may only have subtle clinical markers of this. Comprehensive clinical assessment maximizes the chance of making the correct diagnosis, and neurologists' specific expertise is in eliciting a neurological history and interpreting neurological examination findings.

While most concussions resolve in an uncomplicated manner within 7–10 days, a proportion of individuals may suffer from prolonged symptoms following concussion (King, 2014). Such symptoms tend to be neurological in nature and commonly include headache, balance disturbances, and cognitive issues. However, the underlying cause of these symptoms can vary between individuals. For instance, exercise-induced symptom recurrence could be due to incompletely resolved brain injury, posttraumatic migraine, or vestibular dysfunction. A thorough assessment by a neurologist can define the likely cause of ongoing symptoms and guide further management. Neurologists have extensive experience of managing the chronic symptoms that can occur following concussion. For example, posttraumatic headache can require the aggressive use of migraine prophylaxis medication or other interventions (greater occipital nerve injection or botulinum toxin therapy) that other clinicians may not feel comfortable using. Given their experience, neurologists can also provide critical information for athletes regarding the likely prognosis of any ongoing symptoms. This can provide reassurance and allows informed decisions to be made regarding career-threatening injuries.

Not all athletes with concussion necessarily need every available investigation. If a neurologist can make a clear diagnosis regarding the acute injury and a hypothesis regarding the cause of any ongoing symptoms, individualized and focused investigations can be arranged. Not only is this method of clinically guided investigation an efficient use of costly resources, but it also reduces the chance of false-positive results. Brain imaging, neuropsychological testing, and vestibular assessment frequently generate "abnormal" results. Incidental findings are common; however, up to 15% of the general population will have an abnormality on MRI brain (Vernooij et al., 2007), and even seemingly relevant results can have multiple interpretations. For example, deficits in neuropsychology testing may reflect the consequences of brain injury but could be due to anxiety or somatic symptoms impacting on attentional abilities. Neurologists understand the limitations of these investigations and are used to interpreting results in the context of the clinical presentation.

3.4 PHYSIOTHERAPIST

The appropriately trained physiotherapist can constitute a useful part of the MDT in facilitating the recovery of an athlete with concussion (Reneker and Cook, 2015). One of the key areas in which physiotherapists can seek to make improvements with an athlete in the clinic setting is through the process of vestibular rehabilitation. Part of this approach may include the provision of exercises, and Alsalaheen et al. (2013) examined the types of vestibular rehabilitation exercises given to individuals who were experiencing concussion symptoms. These authors suggested that the most commonly prescribed exercises were eye–head coordination exercises, followed by standing static balance exercises, and ambulation exercises. Although there is limited evidence to support vestibular rehabilitation for concussion at present, a recent systematic review suggested that this approach has promise (Murray et al., 2017). Encouraging outcomes were also found in randomized controlled trials for vestibular rehabilitation in India (Sisodia et al., 2015), a blended cervical and vestibular rehabilitation approach in Canada (Schneider et al., 2014), and a vestibular rehabilitation and aerobic training approach in the USA (Moore et al., 2016).

Manual therapy of the cervical spine is another potential means whereby a physiotherapist can take a proactive role in facilitating the recovery of the athlete with a concussion. There is a growing understanding relating to the role of the cervical spine in concussion-related injuries (Marshall et al., 2015), and it is suggested that some of the symptoms associated with chronic concussion may be attributable to cervical spine dysfunction. Such cervical spine disorders have been shown to respond well to appropriate manual therapy interventions (Reid et al., 2008), and with careful selection of both patient and procedure, this is a valuable method of reducing symptoms. Given the often traumatic mechanism of injury involved in sustaining a concussion, it is reasonable to infer that the cervical spine may be an additional management consideration (Morin et al., 2016) especially as symptoms after head injury may overlap between both a concussive injury and a cervical injury (Leddy et al., 2015).

An additional input that the physiotherapist can have in the clinic is related to functional rehabilitation. By working in close association with the clinic sports physician and the athlete's club medical staff, the clinic physiotherapist can help to develop a tailored RTP programme in keeping with the best-practice RTP guidelines detailed in the current consensus statement (McCrory et al., 2013). Given the unique skill set required from a physiotherapist in a concussion clinic, specialized training which extends beyond the typical scope of a physiotherapist would be required. There are several courses which offer this opportunity for specific concussion rehabilitation for physical therapists in the USA (Evidence in Motion, 2017), and it is likely in the near future there will be a proliferation of similar courses in the UK.

3.5 PSYCHOLOGIST AND PSYCHIATRIST

Psychologists and psychiatrists in sport are involved in a pathway of mental health care from "upstream" prevention, screening, and early detection of mental distress, to "downstream" assessment, treatment, rehabilitation, and recovery from mental illness. At each stage on this pathway there is a broad variety of biological, psychological, and social factors relevant to each athlete's presentation and eventual recovery

(Currie and Johnston, 2016). "Upstream," the development of psychological resilience (Fletcher and Sarkar, 2012), is described along with the concept of mental "prehabilitation" (the prevention of physical injury). Players are taught preventative measures: that early detection is improved by education; mental health awareness; the use of effective mental health screening tools; the establishment of referral systems. Failure to prevent common mental health conditions in athletes (such as anxiety and affective disorders) is likely to increase the concussion risk of the athlete, because the symptoms of these syndromes include impairments in attention, concentration, fatigue, and psychomotor changes.

"Downstream," the ready access to psychiatric and psychological expertise and good collaboration with the world of sports, can reduce stigma, improve access to treatment, delivery of that treatment, facilitate rehabilitation, and assist the eventual RTP following recovery (Currie and Johnston, 2016). The occurrence of an injury of any type is recognized as a precipitant and a risk factor for diminished well-being and also psychiatric illness in athletes. This comprises a major part of the workload of a sport psychiatrist. Depression, fatigue, irritability, confusion, and general mood disturbance are frequently reported after concussion in sport (Mainwaring et al., 2012). The psychological context of concussion as a cause of mental state changes is important, and concussion has been described an "invisible injury" (Bloom et al., 2004). The absence of swelling, stitches, or other visual signs of the injury can have various problematic effects including: less frequent offers of social support from the casual observer; a seemingly "unexplained" loss of fitness or form; and lack of acceptance or understanding of the need for necessary care and rehabilitation (which is especially problematic if arising in coaches, selectors, or the media). Where the symptoms of concussion are mild, the effects (including psychological effects) can be subtle, can often go unnoticed, or be misdiagnosed (Moser, 2007). Concussion clinics can offer an advantageous role in such situations by providing a detailed psychological and psychiatric assessment.

In a concussion clinic neuropsychological assessments are undertaken to evaluate cognitive functioning postinjury. One important task is to separate any impairment due to persistent psychiatric or psychological symptoms from that considered to relate to the index neurological injury. This judgment can be complex and can in some cases only become apparent over a longer period of time. Where indicated, psychotropic medications may also be used in players who have sustained a concussion. Generally these medications are not of a nature that require therapeutic use exemptions for elite athletes. Depending upon factors such as the severity of the concussion, a clinical psychologist's intervention approach is adapted by individual player need. Therapeutic methods and techniques can include cognitive behavioral therapy, mindfulness, applied relaxation techniques, and acceptance and commitment therapy. A key element is focusing on enhancing self-efficacy and confidence. Guidance and support to address other relevant biopsychosocial factors, for example, those arising within the player's family system or relevant organization (such as issues within a professional club) are also important in achieving optimum recovery.

4 CONCLUSIONS

Despite hyperbolic suggestions that the future of some contact sports are in jeopardy due to concussion (Sports Illustrated, 2016), at present the majority opinion across society is that the participation in these sports outweighs the risks (Mannix et al., 2016). The preceding sections of this chapter have highlighted the myriad of ways in which the MDT can collaborate in a concussion clinic to best serve the needs of the patient. The relatively new emergence of concussion clinics in the UK mirrors concussion itself; just as the medical community are only just beginning to understand the diagnosis, short-term, and long-term implications of concussion, they are also beginning to understand more about the role that dedicated clinics can play in the management of concussion. In order for concussion clinics in the UK to best serve the needs of their population, there are several key areas for development we have highlighted. The underpinning factor across all of these areas is the need for collaboration and cooperation across concussion clinics in the UK and worldwide.

4.1 DIAGNOSIS

An emerging area of interest is the role that biomarkers have in relation to the diagnosis of concussion. There have been several recent studies exploring this concept, including the work of Papa et al. (2015) which proposed that over 11 different biomarkers (including tau, creatine kinase, and cortisol) have a link to concussion. The establishment of valid and reliable biomarkers which could be used to assist concussion diagnosis would be an exceptionally useful tool to aid the safe management of athletes. Once this step has been reached, then these biomarkers could also be used in a prognostic manner by providing feedback throughout the recovery of an athlete.

4.2 IMAGING

As technology related to imaging develops, the sophistication that this technology allows will enable concussion clinics to provide an enhanced level of service to their patients to aid in the prognosis and management of their recovery. Kutcher and colleagues published a review of emerging technologies to diagnose sports concussion (Kutcher et al., 2013). A multitude of technologies were mentioned in this study and included: quantitative electroencephalography (qEEG), fMRI, and single-photon emission CT. Although the clinical utility of these tests was not fully established, some (in particular qEEG and fMRI) showed some promise. By keeping abreast of developments related to these technologies and adopting those that are shown to be efficacious, concussion clinics in the UK will be able to provide the best services to their athletes.

4.3 MULTIMEDIA TECHNOLOGIES

The explosion of social media and smartphones in everyday life generates several implications for concussion management. At a community level, smartphone apps could constitute a valuable resource to assist both the diagnosis and management of this condition (Lee et al., 2015). A greater understanding of how these concussion apps are being used will assist concussion clinics in guiding patients to employ them in a more informed and controlled manner. Examples are being reported of social media innovations being used as an adjunct to traditional concussion management (Ahmed et al., 2017), as well as concussion assessment and management being done remotely (Vargas et al., 2012). Utilizing technologies which allow the ready dissemination of information and advice (incorporating the functionality of smartphones) is likely to provide significant and beneficial advances in patient care.

4.4 EDUCATION/DISSEMINATION

Despite the best attempts of several of the UK's professional sporting organizations, gaps persist in the concussion-related knowledge of athletes in sports including football (Williams et al., 2016). Until athletes are comprehensively aware of the signs/symptoms of concussion and the short-, medium-, and long-term effects that this can have on them, it is unrealistic to expect them to consistently self-report their symptoms and resist pressure from coaches, teammates, and others to RTP (Kroshus et al., 2015b). Continued and unified effort is required from the major sporting organizations in the UK to better educate those playing their sports at all levels, and to use innovative educational approaches to assist with this process (Kroshus et al., 2015c).

4.5 CONCUSSION CLINICS FOR ALL

To become established within the NHS, dedicated concussion clinics would have to prove value for money to the Clinical Commissioning Groups (CCG) new "Sustainability and Transformation Plans," which form part of NHS England's Five Year Forward View (NHS England, 2016). As the awareness of concussion grows in the UK, it is possible that public demand for these services will rise and lead to the commissioning of NHS concussion clinics. These would form a similar service (albeit on a different scale) to the existing private concussion clinics, by facilitating easy access for concussed patients for diagnosis, treatment, and if necessary onward referral onto suitable care pathways.

In conclusion, there is a need for increased knowledge of the correct management and advice for concussion among professionals and the public alike. The areas highlighted in this chapter have outlined the key aspects where a well-equipped MDT can provide a comprehensive and streamlined service to the athlete with concussion in UK clinic setting. Given the current constraints on both the NHS and the practitioners working in the NHS, it is unrealistic to expect front-line clinicians in the NHS to provide best-practice RTP advice and management for the athlete (Ahmed and Weiler, 2016). In order for the UK to deliver concussion care which is of a global quality, the creation of concussion-specific clinics (both in the NHS and privately) will allow a vastly enhanced and tailored standard of concussion care to be delivered.

REFERENCES

Ahmed, O.H., Weiler, R., 2016. Optimising concussion care in the United Kingdom: a rethink in the management strategy for sports concussion. Phys. Ther. Sport 21, 107–108.

Ahmed, O.H., Lee, H., Schneiders, A.G., McCrory, P., Sullivan, S.J., 2014. Concussion and comedy: no laughing matter? PM&R 6 (12), 1071–1072.

Ahmed, O.H., Schneiders, A.G., McCrory, P., Sullivan, S., 2017. Sports concussion management using Facebook: a feasibility study of an innovative adjunct "iCon" J. Athl. Train. 52 (4), 339–349.

Alsalaheen, B.A., Whitney, S.L., Mucha, A., Morris, L.O., Furman, J.M., Sparto, P.J., 2013. Exercise prescription patterns in patients treated with vestibular rehabilitation after concussion. Physiother. Res. Int. 18 (2), 100–108.

Alsalaheen, B., Stockdale, K., Pechumer, D., Broglio, S.P., 2016. Validity of the immediate post concussion assessment and cognitive testing (ImPACT). Sports Med. 46, 1487–1501.

BBC News, 2016. Should You Let Someone With Concussion Fall Asleep? Retrieved 8th February, 2017. Available at: http://www.bbc.com/future/story/20160616-should-you-let-someone-with-concussion-fall-asleep.

Bloom, G.A., Horton, A.S., McCrory, P., Johnston, K.M., 2004. Sport psychology and concussion: new impacts to explore. Br. J. Sports Med. 38, 519–521.

Burman, E., Lysholm, J., Shahim, P., Malm, C., Tegner, Y., 2016. Concussed athletes are more prone to injury both before and after their index concussion: a database analysis of 699 concussed contact sports athletes. BMJ Open Sport Exerc. Med. 2 (1), e000092.

Care Quality Commission, 2015. Regulations for Service Providers and Managers. Retrieved 24th January 2017. Available at: www.cqc.org.uk/content/regulations-service-providers-and-managers.

CBC News—Health, 2016. Private concussion clinics called a 'Wild West' of unregulated treatment. Retrieved 1st February 2017. Available at: http://www.cbc.ca/news/health/concussion-hotline-baseline-testing-treatment-industry-for-profit-unregulated-1.3833158.

Concussion Clinic, 2016. Concussion Clinic. Retrieved 8th February 2017. Available at: http://www.concussionclinic.co.uk/3_services.html.

Currie, A., Johnston, A., 2016. Psychiatric disorders: the psychiatrist's contribution to sport. Int. Rev. Psychiatry 28, 587–594.

Ellis, M., Ritchie, L., Selci, E., Chu, S., McDonald, P., Russell, K., 2017. Googling concussion care: a critical appraisal of online concussion healthcare providers and practices in Canada. Clin. J. Sport Med. 27 (2), 179–182.

Evidence in Motion, 2017. Concussion Management Certification. Retrieved 8th February 2017. Available from: http://www.evidenceinmotion.com/educational-offerings/course/concussion-management-certification/.

Finkbeiner, N.W., Max, J.E., Longman, S., Debert, C., 2016. Knowing what we don't know: long-term psychiatric outcomes following adult concussion in sports. Can. J. Psychiatry 61 (5), 270–276.

Fletcher, D., Sarkar, M., 2012. A grounded theory of psychological resilience in Olympic champions. Psychol. Sport Exerc. 13 (5), 669–678.

Gonzalez, P.G., Walker, M.T., 2011. Imaging modalities in mild traumatic brain injury and sports concussion. PM&R 3 (10), S413–S424.

Holshouser, B.A., Tong, K.A., Ashwal, S., 2005. Proton MR spectroscopic imaging depicts diffuse axonal injury in children with traumatic brain injury. Am. J. Neuroradiol. 26, 1276–1285.

Hudson, R., Spradley, B., 2016. Concussions: a sport ethics commentary. Sport J.

International Federation of Sports Medicine, 2017. Code of Ethics. Retrieved 7th February 2017. Available at: http://www.fims.org/about/code-ethics/.

King, N.S., 2014. A systematic review of age and gender factors in prolonged post-concussion symptoms after mild head injury. Brain Inj. 28 (13–14), 1639–1645.

Kou, Z., Wu, Z., Tong, K.A., Holshouser, B., Benson, R.R., Hu, J., Haacke, E.M., 2010. The role of advanced MR imaging findings as biomarkers of traumatic brain injury. J. Head Trauma Rehabil. 25, 267–282.

Kroshus, E., Baugh, C.M., Daneshvar, D.H., Stamm, J.M., Laursen, R.M., Austin, S.B., 2015a. Pressure on sports medicine clinicians to prematurely return collegiate athletes to play after concussion. J. Athl. Train. 50 (9), 944–951.

Kroshus, E., Garnett, B., Hawrilenko, M., Baugh, C.M., Calzoh, J.P., 2015b. Concussion under-reporting and pressure from coaches, teammates, fans, and parents. Soc. Sci. Med. 134, 66–75.

Kroshus, E., Garnett, B.R., Baugh, C.M., Calzo, J.P., 2015c. Social norms theory and concussion education. Health Educ. Res. 30 (6), 1004–1013.

Kutcher, J.S., McCrory, P., Davis, G., Pitto, A., Meeuwisse, W.H., Broglio, S.P., 2013. What evidence exists for new strategies or technologies in the diagnosis of sports concussion and assessment of recovery? Br. J. Sports Med. 47 (5), 299–303.

Lancaster, M.A., Olson, D.V., McCrea, M.A., Nelson, L.D., LaRoche, A.A., Muftuler, L.T., 2016. Acute white matter changes following sport-related concussion: a serial diffusion tensor and diffusion kurtosis tensor imaging study. Hum. Brain Mapp. 37, 3821–3834.

Leddy, J.J., Baker, J.G., Merchant, A., Picano, J., Gaile, D., Matuszak, J., Willer, B., 2015. Brain or strain? Symptoms alone do not distinguish physiologic concussion from cervical/vestibular injury. Clin. J. Sport Med. 25 (3), 237–242.

Lee, H., Sullivan, S.J., Schneiders, A.G., Ahmed, O.H., Balasundaram, A.P., Williams, D., Meeuwisse, W.H., McCrory, P., 2015. Smartphone and tablet apps for concussion road warriors (team clinicians): a systematic review for practical users. Br. J. Sports Med. 49, 499–505.

Liu, A.Y., Maldjian, J.A., Bagley, L.J., Sinson, G.P., Grossman, R.I., 1999. Traumatic brain injury: diffusion-weighted MR imaging findings. Am. J. Neuroradiol. 20, 1636–1641.

Mainwaring, L., Hutchison, M., Camper, P., Richards, R., 2012. Examining emotional sequelae of sport concussion. J. Clin. Sport Psychol. 6 (3), 247–274.

Mannix, R., Meehan, W.P., Pascual-Leone, A., 2016. Sports-related concussions—media, science and policy. Nat. Rev. Neurol. 12, 486–490.

Marshall, C.M., Vernon, H., Leddy, J.J., Baldwin, B.A., 2015. The role of the cervical spine in post-concussion syndrome. Phys. Sportsmed. 43 (3), 274–284.

Mayer, A.R., Ling, J., Mannell, M.V., Gasparovic, C., Phillips, J.P., Doezema, D., Reichard, R., Yeo, R.A., 2010. A prospective diffusion tensor imaging study in mild traumatic brain injury. Neurology 74, 643–650.

McCrory, P., Meeuwisse, M., Aubry, M., Cantu, B., Dvorák, J., Echemendia, R.J., Engebretsen, L., Johnston, K., Kutcher, J.S., Raftery, M., Sills, A., Benson, B.W., Davis, G.A., Ellenbogen, R.G., Guskiewicz, K., Herring, S.A., Iverson, G.L., Jordan, B.D., Kissick, J., McCrea, M., McIntosh, A.S., Maddocks, D., Makdissi, M., Purcell, L., Putukian, M., Schneider, K., Tator, C.H., Turner, M., 2013. Consensus statement on concussion in sport: the 4th international conference on concussion in sport held in Zurich, November 2012. Br. J. Sports Med. 47 (5), 250–258.

McNamee, M.J., Partridge, B., Anderson, L., 2016. Concussion ethics and sports medicine. Clin. Sports Med. 35 (2), 257–267.

Mihalik, J., De Maio, V.J., Tibbo-Valeriote, H., Lanier, B., Jackson, T., Wooten, J.D., 2013. Epidemiology and age comparisons of patients presenting to a dedicated community concussion clinic. Br. J. Sports Med. 47 (5), e1.

Moore, B.M., Adams, J.T., Barakatt, E., 2016. Outcomes following a vestibular rehabilitation and aerobic training program to address persistent post-concussion symptoms. J. Allied Health 45 (4), e59–e68.

Morin, M., Langevin, P., Fait, P., 2016. Cervical spine involvement in mild traumatic brain injury: a review. J. Sports Med. http://dx.doi.org/10.1155/2016/1590161.

Moser, R.S., 2007. The growing public health concern of sports concussion: the new psychology practice frontier. Prof. Psychol. Res. Pract. 38 (6), 699–704.

Murray, D.A., Meldrum, D., Lennon, O., 2017. Can vestibular rehabilitation exercises help patients with concussion? A systematic review of efficacy, prescription and progression patterns. Br. J. Sports Med. 51, 442–451. http://dx.doi.org/10.1136/bjsports-2016-096081.

National Football League, 2013. NFL, Ex-Players Agree to $765M Settlement in Concussions Suit. Retrieved 7th February 2017. Available at: http://www.nfl.com/news/story/0ap1000000235494/article/nfl-explayers-agree-to-765m-settlement-in-concussions-suit.

National Institute for Health and Care Excellence, 2014. Head Injury. Retrieved 8th February 2017. Available at: https://www.nice.org.uk/guidance/cg176/resources/imaging-algorithm-498950893.

Neurocog Concussion Clinic, 2016. Taking the Guesswork Out of Concussion. Retrieved 8th February 2017. Available at: http://www.neurocogconcussion.co.uk/.

Neurology Central, 2016. UK Experts Meet to Discuss Concussion in Sport. Retrieved 6th February 2017. Available at: https://www.neurology-central.com/2016/11/22/uk-experts-meet-discuss-concussion-sport/.

NHS England, 2016. Sustainability and Transformation Plans September 2016. Retrieved 24th January 2017. Available from: www.england.nhs.uk/ourwork/futurenhs/deliver.

Pabian, P.S., Oliveira, L., Tucker, J., Beato, M., Gual, C., 2016. Interprofessional management of concussion in sport. Phys. Ther. Sport 23, 123–132.

Papa, L., Ramia, M.M., Edwards, D., Johnson, B.D., Slobounov, S.M., 2015. Systematic review of clinical studies examining biomarkers of brain injury in athletes after sports-related concussion. J. Neurotrauma 32 (10), 661–673.

Parizel, P.M., Van Goethem, J.W., Ozsarlak, O., Maes, M., Phillips, C.D., 2005. New developments in the neuroradiological diagnosis of craniocerebral trauma. Eur. J. Radiol. 15, 569–581.

Phillips, T., Guy, K., Tompkinson, A., Martin, R., Phillips, D., Sloan, N., 2016. Concussion in the emergency department; unconsciously incompetent? Emerg. Med. J. 33, 905–906.

Provvidenza, C., Engebretsen, L., Tator, C., Kissick, J., McCrory, P., Sills, A., Johnston, K.M., 2013. From consensus to action: knowledge transfer, education and influencing policy on sports concussion. Br. J. Sports Med. 47, 332–338.

Reid, S.A., Rivett, D.A., Katekar, M.G., Callister, R., 2008. Sustained natural apophyseal glides (SNAGs) are an effective treatment for cervicogenic dizziness. Man. Ther. 13 (4), 357–366.

Reneker, J.C., Cook, C.E., 2015. Dizziness after sports-related concussion: can physiotherapists offer better treatment than just 'physical and cognitive rest'? Br. J. Sports Med. 49, 491–492.

Ross, B.D., Ernst, T., Kreis, R., Haseler, L.J., Bayer, S., Danielsen, E., Blüml, S., Shonk, T., Mandigo, J.C., Caton, W., Clark, C., Jensen, S.W., Lehman, N.L., Arcinue, E., Pudenz, R., Shelden, C.H., 1998. 1H MRS in acute traumatic brain injury. J. Magn. Reson. Imaging 8, 829–840.

Rugby Football Union, 2016. Concussion-Headcase. Retrieved 8th February 2017. Available from: http://www.englandrugby.com/my-rugby/players/player-health/concussion-headcase/.

Rutgers, D.R., Fillard, P., Paradot, G., Tadié, M., Lasjaunias, P., Ducreux, D., 2008. Diffusion tensor imaging characteristics of the corpus callosum in mild, moderate, and severe traumatic brain injury. Am. J. Neuroradiol. 29, 1730–1735.

Sarmiento, K., Hoffman, R., Dmitrovsky, Z., Lee, R., 2014. A 10-year review of the centers for disease control and prevention's heads up initiatives: bringing concussion awareness to the forefront. J. Safety Res. 50, 143–147.

Schneider, K.J., Meeuwisse, W.H., Nettel-Aguirre, A., Barlow, K., Boyd, L., Kang, J., Emery, C.A., 2014. Cervicovestibular rehabilitation in sport-related concussion: a randomised controlled trial. Br. J. Sports Med. 48, 1294–1298.

Sharp, D.J., Jenkins, P.O., 2015. Concussion is confusing us all. Pract. Neurol. 15 (3), 172–186.

Singh, R., Savitz, J., Teague, T.K., Polanski, D.W., Mayer, A.R., Bellgowan, P.S., Meier, T.B., 2016. Mood symptoms correlate with kynurenine pathway metabolites following sports-related concussion. J. Neurol. Neurosurg. Psychiatry 87 (6), 670–675.

Sisodia, V., Hamid, J., Guru, K., 2015. Efficacy of vestibular rehabilitation in management of balance deficit in Indian collegiate football players, with sport-related concussion—randomized clinical trial. Physiotherapy 101 (Suppl. 1), e1403–e1404.

Smith, D.H., Stewart, W., 2016. Tackling concussion, beyond Hollywood. Lancet Neurol. 15 (7), 662–663.

Sports Illustrated, 2016. Football's Endgame: What Would Happen If America's Pastime Just ... Died? Retrieved 5th February 2017. Available at: http://www.si.com/nfl/2016/08/31/future-of-football-nfl-safety-concussions.

STAT, 2015. Concussion, Inc.: The Big Business of Treating Brain Injuries. Retrieved 8th February 2017. Available at: https://www.statnews.com/2015/12/16/concussion-brain-big-business/.

Stewart, G.W., McQueen-Borden, E., Bell, R.A., Barr, T., Juengling, J., 2012. Comprehensive assessment and management of athletes with sport concussion. Int. J. Sports Physiol. Therapy 7 (4), 433–447.

2014. Tackling the sports-related concussion crisis [Editorial]. Lancet Neurol. 13 (8), 747.

The Institute of Sport Exercise & Health, 2016. Concussion Clinic. Retrieved 6th February 2017. Available at: http://iseh.co.uk/patients/clinical-services/concussion-clinic.

The New York Times, 2013. A New Way to Care for Young Brains. Retrieved 8th February 2017. Available at: http://www.nytimes.com/2013/05/06/sports/concussion-fears-lead-to-growth-in-specialized-clinics-for-young-athletes.html.

Vargas, B.B., Channer, D.D., Dodick, D.W., Demaerschalk, B.M., 2012. Teleconcussion: an innovative approach to screening, diagnosis, and management of mild traumatic brain injury. Telemed. J. E Health 18 (10), 803–806.

Vernooij, M.W., Ikram, M.A., Tanghe, H.L., Vincent, A.J., Hofman, A., Krestin, G.P., Niessen, W.J., Breteler, M.M., van der Lugt, A., 2007. Incidental findings on brain MRI in the general population. N. Engl. J. Med. 357 (18), 1821–1828.

Wade, D.T., King, N.S., Wenden, F.J., Crawford, S., Caldwell, F.E., 1998. Routine follow up after head injury: a second randomised controlled trial. J. Neurol. Neurosurg. Psychiatry 65 (2), 177–183.

Weber, M., Edwards, MG., 2012. Sport concussion knowledge in the UK general public. Archives of Clinical Neuropsychology 27 (3), 355–361.

Wilde, E.A., McCauley, S.R., Hunter, J.V., Bigler, E.D., Chu, Z., Wang, Z.J., Hanten, G.R., Troyanskaya, M., Yallampalli, R., Li, X., Chia, J., Levin, H.S., 2008. Diffusion tensor imaging of acute mild traumatic brain injury in adolescents. Neurology 70, 948–955.

Williams, V.B., Danan, I.J., 2016. A historical perspective on sports concussion: where we have been and where we are going. Curr. Pain Headache Rep. 20 (6), 43.

Williams, J.M., Langdon, J.L., McMillan, J.L., Buckley, T.A., 2016. English professional football players concussion knowledge and attitude. J. Sport Health Sci. 5 (2), 197–204.

World Rugby, 2015. Head injury assessment adopted into law. Retrieved 2nd February 2017. Available at: http://www.worldrugby.org/news/70796.

CHAPTER 12

Neurocognitive mechanisms of the flow state

David J. Harris[1], Samuel J. Vine, Mark R. Wilson

School of Sport and Health Sciences, University of Exeter, Exeter, United Kingdom
[1]*Corresponding author: Tel.: +44-1392-722891, e-mail address: d.j.harris@exeter.ac.uk*

Abstract

While the experience of flow is often described in attentional terms—focused concentration or task absorption—specific cognitive mechanisms have received limited interest. We propose that an attentional explanation provides the best way to advance theoretical models and produce practical applications, as well as providing potential solutions to core issues such as how an objectively difficult task can be subjectively effortless. Recent research has begun to utilize brain-imaging techniques to investigate neurocognitive changes during flow, which enables attentional mechanisms to be understood in greater detail. Some tensions within flow research are discussed; including the dissociation between psychophysiological and experiential measures, and the equivocal neuroimaging findings supporting prominent accounts of hypofrontality. While flow has received only preliminary investigation from a neuroscientific perspective, findings already provide important insights into the crucial role played by higher-order attentional networks, and clear indications of reduced activity in brain regions linked to self-referential processing. The manner in which these processes may benefit sporting performance are discussed.

Keywords

Attention, Neuroscience, Mechanisms, Peak performance, Zone

> *The consciousness of self is the greatest hindrance to the proper execution of all physical action*
> **Bruce Lee (1975)**

1 INTRODUCTION

The crucial role that attention plays in shaping the learning of sporting skills, performing under pressure, and facilitating expertise is well established (Mann et al., 2007; Moore et al., 2012). However, the state of flow, something of an attentional anomaly (Bruya, 2010), provides an alternate perspective from which to examine attention during peak performance. During flow increased demands are paradoxically met with no apparent increase in effort (Csikszentmihalyi, 1975), and athletes report a laser-like task focus in the face of distractions (Jackson and Csikszentmihalyi, 1999). Flow is frequently linked to peak performances (Jackson et al., 2001; Koehn and Morris, 2012) as well as heightened enjoyment (Privette, 1983). Therefore understanding the cognitive processes responsible for this state may inform both attentional models and practical endeavors for finding peak focus during sporting performance.

Flow, or "the zone," is a phenomenological state where an individual finds an effortless involvement and deep, task-related focus in the current activity. It is accompanied by a feeling of control and a loss of self, creating an intrinsically motivated, optimal experience (Csikszentmihalyi, 1975, 1990, see Fig. 1). Athletes in flow often report reaching the apex of their abilities and levels of fulfillment that are unrivaled in the rest of their lives (Jackson and Csikszentmihalyi, 1999). While research has described the nature of the experience and it's antecedents in sport (Jackson, 1996; Swann et al., 2016), the underlying mechanisms of flow are not well understood. We propose that a deeper consideration of attentional processes may be crucial to this understanding. Therefore we review current findings related to the role of attention, and the ways in which direct brain recording can inform this understanding, with the aim of proposing attentional changes as the fundamental mechanism for creating the state of flow.

In order to understand flow in the context of sporting performance (and the potential benefits it provides), findings from the sporting literature will be considered alongside neuroimaging research. Imaging findings are well established across the wider field of attention (Corbetta and Shulman, 2002; Fan et al., 2005) and will be used to highlight how attentional changes may provide beneficial outcomes for sport. We do not attempt an exhaustive review of either sporting flow or neuroimaging studies, but endeavor to give an overview of findings central to understanding attentional processes during flow.

1.1 THE FLOW STATE

Flow, the mental state of complete absorption in the present task, is traditionally described along Csikszentmihalyi's (1990) nine dimensions. Three of these dimensions are described as setting the conditions for flow; a balance of challenges and skills, clear goals, and immediate feedback (Kawabata and Mallett, 2011; Nakamura and Csikszentmihalyi, 2009). Csikszentmihalyi suggests that flow occurs when challenges and skills are in balance, in a channel between the states of boredom

1 Introduction

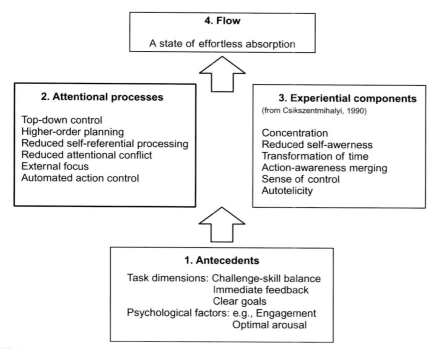

FIG. 1

Schematic representation of the focus of the chapter. While the experience of flow (box 3) and it's antecedents (box 1) have been well documented, the focus of our discussion is the attentional processes (box 2) that happen alongside, and may be responsible for, the experiential components.

and anxiety. The remaining six dimensions are experiential components (Nakamura and Csikszentmihalyi, 2009); intense concentration, merging of action and awareness, loss of self-consciousness, a sense of control, distortion of time, and intrinsic rewards (autotelicity). This conceptualization of flow has received robust support across a range of domains; including art and science (Csikszentmihalyi, 1996), sport (Jackson, 1996), and literary writing (Perry, 1999).

1.2 FLOW IN SPORT

The sport setting is ideally suited for achieving flow, as sporting activity can provide the three basic conditions Csikszentmihalyi suggests are necessary; clear goals, immediate feedback, and a balance between challenge and skill. The added effect of physical exertion may further enhance the experience of total absorption (Dietrich, 2006). Finding flow is highly desirable, as athletes report effortlessness and fluency of performance when in the state (Swann et al., 2016). As there is reasonable agreement over what characterizes a flow experience (Swann et al., 2012) research in sport has mainly focused on its antecedents. For example, a systematic review by Swann et al. (2012) found the most common antecedents to be focus, preparation, motivation, arousal, and positive thoughts. Unfortunately, research in sport is yet to identify important mechanisms that underpin the experience and has made limited use of experimental approaches to studying flow. As such, findings from other domains applying neuroscientific approaches will be discussed as they provide both methods that can drive flow research forwards, and important implications for how attention during flow may enable superior performance.

2 ATTENTION

Attention is the cognitive process of selecting discrete aspects of information for further processing; it allows an animal to behave adaptively in a world containing an overabundance of information (Knudsen, 2007). Attention therefore shapes our experience but is also key to selecting our responses to it (Allport, 1989) and is a viable candidate for facilitating beneficial effects of flow given that effective attentional control has been shown to have significant benefits for sporting performance (Abernethy et al., 2007). For example, optimizing the target and timing of visual attention can enhance psychophysiological efficiency, visuomotor coordination, and movement kinematics (Moore et al., 2012).

As four of Csikszentmihalyi's (1990) six experiential components of the flow experience (intense concentration, merging of action and awareness, loss of self-consciousness, and distortion of time) allude to changes in attentional processes, these mechanisms deserve further clarification. Within the sporting literature athletes report focus as being crucial for finding flow (Jackson, 1992). Swann et al. (2012) also found the most reported aspects of the experience to be concentration and action-awareness merging, again suggesting a fundamental role for attention. In order to build an initial picture of attention during flow, we focus on four topics that emerge from current literature: the degree of automaticity, the role of effective attentional control, the extent to which flow requires effortful attention, and the experience of reduced self-awareness.

2.1 AUTOMATICITY

A prominent theoretical approach is to describe flow as a state of enhanced automaticity, essentially an absence of controlled attention. Automated control is generally, though not always, thought to provide benefits for action in terms of efficiency, fluency, and reduced resource allocation (see Toner et al., 2015). Automatic or implicit processes are fast and efficient, and circumvent conscious awareness of effort (Moors and De Houwer, 2006), and as such may contribute to the frequent perception of ease during flow. As automaticity involves reduced executive activity, it is characterized by reduced frontal activation (esp. lateral and dorsolateral PFC), as regions of basal ganglia acquire motor sequence knowledge (Poldrack et al., 2005). A model of the flow experience centered on a reduction in prefrontal control is proposed by Dietrich (2003).

2.1.1 Transient Hypofrontality Theory (THT)

Dietrich (2003) suggests that the core components of the flow experience, and other altered states of consciousness, can be explained by reductions in processing by the prefrontal cortex (PFC). Higher processes like abstract thinking, self-reflective consciousness, and working memory are lost (known as phenomenological subtraction), creating states of reduced function (common in running, meditation, or hypnosis). These states are suggested to be linked to flow due to a common mechanism of hypofrontality (i.e., reduced frontal activity and executive functioning).

Dietrich and Stoll (2010) argue that sports are particularly good at engendering flow states for two reasons. First, bodily motion is extremely complex in computational terms and as such uses considerable mental resources (demonstrated by Stoll and Pithan, 2016). When prolonged, this is sufficient to divert processing away from noncritical cognitive processing (i.e., higher-order function in the PFC). Second, Dietrich suggests that as an action becomes more automatic, and under control of the basal ganglia (Poldrack et al., 2005), it is easier for frontal function to decrease and for this altered state of consciousness to arise. Dietrich's suggestion that action in flow is more automatic also accounts for the effortless experience often associated with flow, as automatic action is controlled through unconscious or implicit processes, and unavailable to conscious awareness (Moors and De Houwer, 2006). In attentional terms, they are not constrained by capacity limitations, excluding them from mental effort (Kahneman, 2011).

Imaging studies have provided some support for a reduction in frontal activity, but results suggest a very specific pattern, that does not constitute a general deactivation. Ulrich et al. (2014, 2016) found reduced activity of the medial PFC, an important structure in self-referential processing. Conversely, however, research by Yoshida et al. (2014) using functional near-infrared spectroscopy (fNIRS) found increased ventrolateral PFC activation during a flow level of a gaming task. Additionally, research by Harmat et al. (2015) again used fNIRS to test prefrontal activation during flow in a widely used paradigm utilizing continual adjustment of difficulty in the game "Tetris." There was found to be no reduction in frontal activity, suggesting that a general mechanism of hypofrontality may be overly simplistic. Support for hypofrontality comes mainly from studies finding reductions in cognitive function as a result of prolonged exercise (see Stoll and Pithan, 2016). Whether this state is representative of a general flow mechanism is questionable. Harmat et al. (2015) suggest that the demands on executive control, even during a relatively simple Tetris task, make a reduction in prefrontal activity unlikely.

2.1.2 Reduced Verbal-Analytic Processing

Initial electroencephalogram (EEG) research does provide support for a move toward automaticity, and the avoidance of deliberative control of motor responses (e.g., *Reinvestment Theory*, Masters and Maxwell, 2008). EEG assesses cortical activity through electrodes placed on the scalp, which give information regarding rhythmic neural oscillations, measured in Hertz (Hz), and may be useful for identifying brain areas that are particularly active or inactive during flow. Of particular interest is the alpha band (8–12 Hz), which reflects relaxed wakefulness and may signpost areas that are inhibited during flow (Pfurtscheller et al., 1996).

Wolf et al. (2015) demonstrated that flow during table tennis imagery was related to a reduced influence of verbal-analytic processing on motor control, which was also related to expertise. EEG recordings showed relative deactivation (higher alpha power) of a left temporal site (T3), associated with verbal-analytic processing, compared to the corresponding right site (T4) which reflects visuospatial processing. Their results suggest that the flow experience may involve a shift away from verbal-analytic influence to a more automatic mode of operating; the same shift that has been found with a progression from novice to expert performance (Deeny et al., 2003), and from more explicit to more implicit forms of motor performance (Zhu et al., 2011). Further success of EEG in identifying markers of flow may provide opportunities for training flow experiences (see Cheron, 2016), due to previous success with EEG in neurofeedback training (Ring et al., 2015).

This conceptualization of flow based on automation is appealing and indeed seems sufficiently entrenched to have permeated areas like philosophy; from the ancient Chinese concept of "wu-wei" (Slingerland, 2014) to Dreyfus' antirepresentationalist account of choking and coping under pressure (see Dreyfus, 2007; Gottlieb, 2015). Automated processes provide benefits for action and are a hallmark of expertise (Fitts and Posner, 1967; Singer, 2002), so while the mechanism of hypofrontality may be questionable (based on neuroimaging and mental effort findings, see later), the benefits of automated action for avoiding conscious disruption of well-learned movement sequences are clear. As such, if flow enables an enhanced degree of implicitness in motor control we can expect to see benefits for sport. Whether this degree of automaticity extends to aspects of performance like decision making, selective attention, creativity,[a] and problem solving seems unlikely (Harmat et al., 2015).

A mixed pattern of automatic and controlled processing has also been found in the study of musical improvisation, a flow-like activity where musicians display high levels of creativity, spontaneously producing complex musical structures (Nisenson, 1995). Limb and Braun (2008) found that during improvisation, in contrast to over-learned sequences, jazz pianists displayed a dissociated pattern of frontal activity with deactivation of lateral prefrontal areas associated with effortful problem solving and monitoring of goal-directed behaviors (Ashby et al., 1999). However, several other studies of improvisation *have* supported the involvement of monitoring and cognitive control processes (Bengtsson et al., 2007) suggesting a greater coupling between creative and controlled (executive) processes (see Beaty, 2015 for

---

[a] Flow is often associated with creativity (Csikszentmihalyi, 1996), which has itself received neuroimaging interest. However findings from this area were not included as methods, definitions, and results are sufficiently diverse as to create additional confusion. For a review, see Dietrich and Kanso (2010).

a review). Losing yourself in the performance may not be as simple as a complete absence of cognitive control, but a more balanced cooperation between spontaneous and deliberative processing.

Similarly, Csikszentmihalyi and Nakamura (2010) postulate that the importance of automaticity during flow lies in allowing automated sequences to take care of themselves, so that *more* attention can be paid to essential aspects of the activity. They suggest that effortless attention is rarely fully automatic, and a person is often more open, alert, and flexible within the structure of the activity. Such a balance between attentional effort and automatic processing perhaps provides the benefits for sporting performance that accompany automated action sequences (Singer, 2002) and the flexibility and problem solving that come with more effortful, controlled processing (Miyake et al., 2000).

2.2 ATTENTIONAL CONTROL

Attentional control involves the ability to direct attention to only those stimuli that are relevant to our current goals, minimizing the extent to which bottom-up influences capture our attention (Corbetta and Shulman, 2002). Sarter et al. (2006) describe top-down control as the "biasing of attentional resources toward the detection and processing of target stimuli" (p. 148) which they link to attentional effort. While flow has not traditionally been associated with controlled processing (Dietrich, 2006), the characterization of flow as an extreme focus with immunity to distraction (Jackson and Csikszentmihalyi, 1999) suggests a strong influence of top-down attentional control. Effective attentional control might explain the performance advantages reported for flow, as limited attentional resources are optimized through becoming highly task focused.

2.2.1 Synchronization Theory

A theory of flow proposed by Weber et al. (2009) suggests that widespread synchronization between neural attention networks may provide the basis for the complete absorption apparent during a flow state. The theory is based on Posner et al.'s (1987) tripartite theory of attention; involving executive, alerting, and orienting networks. The alerting network is responsible for initiating and maintaining attentiveness, while the orienting network directs attention to a stimulus. The executive network modulates both and plays a crucial role in top-down control. Optimal attentional control depends upon directing attention to relevant stimuli, and if alerting and orienting networks were optimally synchronized with executive control we would expect attention to be highly goal-directed (Petersen and Posner, 2012).

Weber and colleagues propose that this organization is achieved through synchronized firing rates of neurons within attentional networks. Neurons fire at varying rates at rest, but oscillatory activity in groups of neurons can arise from feedback connections between them, particularly in the alpha (8–12 Hz) and theta (4–7 Hz) bands. Synchronized neural activity is proposed as a solution to the binding problem of consciousness (Crick and Koch, 1990), as groups of neurons that have become synchronized may coordinate their processing. Weber et al. suggest that coordinated firing of alerting, orienting, and executive attention networks, alongside reward networks (Schultz, 2006), gives rise to the experience of flow, as all attention and reward systems are working harmoniously. This optimal organization (which Csikszentmihalyi terms *negentropy*) is highly efficient, creating mental ease, which is often reported during flow (Bruya, 2010). In effect, this theory suggests that flow enables an optimal, highly efficient organization of attention, with an important role for executive function and top-down maintenance of goal-directed attention.

2.2.2 Top-Down Attention Networks

The importance of higher attentional processes during the flow experience is supported by neuroimaging research. Ulrich et al. (2016) have identified activation of brain areas associated with the multiple demand (MD) network during flow in an arithmetic task. The MD system (Duncan, 2010) reflects activity of a functionally general network of brain regions associated with a variety of cognitive challenges. The MD network involves areas of the prefrontal and parietal cortex, including: the inferior frontal sulcus, anterior insula, presupplementary motor area, and in and around the intraparietal sulcus. Duncan (2010, 2013) link the MD system to goal-directed behavior, fluid intelligence, and selective visual attention. In particular it plays a function in organizing multistep behavior. Interestingly, Ulrich et al. found brain areas related to the MD network to be activated *more* during a flow level of difficulty than a harder level of difficulty, with activation associated with performance.

Alternatively, the areas identified by Ulrich et al. (2016) could also be interpreted as reflecting increased activity of the dorsal stream of attention (Corbetta and Shulman, 2002, see Fig. 2). The MD system depends on areas of parietal and frontal cortex which are identified by Corbetta and Shulman (2002) as part of the frontoparietal dorsal stream. Indeed, Duncan (2013) notes the overlap between MD and dorsal networks, and therefore the findings of Ulrich et al. may also be instructive of the important role played by the dorsal stream in promoting goal-directed control of attention during flow. Either way, these findings are indicative of top-down control of attention, as the MD system serves to coordinate a series of multistep behaviors, guide selective focus to task-relevant information, and provide cognitive control. These findings strongly challenge views of flow as a state of automaticity, with reduced frontal influence (Dietrich, 2003), and highlight that during flow supervisory attentional and cognitive control systems of the brain are highly active.

2.2.3 Dopamine Pathways

Attentional processes during flow may also be affected by neurotransmitter activity, as findings highlight a role for dopamine. Dopamine pathways are primarily associated with reward networks in the brain (Schultz, 2006), but also modulate attentional focusing (Nieoullon, 2002), error monitoring (Holroyd and Coles, 2002; Ridderinkhof et al., 2004), and response inhibition (Chambers et al., 2009; Congdon et al., 2008). de Manzano et al. (2013) have demonstrated that flow prone

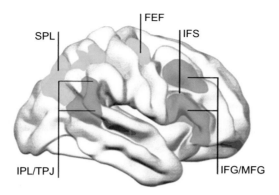

FIG. 2

The ventral network (*blue*), responsible for reorienting attention to salient stimuli, projects from the temporoparietal junction (TPJ) toward inferior frontal gyrus (IFG) and middle frontal gyrus (MFG). The dorsal network (*orange*), responsible for top-down voluntary allocation of attention, projects from the superior parietal lobe (SPL) toward the frontal eye fields (FEF). The MD system includes overlapping frontoparietal areas, from the SPL to the premotor cortex and inferior frontal sulcus (IFS).

Figure reproduced from Aboitiz, F., Ossandón, T., Zamorano, F., Palma, B., Carrasco, X., 2014. Irrelevant stimulus processing in ADHD: catecholamine dynamics and attentional networks. Front. Psychol. 5, 183.

individuals have increased availability of dopamine D2 receptors in the striatum, which is functionally related to selective attention (Nieoullon, 2002). Gyurkovics et al. (2016) have identified flow proneness to be related to a D2 receptor coding gene and suggest that the relationship between dopamine and flow may be through reduced impulsivity and more effective response inhibition.

Reduced dopamine action is associated with impulsive behavior (Dalley and Roiser, 2012) evidenced by its therapeutic effects on impulsivity in ADHD (Kollins and March, 2007), so individuals with enhanced D2 availability are predisposed to the behavioral control and monitoring benefits of dopamine. The findings related to dopamine fit with our wider discussion of attentional control, which requires response inhibition and impulse control (Miyake et al., 2000), which are modulated by dopamine action (Dalley and Roiser, 2012; Nieoullon, 2002). These findings also have important implications for individual differences in flow and the "autotelic personality" (Csikszentmihalyi, 1975), which may therefore have a biological basis.

The findings discussed here point strongly to an important role for higher-order attention control mechanisms during flow, although this is perhaps not the dominant approach within the flow literature (Dietrich, 2006; Jackson, 1996; Swann et al., 2016). Effective control of attention is established as crucial for optimal sporting performance. For example, optimal attention control, as indexed by the quiet eye (Vickers, 1996; see Chapter "Quiet eye training improves accuracy in basketball field goal shooting" by Joan N. Vickers et al.), provides significant inter- and intraindividual performance benefits (see metaanalysis by Lebeau et al., 2016). The quiet eye is a final fixation to a target prior to movement execution, which provides task focus and allows organization of neural networks for controlling movement (Vickers, 2009). The importance of optimal control of visual attention is highlighted by its negative corollary; as disrupted attentional control under pressure has been shown to impair motor performance (see Eysenck and Wilson, 2016 for a review). For example, there is ample evidence that conditions designed to increase anxiety and reduce flow can impair attention (indexed via quiet eye) and subsequent performance in a range of sport skills; including basketball free-throw shooting (Wilson et al., 2009), soccer penalties (Wood and Wilson, 2010), golf putting (Vine et al., 2013), and biathlon (Vickers and Williams, 2007). The use of technology, such as mobile eye trackers, to measure objective, task-relevant indices of attentional control (e.g., quiet eye) may therefore provide useful insights into the flow process while it unfolds.

2.3 ATTENTIONAL EFFORT

Traditional models of attention suggest that as a task becomes more difficult, requiring cognitive or executive control, more mental energy (or effort) is required to meet demands (McGuire and Botvinick, 2010). Mental effort, the motivated activation of attentional systems (Sarter et al., 2006), has a physical basis, leading to changes in both central (neural) and peripheral (e.g., cardiovascular) psychophysiological indicators (Berntson et al., 1997). However flow presents a challenge for attention researchers, as difficult tasks are met with a perceived *decrease* in felt effort (Bruya, 2010). This type of effortless attention has been referred to as "postvoluntary": neither voluntary (effortful) nor involuntary (automatic), but captured by an absorbing activity (Dobrynin, 1966).

During flow, individuals report that task control becomes effortless as they become fully absorbed. Therefore either the task has become easier, and does not require mental effort, or the subjective experience has become dissociated from the objective level of mental work. Several studies utilizing objective markers of mental effort during flow, would suggest the latter. For example, Gaggioli et al. (2013) revealed that everyday flow experiences were indexed by increased sympathetic activity (increased heart rate), suggesting that the perception of ease is not reliably reflected in physiological activity.

2.3.1 Psychophysiological Measures of Attentional Effort

Heart rate variability (HRV), the fluctuation in beat-to-beat interval, provides an objective measure of mental effort (Berntson et al., 1997) and attention (Backs, 1997). Mental effort reduces HRV (particularly in the low frequency, 0.04–0.15 Hz band) due to increased sympathetic regulation and decreased influence of the baroreflex (Berntson et al., 1997). Keller et al. (2011) found that a balance between challenge and skills (a precursor to flow) produced positive subjective reports, but decreased HRV, indicating increased mental effort. Likewise, Tozman et al. (2015) found that a flow-task reduced HRV in the low frequency component (i.e., required more mental effort) compared to an objectively easy task. Finally, Peifer et al. (2014) found an inverted-U relationship between flow and low frequency HRV; suggesting that once a moderate level of mental effort is reached, additional mental load is detrimental.

2.3.2 Conflict Monitoring

The psychophysiological research provides a consistent picture that flow does demand attentional effort, and hence cannot be considered "automatic." A possible explanation for the lack of felt effort can be found in the work of Botvinick and colleagues on the conflict-monitoring hypothesis (Botvinick et al., 2001), which indicates how a task may feel easier during flow. Within the conflict monitoring hypothesis, subjective mental effort arises from a change in cognitive control, when the demands of the situation are appraised as not being met. During flow, challenges, and skills balance, performance is fluent and no change in cognitive control is required (McGuire and Botvinick, 2010). As highlighted by Dobrynin (1966), effortless attention occurs when the task is fully absorbing, and hence cognitive control is not required to maintain focus. The anterior cingulate cortex (ACC, see Fig. 3) is proposed as a key structure within cognitive control and conflict monitoring (Botvinick et al., 2001, 2004; Van Veen and Carter, 2002), serving to monitor processing and determine when attention needs to be refocused or redirected. The importance of the ACC in the perception of effort is demonstrated by Naccache et al. (2005), who found a patient with ACC lesion did not experience effort in a Stroop task, in spite of normal executive control. In addition, Posner et al. (2010) suggest that early stages of meditation may depend on ACC to exert control, with activity receding as it becomes more effortless.

Neuroimaging studies provide evidence that the same process may be occurring during flow. Klasen et al. (2011) found reductions in rostral ACC activity associated with increased focus, and reductions in dorsal ACC activity associated with clear goals in the task, in participants experiencing flow while playing a video game. Ulrich et al. (2016) also found reduced neural activation in the right anterior cingulate during a flow level of an arithmetic task. These findings indicate how the nature of a flow task, providing clear goals, and holding an individual's focus, may allow reductions in the need for cognitive control through reduced activity of the ACC, and subsequent reductions in felt effort.

The predictions of the CMH identify how felt effort may dissociate from the physiological markers of attentional effort discussed previously. This dissociation has been demonstrated experimentally by Harris et al. (2017), who directly tested opposing predictions for felt and physiological effort in a simulated driving task. While greatest physiological effort was found during a matched-to-skills (flow) condition, subjects reported effort to be the highest in a difficult drive and only moderate in the matched level. Harris et al. (2017) propose a model of effort during flow whereby (low) felt effort relates to the conflict monitoring mechanism, but invested physiological effort is greatest under optimal challenge (based on the predictions of motivational intensity theory, Wright, 1996). These findings indicate that despite greater investment of resources, flow felt comparatively easy, as conflict was minimized. This effect may help to explain the *effort paradox* during flow.

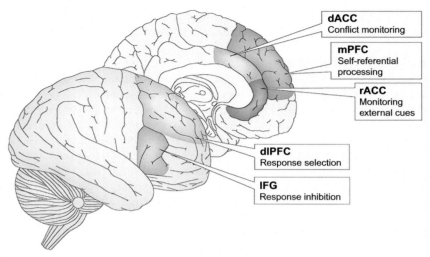

FIG. 3

Key prefrontal areas and some of their functions. The ACC may contribute to perceptions of effort during flow, while imaging has suggested reduced mPFC activity (Ulrich et al., 2016) but increased activity within IFG (Ulrich et al., 2014) and dlPFC (Yoshida et al., 2014).

Reprinted by permission from Macmillan Publishers Ltd.: Nature Reviews. Neuroscience (Amodio***, 2014), copyright (2014) www.nature.com.

The research discussed demonstrates that attentional effort is required during flow in order to meet the demands of a challenging task, which fits with our preceding consideration of top-down attention. It has been illustrated that while these processes can be highly effortful, during flow this is not necessarily the case, as when goals direct attention there is little psychological cost (Schmeichel and Baumeister, 2010) and goal-directed control may be maintained with little effort (Land, 2006). Optimal attentional control during flow is perhaps more dependent upon the absence of stimulus-driven disruption and monitoring processes, than of effortful top-down control. Subsequent reductions in cognitive control may be beneficial for sporting performance through avoidance of deliberative control of implicit processes which can lead to reinvestment and disrupted performance (Masters, 1992; Masters and Maxwell, 2008). Feelings of ease may further contribute to performance through promoting confidence and positive emotions. While potential explanations have been outlined here, this paradoxical aspect of attentional effort during flow requires further consideration and should be considered a significant challenge for attention researchers (Bruya, 2010).

2.4 SELF-AWARENESS

While there appears to be an increased deployment of attentional resources during flow, they appear to be largely directed away from the self (Csikszentmihalyi, 1975, 1990). When fully absorbed in a flow inducing activity, attention is directed toward the goal and the self recedes; bodily actions are reported to feel as though they are moving on their own, without conscious willing and where self-awareness does exist, it seems to be prereflective (see Toner et al., 2016 for discussion). Phenomenologist Sartre (1957) recognized the reduction in awareness of the self that comes from complete absorption in an activity. He noted that "When I run after a streetcar, when I look at the time, when I am absorbed in contemplating a portrait, there is no I" (p. 48). The role of self-awareness links strongly to the preceding discussion of automaticity, as a switch from internal to external attentional focus provides performance benefits through enhanced automaticity (Zachry et al., 2005). This reduction in self-consciousness during flow is also strongly supported by recent neuroimaging findings.

For instance, Ulrich et al. (2014) utilized magnetic resonance perfusion imaging to investigate flow during differing levels of arithmetic challenge. Reduced relative cerebral blood flow was found in the medial PFC (see Fig. 4), an area strongly linked to self-referential processing (Jenkins and Mitchell, 2011; Northoff et al., 2006) and an important part of the brains default mode network (DMN, Buckner et al., 2008; Raichle et al., 2001). The DMN is an interacting system of brain regions including the posterior cingulate cortex, medial PFC, and angular gyrus, which are active during passive states. Activity in the DMN is associated with mind-wandering and thinking about the self, past, and future, and is known to be reduced during goal-directed behaviors (Raichle et al., 2001). As well as medial PFC, Ulrich et al. found reduced activity in other regions associated with the DMN (e.g., angular gyrus, supramarginal gyrus, and parahippocampal cortex). While decreased activity in this network is not unique to flow, it highlights the importance of reductions in self-awareness and internal focus during the state.

Further research by Ulrich et al. (2016) using fMRI again found reductions in medial prefrontal areas and brain regions associated with the DMN. Goldberg et al. (2006) found reduced activity in self-related structures during sensory processing, likening it to "losing yourself in the act," which is strongly suggestive of Csikszentmihalyi's (1990) dimension of action awareness merging. Ulrich et al. (2016) suggest that this reduced self-awareness may also contribute to another facet of flow; the positive nature of the experience. As self-referential processing is associated with negative affectivity (Lemogne et al., 2011), reductions in activity of the medial PFC and DMN may contribute to a positive flow experience. Furthermore, an fMRI study by Garrison et al. (2013) applying real-time neurofeedback in experienced meditators found reductions in areas of the DMN to link to effortless doing and contentment.

Reduced awareness of the self may contribute, not just to the experience of flow, but to its benefits for performance. In sport, awareness of, and focus on, the self is associated with impaired skill learning and performance (Baumeister, 1984; Beilock and Carr, 2001; Masters, 1992). Wulf and Lewthwaite (2010) outline the self-invoking trigger hypothesis to describe how activation of the self-schema by environmental triggers (e.g., instructions, presence of others) can account for a variety of findings on motor learning and performance. Wulf and Lewthwaite (2010) link this self-schema system to the functional network of cortical midline structures found to be inactive during flow. As such, during flow the athlete may be resistant to

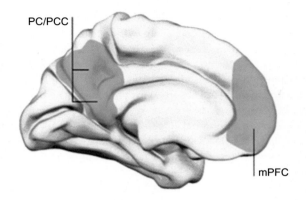

FIG. 4

Medial areas of the default mode network. Medial prefrontal cortex (mPFC), posterior cingulate cortex (PCC), and precuneus (PC) are active when the individual is engaged in mind wandering and thoughts about the self. The DMN also includes lateral parietal and medial temporal areas.

Figure reproduced from Aboitiz, F., Ossandón, T., Zamorano, F., Palma, B., Carrasco, X., 2014. Irrelevant stimulus processing in ADHD: catecholamine dynamics and attentional networks. Front. Psychol. 5, 183.

the self-related triggers and their negative consequences. The reduced activity of the DMN also points to a reduction in internal focus, in favor of focusing externally, on the goal (Nideffer, 1976). Wulf (2013) reports how external focus has been shown to enhance effective and efficient movement, something that is often described in flow, but is yet to be measured empirically. Training target related, external focus is therefore a potential route for enhancing flow (e.g., Moore et al., 2012). As such, a reduction in activity of self-referential neural structures and the DMN may be highly beneficial for sporting performance, as well as underlying key features of the flow experience.

2.5 ATTENTION SUMMARY

We have considered four issues related to attention: attention control, mental effort, automaticity, and self-awareness that have implications for understanding the experience and maintenance of flow. While there are certainly tensions, for example between automaticity and mental effort, all these features seem important in providing a full picture of the flow state. A combination of efficient attention and automated action control would seem to account for many of the features identified by Csikszentmihalyi (1990) as well as providing benefits for sporting performance. Based on flow reports we might expect attention during flow to be more external, less self-conscious, less prone to distraction and more task directed, thus leading to improved performance (see Fig. 1). Further empirical testing is needed in order to understand the extent of these attentional benefits.

3 DISCUSSION

This chapter set out to use recent neuroimaging findings to illustrate how attentional processes may provide the best way to understand the mechanisms behind the flow experience. The studies discussed support our contention that attentional processes can explain the key features of flow (e.g., Gyurkovics et al., 2016; Klasen et al., 2011; Ulrich et al., 2016). Findings suggest a reduction in self-awareness, through reductions in medial prefrontal areas and the DMN (Ulrich et al., 2014); improvements in impulse control related to dopamine activity (de Manzano et al., 2013); and considerable activity in networks related to higher-order attentional processing in the MD system or dorsal stream of attention (Ulrich et al., 2016). Together these findings account for many of the key features of the flow experience (Csikszentmihalyi, 2000). In addition, these features have implications for sporting performance. For example, the activation of higher attentional processes like the MD network support selective and goal-directed attention, which have beneficial performance effects in many sports (Abernethy et al., 2007; Vine and Wilson, 2011; Williams and Davids, 1998). Additionally, reduced verbal-analytic influence (Wolf et al., 2015) and automated performance (Dietrich, 2003) have been linked to sporting expertise (Beilock et al., 2002). Overall, these findings suggest a mental state where improved performance is almost inevitable.

These results also have implications for theories of flow. Dietrich's (2003) hypofrontality theory has received mixed support, suggestive of a more nuanced pattern of activity across the PFC. While unnecessary or even detrimental activity such as self-awareness may be limited, areas of dorsolateral and ventrolateral PFC involved in response selection and inhibition are highly active. As such, current findings suggest that Weber et al.'s (2009) synchronization theory may more accurately reflect neural activity during flow, as areas related to higher-order attentional networks appear to be crucial. The overall pattern of activity during flow suggests a highly efficient "switching on" of networks for goal-directed activity and a "switching off" of areas related to the self and conscious control of movement. Additionally, models of effort derived from the conflict monitoring hypothesis (Botvinick et al., 2001) such as that of Harris et al. (2017) have received initial support in describing the effort paradox during flow by addressing mechanisms for subjective and objective effort separately.

The important role of top-down focusing and higher attentional networks in flow that has been highlighted here has clear practical implications. To achieve an optimal task focus, the engaging and challenging nature of the task is clearly crucial (Csikszentmihalyi, 1990), but enhancing personal skills and capacities in attention control may also provide substantial benefits. Attention and self-regulation are trainable (Tang and Posner, 2009) and the efficacy of attentional training for sporting skills has recently been demonstrated (Ducrocq et al., 2016). While the transferability of the trained attention skills may require further validation (Shipstead et al., 2012), training sport specific attention skills through methods like quiet eye training may enable more frequent flow experience in the given task. As discussed, the quiet eye is an instance of optimal attention control in an aiming task (Vickers, 1996), but is also trainable, promoting external, goal-directed attention with subsequent performance benefits. Flow experiences are known to be fleeting and are not a prerequisite for performing well, but given the important role of attentional mechanisms, developing the ability to control attention may be the best way to find flow more often.

Significant research questions lie ahead in understanding the role of attention in creating and maintaining flow. For example, do those in flow experience enhanced perceptual abilities? Do they always demonstrate an external focus of attention? And does flow prevent skill disruption through reinvestment (Swann et al., 2017)? Future studies may wish to examine attentional benefits, like immunity to distraction or improved focus on task relevant stimuli. Initially inquiry may have to rely on creative laboratory research, where internal control is high and objective measures can be reliably assessed, even if this is unrepresentative of real sport environments. However, the need to disentangle cause and effect makes such an approach a crucial step in furthering our understanding of this complex phenomenon. Some of these questions can be answered by utilizing eye tracking techniques which provide an objective measure of how attention is being directed and controlled (Corbetta, 1998).

For example, improvements in attentional control can be easily measured through assessing the extent to which individuals direct attention to goal-relevant over irrelevant stimuli in the environment. Attentional probes, attentional bias, and unexpected events can also be utilized to see how strongly visual attention is held by the goals of the current task. Brain measurement should continue to be utilized, while psychophysiological methods may prove a useful, more accessible first step in sport, due to the difficulties of direct brain recordings during sporting performance. Intermediate steps like movement-controlled gaming (see Thin et al., 2011), simulators (Harris et al., 2017; Tozman et al., 2015), and virtual reality may be important in applying imaging techniques to tasks representative of sport.

4 CONCLUSIONS

In this chapter, we hoped to provide an overview of findings from direct measurement techniques regarding the nature of the flow experience and its underlying attentional mechanisms. Imaging findings have indicated that flow is a state where attention is focused, self-awareness is reduced, positive emotions are elicited, and automatic actions are allowed to take control. A greater focus on the specific attentional processes and networks discussed will allow clearer theoretical predictions that will drive flow research forward, as well as making use of flow as an attentional anomaly that can inform wider theories of attention.

Funding: The authors have received no funding for the preparation of this chapter.

Conflict of interest: D.H., S.V., and M.W. declare that they have no conflicts of interest relevant to the content of this review.

REFERENCES

Abernethy, B., Maxwell, J.P., Masters, R.S., van der Kamp, J., Jackson, R.C., 2007. Attentional processes in skill learning and expert performance. Handb. Sport Psychol. 3, 245–263.

Allport, A., 1989. Visual attention. In: Posner, M.I. (Ed.), Foundations of Cognitive Science. MIT Press, Massachusetts, pp. 631–682.

Ashby, F.G., Isen, A.M., Turken, A.U., 1999. A neuropsychological theory of positive affect and its influence on cognition. Psychol. Rev. 106, 529–550.

Backs, R.W., 1997. Psychophysiological aspects of selective and divided attention during continuous manual tracking. Acta Psychol. (Amst) 96 (3), 167–191.

Baumeister, R.F., 1984. Choking under pressure: self-consciousness and paradoxical objects of incentives on skillful performance. J. Pers. Soc. Psychol. 46, 610–620.

Beaty, R.E., 2015. The neuroscience of musical improvisation. Neurosci. Biobehav. Rev. 51, 108–117.

Beilock, S.L., Carr, T.H., 2001. On the fragility of skilled performance: what governs choking under pressure? J. Exp. Psychol. Gen. 130 (4), 701.

Beilock, S.L., Carr, T.H., MacMahon, C., Starkes, J.L., 2002. When paying attention becomes counterproductive: impact of divided versus skill-focused attention on novice and experienced performance of sensorimotor skills. J. Exp. Psychol. Appl. 8 (1), 6.

Bengtsson, S.L., Csikszentmihalyi, M., Ullen, F., 2007. Cortical regions involved in the generation of musical structures during improvisation in pianists. J. Cogn. Neurosci. 19, 830–842.

Berntson, G.G., Bigger, J.T., Eckberg, D.L., Grossman, P., Kaufmann, P.G., Malik, M., Van Der Molen, M.W., 1997. Heart rate variability: origins, methods, and interpretive caveats. Psychophysiology 34 (6), 623–648.

Botvinick, M.M., Braver, T.S., Barch, D.M., Carter, C.S., Cohen, J.D., 2001. Conflict monitoring and cognitive control. Psychol. Rev. 108 (3), 624.

Botvinick, M.M., Cohen, J.D., Carter, C.S., 2004. Conflict monitoring and anterior cingulate cortex: an update. Trends Cogn. Sci. 8 (12), 539–546.

Bruya, B., 2010. Effortless Attention: A New Perspective in the Cognitive Science of Attention and Action. MIT Press, Massachusetts, pp. 1–28.

Buckner, R.L., Andrews-Hanna, J.R., Schacter, D.L., 2008. The brain's default network. Ann. N. Y. Acad. Sci. 1124 (1), 1–38.

Chambers, C.D., Garavan, H., Bellgrove, M.A., 2009. Insights into the neural basis of response inhibition from cognitive and clinical neuroscience. Neurosci. Biobehav. Rev. 33 (5), 631–646.

Cheron, G., 2016. How to measure the psychological 'flow'? A neuroscience perspective. Front. Psychol. 7, 1823.

Congdon, E., Lesch, K.P., Canli, T., 2008. Analysis of DRD4 and DAT polymorphisms and behavioral inhibition in healthy adults: implications for impulsivity. Am. J. Med. Genet. B Neuropsychiatr. Genet. 147 (1), 27–32.

Corbetta, M., 1998. Frontoparietal cortical networks for directing attention and the eye to visual locations: identical, independent, or overlapping neural systems? Proc. Natl. Acad. Sci. U.S.A. 95 (3), 831–838.

Corbetta, M., Shulman, G.L., 2002. Control of goal-directed and stimulus-driven attention in the brain. Nat. Rev. Neurosci. 3 (3), 201–215.

Crick, F., Koch, C., 1990. Towards a neurobiological theory of consciousness. Semin. Neurosci. 2, 263–275.
Csikszentmihalyi, M., 1975. Beyond Boredom and Anxiety. Jossey-Bas, San Francisco.
Csikszentmihalyi, M., 1990. Flow: The Psychology of Optimal Experience. Harper and Row, New York.
Csikszentmihalyi, M., 1996. Creativity. Harper Collins, New York.
Csikszentmihalyi, M., Nakamura, J., 2010. Effortless attention in everyday life: a systematic phenomenology. In: Bruya, B. (Ed.), Effortless Attention: A New Perspective in the Cognitive Science of Attention and Action. MIT Press, Massachusetts, pp. 179–189.
Dalley, J.W., Roiser, J.P., 2012. Dopamine, serotonin and impulsivity. Neuroscience 215, 42–58.
de Manzano, Ö., Cervenka, S., Jucaite, A., Hellenäs, O., Farde, L., Ullén, F., 2013. Individual differences in the proneness to have flow experiences are linked to dopamine D2-receptor availability in the dorsal striatum. Neuroimage 67, 1–6.
Deeny, S.P., Hillman, C.H., Janelle, C.M., Hatfield, B.D., 2003. Cortico-cortical communication and superior performance in skilled marksmen: an EEG coherence analysis. J. Sport Exerc. Psychol. 25 (2), 188–204.
Dietrich, A., 2003. Functional neuroanatomy of altered states of consciousness: the transient hypofrontality hypothesis. Conscious. Cogn. 12 (2), 231–256.
Dietrich, A., 2006. Transient hypofrontality as a mechanism for the psychological effects of exercise. Psychiatry Res. 145 (1), 79–83.
Dietrich, A., Kanso, R., 2010. A review of EEG, ERP, and neuroimaging studies of creativity and insight. Psychol. Bull. 136 (5), 822.
Dietrich, A., Stoll, O., 2010. Effortless attention in sports performance. In: Bruya, B. (Ed.), Effortless Attention: A New Perspective in the Cognitive Science of Attention and Action. MIT Press, Massachusetts, pp. 159–178.
Dobrynin, N., 1966. Basic Problems of the Psychology of Attention. Department of Commerce, Washington.
Dreyfus, H.L., 2007. Response to McDowell. Inquiry 50 (4), 371–377.
Ducrocq, E., Wilson, M., Vine, S., Derakshan, N., 2016. Training attentional control improves cognitive and motor task performance. J. Sport Exerc. Psychol. 38, 521–533.
Duncan, J., 2010. The multiple-demand (MD) system of the primate brain: mental programs for intelligent behaviour. Trends Cogn. Sci. 14 (4), 172–179.
Duncan, J., 2013. The structure of cognition: attentional episodes in mind and brain. Neuron 80 (1), 35–50.
Eysenck, M.W., Wilson, M.R., 2016. Pressure and sport performance: a cognitive approach. Introducing attentional control theory: sport. In: Groome, D., Eysenck, M. (Eds.), Applied Cognitive Psychology. Psychology Press, Hove, UK.
Fan, J., McCandliss, B.D., Fossella, J., Flombaum, J.I., Posner, M.I., 2005. The activation of attentional networks. Neuroimage 26 (2), 471–479.
Fitts, P.M., Posner, M.I., 1967. Human Performance. Cole, Belmont, CA, Brooks.
Gaggioli, A., Cipresso, P., Serino, S., Riva, G., 2013. Psychophysiological correlates of flow during daily activities. Stud. Health Technol. Inform. 191, 65–69.
Garrison, K., Santoyo, J., Davis, J., Thornhill, T., Kerr, C., Brewer, J., 2013. Effortless awareness: using real time neurofeedback to investigate correlates of posterior cingulate cortex activity in meditators' self-report. Front. Hum. Neurosci. 7, 440.
Goldberg, I.I., Harel, M., Malach, R., 2006. When the brain loses its self: prefrontal inactivation during sensorimotor processing. Neuron 50 (2), 329–339.

Gottlieb, G., 2015. Know-how, procedural knowledge, and choking under pressure. Phenomenol. Cogn. Sci. 14 (2), 361–378.

Gyurkovics, M., Kotyuk, E., Katonai, E.R., Horvath, E.Z., Vereczkei, A., Szekely, A., 2016. Individual differences in flow proneness are linked to a dopamine D2 receptor gene variant. Conscious. Cogn. 42, 1–8.

Harmat, L., de Manzano, Ö., Theorell, T., Högman, L., Fischer, H., Ullén, F., 2015. Physiological correlates of the flow experience during computer game playing. Int. J. Psychophysiol. 97, 1–7.

Harris, D.J., Vine, S.J., Wilson, M.R., 2017. Is flow really effortless? The complex role of effortful attention. Sport Exerc. Perform. Psychol. 6 (1), 103.

Holroyd, C.B., Coles, M.G.H., 2002. The neural basis of human error processing: reinforcement learning, dopamine, and the error related negativity. Psychol. Rev. 109, 679–709.

Jackson, S.A., 1992. Athletes in flow: a qualitative investigation of flow states in elite figure skaters. J. Appl. Sport Psychol. 4 (2), 161–180.

Jackson, S.A., 1996. Toward a conceptual understanding of the flow experience in elite athletes. Res. Q. Exerc. Sport 67 (1), 76–90.

Jackson, S.A., Csikszentmihalyi, M., 1999. Flow in Sports. Human Kinetics, Illinois.

Jackson, S.A., Thomas, P.R., Marsh, H.W., Smethurst, C.J., 2001. Relationships between flow, self-concept, psychological skills, and performance. J. Appl. Sport Psychol. 13 (2), 129–153.

Jenkins, A.C., Mitchell, J.P., 2011. Medial prefrontal cortex subserves diverse forms of self-reflection. Soc. Neurosci. 6 (3), 211–218.

Kahneman, D., 2011. Thinking, Fast and Slow. Macmillan, London.

Kawabata, M., Mallett, C.J., 2011. Flow experience in physical activity: examination of the internal structure of flow from a process-related perspective. Motiv. Emot. 35 (4), 393–402.

Keller, J., Bless, H., Blomann, F., Kleinböhl, D., 2011. Physiological aspects of flow experiences: skills-demand-compatibility effects on heart rate variability and salivary cortisol. J. Exp. Soc. Psychol. 47 (4), 849–852.

Klasen, M., Weber, R., Kircher, T.T., Mathiak, K.A., Mathiak, K., 2011. Neural contributions to flow experience during video game playing. Soc. Cogn. Affect. Neurosci. 7 (4), 485–495.

Knudsen, E.I., 2007. Fundamental components of attention. Annu. Rev. Neurosci. 30, 57–78.

Koehn, S., Morris, T., 2012. The relationship between performance and flow state in tennis competition. J. Sports Med. Phys. Fitness 52 (4), 437–447.

Kollins, S.H., March, J.S., 2007. Advances in the pharmacotherapy of attention-deficit/hyperactivity disorder. Biol. Psychiatry 62 (9), 951–953.

Land, M.F., 2006. Eye movements and the control of actions in everyday life. Prog. Retin. Eye Res. 25 (3), 296–324.

Lebeau, J.C., Liu, S., Sáenz-Moncaleano, C., Sanduvete-Chaves, S., Chacón-Moscoso, S., Becker, B.J., Tenenbaum, G., 2016. Quiet eye and performance in sport: a meta-analysis. J. Sport Exerc. Psychol. 38, 441–457.

Lee, B., 1975. The Tao of Jeet Kune Do. Turtleback, Santa Clarita.

Lemogne, C., Gorwood, P., Bergouignan, L., Pélissolo, A., Lehéricy, S., Fossati, P., 2011. Negative affectivity, self-referential processing and the cortical midline structures. Soc. Cogn. Affect. Neurosci. 6 (4), 426–433.

Limb, C.J., Braun, A.R., 2008. Neural substrates of spontaneous musical performance: an FMRI study of jazz improvisation. PLoS One 3, e1679.

Mann, D.T., Williams, A.M., Ward, P., Janelle, C.M., 2007. Perceptual-cognitive expertise in sport: a meta-analysis. J. Sport Exerc. Psychol. 29 (4), 457.

Masters, R.S., 1992. Knowledge, knerves and know-how: the role of explicit versus implicit knowledge in the breakdown of a complex motor skill under pressure. Br. J. Psychol. 83 (3), 343–358.

Masters, R., Maxwell, J., 2008. The theory of reinvestment. Int. Rev. Sport Exerc. Psychol. 1 (2), 160–183.

McGuire, J.T., Botvinick, M.M., 2010. The impact of anticipated cognitive demand on attention and behavioral choice. In: Bruya, B. (Ed.), Effortless Attention: A New Perspective in the Cognitive Science of Attention and Action. MIT Press, Massachusetts, pp. 103–120.

Miyake, A., Friedman, N.P., Emerson, M.J., Witzki, A.H., Howerter, A., Wager, T.D., 2000. The unity and diversity of executive functions and their contributions to complex "frontal lobe" tasks: a latent variable analysis. Cogn. Psychol. 41 (1), 49–100.

Moore, L.J., Vine, S.J., Cooke, A., Ring, C., Wilson, M.R., 2012. Quiet eye training expedites motor learning and aids performance under heightened anxiety: the roles of response programming and external attention. Psychophysiology 49 (7), 1005–1015.

Moors, A., De Houwer, J., 2006. Automaticity: a theoretical and conceptual analysis. Psychol. Bull. 132 (2), 297.

Naccache, L., Dehaene, S., Cohen, L., Habert, M.O., Guichart-Gomez, E., Galanaud, D., Willer, J.C., 2005. Effortless control: executive attention and conscious feeling of mental effort are dissociable. Neuropsychologia 43 (9), 1318–1328.

Nakamura, J., Csikszentmihalyi, M., 2009. Flow theory and research. In: Snyder, C.R., Lopez, S.J. (Eds.), Oxford Handbook of Positive Psychology, second ed. Oxford University Press, Oxford, UK, pp. 195–206.

Nideffer, R.M., 1976. Test of attentional and interpersonal style. J. Pers. Soc. Psychol. 34 (3), 394.

Nieoullon, A., 2002. Dopamine and the regulation of cognition and attention. Prog. Neurobiol. 67 (1), 53–83.

Nisenson, E., 1995. Ascension: John Coltrane and His Quest. Da Capo Press, New York.

Northoff, G., Heinzel, A., de Greck, M., Bermpohl, F., Dobrowolny, H., Panksepp, J., 2006. Self-referential processing in our brain—a meta-analysis of imaging studies on the self. Neuroimage 31, 440–457.

Peifer, C., Schulz, A., Schächinger, H., Baumann, N., Antoni, C.H., 2014. The relation of flow-experience and physiological arousal under stress—can u shape it? J. Exp. Soc. Psychol. 53, 62–69.

Perry, S.K., 1999. Writing in Flow. Writer's Digest Books, Cincinnati, OH.

Petersen, S.E., Posner, M.I., 2012. The attention system of the human brain: 20 years after. Ann. Rev. Neurosci. 35, 73–89.

Pfurtscheller, G., Stancak, A., Neuper, C., 1996. Event-related synchronization (ERS) in the alpha band—an electrophysiological correlate of cortical idling: a review. Int. J. Psychophysiol. 24 (1), 39–46.

Poldrack, R.A., Sabb, F.W., Foerde, K., Tom, S.M., Asarnow, R.F., Bookheimer, S.Y., Knowlton, B.J., 2005. The neural correlates of motor skill automaticity. J. Neurosci. 25 (22), 5356–5364.

Posner, M.I., Inhoff, A.W., Friedrich, F.J., Cohen, A., 1987. Isolating attentional systems: a cognitive-anatomical analysis. Psychobiology 15, 107–121.

Posner, M.I., Rothbart, M.K., Rueda, M.R., Tang, Y., 2010. Training effortless attention. In: Bruya, B. (Ed.), Effortless Attention: A New Perspective in the Cognitive Science of Attention and Action. MIT Press, Massachusetts, pp. 409–424.

Privette, G., 1983. Peak experience, peak performance, and flow: a comparative analysis of positive human experiences. J. Personal Soc. Psychol. 45 (6), 1361.

Raichle, M.E., MacLeod, A.M., Snyder, A.Z., Powers, W.J., Gusnard, D.A., Shulman, G.L., 2001. A default mode of brain function. Proc. Natl. Acad. Sci. U.S.A. 98 (2), 676–682.

Ridderinkhof, K.R., van den Wildenberg, W.P., Segalowitz, S.J., Carter, C.S., 2004. Neurocognitive mechanisms of cognitive control: the role of prefrontal cortex in action selection, response inhibition, performance monitoring, and reward-based learning. Brain Cogn. 56 (2), 129–140.

Ring, C., Cooke, A., Kavussanu, M., McIntyre, D., Masters, R., 2015. Investigating the efficacy of neurofeedback training for expediting expertise and excellence in sport. Psychol. Sport Exerc. 31 (16), 118–127.

Sarter, M., Gehring, W.J., Kozak, R., 2006. More attention must be paid: the neurobiology of attentional effort. Brain Res. Rev. 51 (2), 145–160.

Sartre, J.P., 1957. The Transcendence of the Ego: An Existentialist Theory of Consciousness (vol. 114). Macmillan; London.

Schmeichel, B.J., Baumeister, R.F., 2010. Effortful attention control. In: Bruya, B. (Ed.), Effortless Attention: A New Perspective in the Cognitive Science of Attention and Action. MIT Press, Massachusetts, pp. 29–50.

Schultz, W., 2006. Behavioral theories and the neurophysiology of reward. Annu. Rev. Psychol. 57, 87–115.

Shipstead, Z., Redick, T.S., Engle, R.W., 2012. Is working memory training effective? Psychol. Bull. 138 (4), 628.

Singer, R.N., 2002. Preperformance state, routines and automaticity: what does it take to realize expertise in self-paced events? J. Sport Exerc. Psychol. 24 (4), 359–375.

Slingerland, E., 2014. Trying Not to Try: Ancient China, Modern Science, and the Power of Spontaneity. Crown, Connecticut.

Stoll, O., Pithan, J.M., 2016. Running and flow: does controlled running lead to flow-states? Testing the transient hypofontality theory. In: Flow Experience. Springer International Publishing, New York, pp. 65–75.

Swann, C., Keegan, R., Crust, L., Piggott, D., 2016. Psychological states underlying excellent performance in professional golfers: "letting it happen" vs. "making it happen". Psychol. Sport Exerc. 23, 101–113.

Swann, C., Crust, L., Vella, S., 2017. New directions in the psychology of optimal performance in sport: flow and clutch states. Curr. Opin. Psychol. 16, 48–53.

Swann, C., Keegan, R.J., Piggott, D., Crust, L., 2012. A systematic review of the experience, occurrence, and controllability of flow states in elite sport. Psychol. Sport Exerc. 13 (6), 807–819.

Tang, Y.Y., Posner, M.I., 2009. Attention training and attention state training. Trends Cogn. Sci. 13 (5), 222–227.

Thin, A., Hansen, L., McEachen, D., 2011. Flow experience and mood states while playing body movement-controlled video games. Game Cult. 6 (5), 414–428.

Toner, J., Montero, B.G., Moran, A., 2015. Considering the role of cognitive control in expert performance. Phenomenol. Cogn. Sci. 14 (4), 1127–1144.

Toner, J., Montero, B.G., Moran, A., 2016. Reflective and prereflective bodily awareness in skilled action. Psychol. Conscious 3 (4), 303.

Tozman, T., Magdas, E.S., MacDougall, H.G., Vollmeyer, R., 2015. Understanding the psychophysiology of flow: a driving simulator experiment to investigate the relationship between flow and heart rate variability. Comput. Hum. Behav. 52, 408–418.

Ulrich, M., Keller, J., Grön, G., 2016. Neural signatures of experimentally induced flow experiences identified in a typical fMRI block design with BOLD imaging. Soc. Cogn. Affect. Neurosci. 11 (3), 496–507.

Ulrich, M., Keller, J., Hoenig, K., Waller, C., Grön, G., 2014. Neural correlates of experimentally induced flow experiences. Neuroimage 86, 194–202.

Van Veen, V., Carter, C.S., 2002. The anterior cingulate as a conflict monitor: fMRI and ERP studies. Physiol. Behav. 77 (4), 477–482.

Vickers, J.N., 1996. Control of visual attention during the basketball free throw. Am. J. Sports Med. 24 (S), 93–97.

Vickers, J.N., 2009. Advances in coupling perception and action: the quiet eye as a bidirectional link between gaze, attention, and action. Prog. Brain Res. 174, 279–288.

Vickers, J.N., Williams, A.M., 2007. Performing under pressure: the effects of physiological arousal, cognitive anxiety, and gaze control in biathlon. J. Mot. Behav. 39 (5), 381–394.

Vine, S.J., Lee, D., Moore, L.J., Wilson, M.R., 2013. Quiet eye and choking: online control breaks down at the point of performance failure. Med. Sci. Sports Exerc. 45 (10), 1988–1994.

Vine, S.J., Wilson, M.R., 2011. The influence of quiet eye training and pressure on attention and visuo-motor control. Acta Psychol. (Amst) 136 (3), 340–346.

Weber, R., Tamborini, R., Westcott-Baker, A., Kantor, B., 2009. Theorizing flow and media enjoyment as cognitive synchronization of attentional and reward networks. Commun. Theory 19 (4), 397–422.

Williams, A.M., Davids, K., 1998. Visual search strategy, selective attention, and expertise in soccer. Res. Q Exerc. Sport 69 (2), 111–128.

Wilson, M.R., Vine, S.J., Wood, G., 2009. The influence of anxiety on visual attentional control in basketball free throw shooting. J. Sport Exerc. Psychol. 31 (2), 152–168.

Wolf, S., Brölz, E., Keune, P.M., Wesa, B., Hautzinger, M., Birbaumer, N., Strehl, U., 2015. Motor skill failure or flow-experience? Functional brain asymmetry and brain connectivity in elite and amateur table tennis players. Biol. Psychol. 105, 95–105.

Wood, G., Wilson, M.R., 2010. A moving goalkeeper distracts penalty takers and impairs shooting accuracy. J. Sports Sci. 28 (9), 937–946.

Wright, R.A., 1996. Brehm's theory of motivation as a model of effort and cardiovascular response. In: Gollwitzer, P.M., Bargh, J.A. (Eds.), The Psychology of Action: Linking Cognition and Motivation to Behavior. Guilford, New York, pp. 424–453.

Wulf, G., 2013. Attentional focus and motor learning: a review of 15 years. Int. Rev. Sport Exerc. Psychol. 6 (1), 77–104.

Wulf, G., Lewthwaite, R., 2010. Effortless motor learning? An external focus of attention enhances movement effectiveness and efficiency. In: Bruya, B.J. (Ed.), Effortless Attention: A New Perspective in the Cognitive Science of Attention and Action. MIT Press, Massachusetts, pp. 75–101.

Yoshida, K., Sawamura, D., Inagaki, Y., Ogawa, K., Ikoma, K., Sakai, S., 2014. Brain activity during the flow experience: a functional near-infrared spectroscopy study. Neurosci. Lett. 573, 30–34.

Zachry, T., Wulf, G., Mercer, J., Bezodis, N., 2005. Increased movement accuracy and reduced EMG activity as the result of adopting an external focus of attention. Brain Res. Bull. 67 (4), 304–309.

Zhu, F.F., Poolton, J.M., Wilson, M.R., Hu, Y., Maxwell, J.P., Masters, R.S., 2011. Implicit motor learning promotes neural efficiency during laparoscopy. Surg. Endosc. 25 (9), 2950–2955.

CHAPTER

13

Discerning measures of conscious brain processes associated with superior early motor performance: Capacity, coactivation, and character

Tina van Duijn*,[1], Tim Buszard[†], Merel C.J. Hoskens*,[‡], Rich S.W. Masters*,[§]

Faculty of Health, Sport and Human Performance, University of Waikato, Hamilton, New Zealand
[†]*Victoria University, Melbourne, VIC, Australia*
[‡]*Vrije Universiteit Amsterdam, Amsterdam, The Netherlands*
[§]*School of Public Health, The University of Hong Kong, Hong Kong, China*
[1]*Corresponding author: e-mail address: tv24@students.waikato.ac.nz*

Abstract

This study explored the relationship between working memory (WM) capacity, corticocortical communication (EEG coherence), and propensity for conscious control of movement during the performance of a complex far-aiming task. We were specifically interested in the role of these variables in predicting motor performance by novices. Forty-eight participants completed (a) an assessment of WM capacity (an adapted Rotation Span task), (b) a questionnaire that assessed the propensity to consciously control movement (the Movement Specific Reinvestment Scale), and (c) a hockey push-pass task. The hockey push-pass task was performed in a single task (movement only) condition and a combined task (movement plus decision) condition. Electroencephalography (EEG) was used to examine brain activity during the single task. WM capacity best predicted single task performance. WM capacity in combination with T8–Fz coherence (between the visuospatial and motor regions of the brain) best predicted combined task performance. We discuss the implied roles of visuospatial information processing capacity, neural coactivation, and propensity for conscious processing during performance of complex motor tasks.

Keywords

EEG coherence, Working memory, Novice performance, Electrophysiology, Information processing

1 INTRODUCTION

Mental processes during sport are not easily assessed. Personality characteristics associated with conscious involvement in performance, for example, are often measured subjectively with questionnaires. Capacity to process information is measured using memory and attention tasks, but efficiency in exploiting this capacity can be represented by neural measures, such as corticocortical communication between different areas of the brain. Seldom are these approaches considered together when examining the role of the brain in successful performance in sport.

Considered in isolation, it is clear that each measure plays an important role in understanding mental processes in sport performance, but evidence of the relationship between the measures is often unclear. Working memory (WM), for instance, provides a "workspace" in which information that is relevant for learning and performance can be manipulated and temporarily stored (MacMahon and Masters, 2002; Maxwell et al., 2003). The amount of information that can be processed is a function of a person's WM capacity.

Associations have been revealed between WM capacity and various aspects of motor performance. For example, studies have reported positive correlations between WM capacity and performance improvements in motor sequence learning using button-pressing tasks (Bo and Seidler, 2009; Bo et al., 2011, 2012). The tasks typically involve pressing buttons on a keyboard, with a fixed sequence embedded within the movements performed. While these studies have shown that participants with higher WM capacity learn faster, other studies have observed no meaningful correlation between WM capacity and motor sequence learning (Feldman et al., 1995; Kaufman et al., 2010; for discussion of the complex relationship between WM capacity and motor sequence learning, see Janacsek and Nemeth, 2013). WM capacity has also been shown to be positively related to performance during decisive (high pressure) sets in tennis matches (Bijleveld and Veling, 2014) and to shooting accuracy under experimentally induced pressure (Wood et al., 2016).

A relationship may exist between WM capacity and movement specific reinvestment, a personality measure of the tendency to consciously process movement-related information (Masters and Maxwell, 2008; Masters et al., 1993). For instance, Buszard et al. (2013) reported a positive relationship in a cohort of children and adults. High capacity to process information may provide greater opportunity for a performer to consciously monitor and control their movements; however, Laborde et al. (2015) found that even under psychological pressure there was no association. Consequently, it is unclear at this stage what the relationship is between WM capacity and movement specific reinvestment.

1 Introduction

Our understanding of the relationship between WM capacity, movement specific reinvestment, and motor performance might therefore improve with the inclusion of neurophysiological measures (for a review on recent research using neurophysiological measures in sport performance, see Cooke, 2013). Notably, neural activity can be monitored on-line during execution of experimental tasks at a precise temporal resolution by electroencephalography (EEG). A measure that has become particularly useful for understanding motor performance and learning is *EEG coherence*—the synchronicity of neural coactivation between different regions in the cerebral cortex. EEG coherence between regions associated with motor planning (Fz, frontal midline) and verbal-analytic processing (T7, left temporal lobe) is thought to indicate the involvement of verbal processes in motor performance (Zhu et al., 2011).[a] Comparatively, coherence between Fz and the region of the brain associated with visuospatial mapping (T8, right temporal lobe) is thought to indicate the involvement of visuospatial processes in motor performance.

The significance of EEG coherence is revealed by studies investigating skilled performers. In particular, experts display lower coherence between the T7 and Fz regions during movement execution compared to novices (Deeny et al., 2003), suggesting that skilled performance is associated with a decrease in explicit, verbal processing of the movements. T7–Fz coherence has also been found to increase in conditions of heightened anxiety (Chen et al., 2005), and Zhu et al. (2011) observed higher T7–Fz coherence in participants with a greater propensity to consciously monitor and control their movements (i.e., movement specific reinvestment). Zhu et al. (2011) also found that participants taught by explicit instructions displayed higher T7–Fz coherence compared to participants taught in such a way that they accrued minimal explicit (conscious) knowledge about their movements (implicit motor learning; Masters, 1992). Zhu et al. (2011) concluded that EEG coherence between T7 and Fz is a useful yardstick of conscious control in motor performance. In an earlier study, however, Zhu et al. (2010) found that increases in T8–Fz coherence accompanied performance improvements in a visuospatial aiming task, so the information processing required by the task is likely to be an important factor in determining the relationship between T8–Fz or T7–Fz coherence and motor performance.

More recently, the relationship between EEG coherence and WM capacity was investigated in a novel tennis task (Buszard et al., 2016). Visuospatial WM capacity was negatively correlated with T7–F3 coherence (see Footnote a), while verbal WM capacity was positively correlated. Buszard et al. (2016) speculated that high verbal WM capacity increases the likelihood of a person learning a motor skill explicitly,

---

[a] In this paper, all labelings are translated to the nomenclature suggested in the American Electroencephalographic Society's "guidelines for standard electrode position nomenclature" (American Electroencephalographic Society, 1994; Chatrian et al., 1985; Jurcak et al., 2007; Klem et al., 1999). The terms T7 and T8 are used for left and right temporal lobes, while Fz, F3, and F4 are used for frontal lobe sites. Odd numbers indicate left hemisphere, even numbers indicate right hemisphere locations.

but high visuospatial WM capacity facilitates nonverbal or implicit motor learning. However, neither WM capacity nor EEG coherence correlated with performance.

The purpose of our study was to examine WM capacity, movement specific reinvestment, and EEG coherence during novice performance of a hockey push-pass. A hockey push-pass is a complex motor skill that requires coordination of many degrees of freedom. We were interested in uncovering whether individuals with high WM capacity were better at performing a complex novel motor task. Early motor learning is dominated by effortful processing of information, much of which is conscious, as suggested by stage models of learning (e.g., Fitts and Posner, 1967). High WM capacity is likely to facilitate conscious control of a novice's movements, so we expected a positive association between WM capacity and performance of the hockey push-pass. We also were interested to know whether WM capacity was associated with the propensity for conscious engagement in performance and/or with concurrent EEG coherence. We expected to find significant correlations between WM capacity and measures of visuospatial and/or verbal–cognitive engagement in the motor task (i.e., T7–Fz and T8–Fz coherence), and between WM capacity and movement specific reinvestment. Performance was also assessed in a more ecologically relevant task in which participants completed the push-pass while making a decision about who should receive the pass (referred to as a *combined task*). The combined task was expected to increase the demands on the participant, heightening the need for efficient information processing. We therefore expected WM capacity and EEG coherence during the movement only (i.e., single task), which potentially is a marker of efficient motor processing (e.g., Hatfield et al., 2004), to be associated with performance during the combined task.

2 METHOD
2.1 PARTICIPANTS
Forty-eight novices to hockey (mean age = 21.31 years, SD = 4.96) with normal or corrected-to-normal vision and no movement impairments participated in the study. Participants were recruited from the institution and were incentivized to participate with cafeteria vouchers (value NZD10). Participants with more than 20 h of experience were excluded from the study. Ethical permission for the study was received from the Faculty Ethical Committee of the institution. Participants all provided informed consent.

2.2 MATERIALS
Standard field hockey sticks of 92.7 cm (36.5 in.) length were used, and hockey balls were replaced with standard Wilson® tennis balls. The laboratory floor was covered with artificial golf turf. A line on the floor marked the starting position of the ball; a red circle on the wall with concentric circles at 10 cm intervals provided a target for the passes. The distance from the starting position to the target

was 340 cm. Performance was measured as distance from the target center and was obtained by manual analysis of video footage from a Sony RX10pi camera focused on the target.

EEG equipment included a wireless EEG/tDCS transmitter, a set of four measuring and two reference electrodes, conducting gel and electrode contact stickers (Neuroelectrics, ESP). The system was connected to a desktop computer and analyzed using Neurosurfer software (Neuroelectrics, ESP). EEG activity was recorded from six silver/silver chloride (Ag/AgCl) electrodes on the scalp positioned using different sized neoprene caps with predefined holes. Two reference electrodes were placed at the earlobe using a clip. If the earlobe was too small to hold the clip, two sticktrodes were attached to the left mastoid. Caps and electrodes were adjusted to be consistent with a subset of the 10–10 system (Chatrian et al., 1985) and were carefully checked by two technicians before the start of the experiment. Signals were amplified at a sampling rate of 500 Hz with 24-bit resolution and 0–125 Hz bandwidth. Measurement noise was under 1 μV RMS.

Prior to commencing the task, impedance needed to reach a satisfactory level (below 15 kΩ). This was achieved by adjusting electrode positions, the participant's hair and the amount of electrolyte gel. The presession criteria were constant across all participants. The presession criteria required participants to (a) wash their hair on the evening/morning before testing and (b) not consume caffeine in the 2 h preceding testing. A thorough check of the impedances ensured good quality contacts at all times. Baseline EEG data were collected prior to commencing the experiment. This took place while the participant was seated and not moving during 120 s. The first 60 s were recorded with the participant's eyes open (blinks were not suppressed), while the second 60 s required the eyes to be closed.

2.3 PROCEDURE

Participants arrived at the laboratory and completed a demographics sheet plus the Movement Specific Reinvestment Scale (MSRS; Masters and Maxwell, 2008; Masters et al., 2005). The MSRS assesses a person's propensity to consciously monitor and control their movements. The scale consists of 10 items and includes items such as "I am self-conscious about the way I look when I am moving" and "I am aware of the way my body works when I am carrying out a movement." Participants indicated to what extent each statement describes them using a six-point Likert-type scale ranging from "strongly disagree" to "strongly agree." The scale has been shown to have a high internal and test–retest reliability (Masters et al., 2005).

After completing the questionnaires, participants were fitted with the Neuroelectrics cap and the EEG electrodes were attached. They then completed a computerized test of WM capacity. The test was an adapted version of the rotation span test (Oswald et al., 2015; Shah and Miyake, 1996). The to-be-remembered stimuli in this task were images of arrows that could be differentiated by (a) their length (long or short) or (b) their angle of rotation. Participants were required to remember and recall the specific arrows that were presented (i.e., the length and the angle of rotation) at the conclusion of each trial. After each arrow was presented, participants had to perform a distraction task during which they had to judge the orientation of a letter on the screen.

Participants then performed a hockey push-pass task in two conditions: single task and combined task, with single task always performed first. Both tasks involved one block of 20 trials. The *single task* required participants to push-pass the ball as accurately as possible to the target on the opposing wall. Prior to completing the task, participants were shown an animation that illustrated how the task was to be completed. For the *combined task*, 20 images of hockey players ($n=3$) standing in different positions were projected onto the opposing wall. Two players were wearing black shirts, and one was wearing an orange shirt. Arrangements of the players varied, with one or two players in the foreground (85% of life size) and the others in the background (70% of life size). Subjects were informed that they were a member of the black team and were to push-pass the ball as quickly and as accurately as possible toward the hockey stick of the player who, in their opinion, was in the best position to receive the ball. Presentation of each image was preceded by a brief countdown on a blank background.

2.4 DEPENDENT VARIABLES AND DATA ANALYSIS
2.4.1 Dependent variables
Measurements for both performance variables were made using video recordings collected at normal speed (30 fps) from the Sony camera focused on the wall. Video clips were played frame by frame in order to determine the time when the ball contacted the wall. *Single task accuracy* was represented by mean contact distance from the target during the single task test (20 trials). *Combined task accuracy* was represented by mean contact distance from the target chosen by the participant. Greater distances therefore represent worse accuracy.

WM capacity measures were calculated via the program "R" using a software script provided by Stone and Towse (2015). The measure *WM capacity* reflects the maximum number of items a participant remembered correctly. Although other measures exist, this variable was considered prior to data collection to be the most informative for our study, as it represents the maximum capacity, rather than general WM ability (Stone and Towse, 2015). Score on the MSRS was calculated by summing the Likert-scale responses, leading to a cumulative range of 10–60 points.

2.4.2 Data reduction
Raw EEG signals were first filtered at 1–30 Hz, and a notch filter was added to exclude 50 Hz line noise. Signals were then resampled at 256 Hz. Data were reduced to 4 s epochs preceding movement initiation in each trial. Epochs were split into 0.25 s trials for artifact removal. Blinks and eye movements are characterized by high potentials (Boudet et al., 2006), so eye artifacts were excluded using an extreme measures approach via EEGLAB. Trials containing signals above 60 mV were discarded.

An average 174.70 (SD=69.33) trials per participant were retained. A fast Fourier transform with a Hamming window taper, 50% overlap with a resolution of 0.49 Hz, was applied. EEG high-alpha (10–12 Hz) power was calculated over the 0.25-s epochs (e.g., Deeny et al., 2003) and averaged for the 4-s epochs preceding each trial.

The EEG analyses in this study focused on the high-alpha (10–12 Hz) band. Activity in this bandwidth indicates medium-range corticocortical communication and represents task-specific attention processes (Smith et al., 1999; for a review, see Klimesch, 1999). Coherence was analyzed for T7–Fz and T8–Fz regional combinations separately, as these areas represent verbal-analytic (T7), visuospatial (T8), and motor planning (Fz) regions (Haufler et al., 2000; Kaufer and Lewis, 1999; Kerick et al., 2001). Matlab scripts were used to calculate EEG outputs. The processing and analysis steps described earlier were implemented with the EEGLAB toolbox (Delorme and Makeig, 2004).

2.4.3 Statistical analyses

Associations between WM capacity, movement specific reinvestment, and EEG coherence were investigated using Pearson's product-moment correlation coefficient. Regression analyses were conducted to predict performance in both the single task and the combined task. The predictor variables included WM capacity, T7–Fz coherence, T8–Fz coherence, and score on the MSRS. A backward elimination approach was chosen for the regression analyses, with the aim of unpicking the relationships between different measures of conscious processes and performance. Values collected during the single task were also used to predict performance in the combined task. Moderated regression analysis was employed to investigate the joint influence of WM capacity and T8–Fz coherence on performance in the combined task. Alpha value for statistical significance was set to 0.05.

3 RESULTS

3.1 CORRELATION BETWEEN EEG COHERENCE AND PERFORMANCE VARIABLES

Means, standard deviations, and Pearson product-moment correlation coefficients for predictive and outcome variables are presented in Table 1. WM capacity correlated significantly with single task accuracy and combined task accuracy. The relationship was negative, suggesting that larger WM capacity was associated with better push-pass accuracy. A significant correlation was also found between WM capacity and T8–Fz coherence, with larger WM capacity associated with higher coherence. However, WM capacity was not associated with T7–Fz coherence. No significant correlations were observed between push-pass accuracy and T7–Fz or T8–Fz coherence. MSRS score was not significantly correlated with any variable.

Table 1 Means, Standard Deviations, and Pearson Product-Moment Correlation Coefficients for Predictive and Outcome Variables

	Variable	M	SD	1.	2.	3.	4.	5.
1.	Single task accuracy (cm)	17.640	7.997	—				
2.	Combined task accuracy (cm)	25.477	8.235	0.259	—			
3.	WM capacity	2.896	0.831	−0.327*	−0.321*	—		
4.	MSRS	38.083	8.225	0.067	0.125	−0.076	—	
5.	T7–Fz	0.457	0.144	−0.033	0.116	0.239	−0.033	—
6.	T8–Fz	0.416	0.149	−0.252	0.139	0.330*	0.065	0.734**

Note: *$P < 0.05$, **$P < 0.01$.

3.2 PREDICTING SINGLE TASK ACCURACY

First, a regression analysis was conducted to determine whether gender influenced performance. Gender was not a significant predictor of single task accuracy, $R^2 = 0.011$, $F(1,47) = 0.531$, $P = 0.470$, and was therefore not included as a predictor variable in the subsequent analyses.

Stepwise regression analysis (backward method) was used to predict single task accuracy. The predictor variables included WM capacity, MSRS score, T7–Fz coherence, and T8–Fz coherence. The threshold value for the predictor variables was $P = 0.05$ for inclusion and $P = 0.10$ for exclusion. Results of the regression analysis are presented in Table 2. Model 1, including all variables, was not significant. Consequently, MSRS, which had a P-value greater than 0.10, was removed to leave a significant model (Model 2). Removal of T7–Fz coherence ($P > 0.10$) further refined the model (Model 3). Finally, removal of T8–Fz coherence ($P > 0.10$) yielded a model (Model 4) in which WM capacity was the only significant predictor of single task accuracy, accounting for 8.8% of the variance (see Fig. 1).

3.3 PREDICTING COMBINED TASK ACCURACY

Stepwise regression analysis was conducted to predict combined task performance using WM capacity, MSRS score, T7–Fz coherence, and T8–Fz coherence as predictor variables. Results of the analysis are presented in Table 3. A first model, including all variables, was not significant. Consequently, T7–Fz coherence and MSRS (P's > 0.10) were eliminated from Models 1 and 2, respectively, to yield a final model (Model 3) in which WM capacity and T8–Fz coherence were an optimal combination of predictors, explaining 13.4% of the variance in combined task performance. Higher WM capacity was associated with more accurate push-pass performance (see Fig. 2). Higher T8–Fz coherence during the single task (i.e., movement only) appeared to be associated with less accurate performance during the combined task (movement plus decision); however, the association fell short of significance (see Fig 3).

Moderation analysis was conducted to investigate whether T8–Fz coherence moderated the effect of WM capacity on combined task performance. Variables were entered into the regression equation in a stepwise manner. In the first step, the predictor variable (WM capacity) and potential moderator variable (T8–Fz coherence) were entered. In the second step, a product term created by multiplying WM capacity by the standardized T8–Fz coherence variable was entered. A significant change in R^2 for the product term would indicate a significant moderator effect. Assumptions for regression analysis were tested and satisfied. Results of the analysis are presented in Table 4. A significant moderation effect was not found (R^2 change = 0.006, $P = 0.568$).

Table 2 Regression Analysis Predicting Single Task Performance

	Model 1		Model 2		Model 3		Model 4	
R^2_{adj}	0.112		0.124		0.092		0.088	
P	0.058		0.032		0.043		0.023	
Variables	β	P	β	P	β	P	β	P
MSRS	0.086	0.541	—	—	—	—	—	—
T7–Fz coherence	0.343	0.100	0.328	0.110	—	—	—	—
T8–Fz coherence	−0.423	0.051	−0.403	0.058	−0.162	0.277	—	—
WM capacity	−0.263	0.080	−0.272	0.066	−0.274	0.070	−0.327	0.023

Beta (β) and P-values for predictor variables, and R^2_{adj} values and P-values for each model.
Abbreviations: MSRS, movement specific reinvestment scale; WM, working memory.

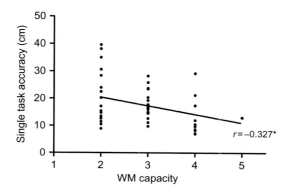

FIG. 1

The relationship of WM capacity with performance in the single task. * indicates that the variable on the x-axis was a significant predictor of the y-variable in the regression analysis.

Table 3 Regression Analysis Predicting Combined Task Accuracy

	Model 1		Model 2		Model 3	
R^2_{adj}	0.100		0.120		0.134	
P	0.073		0.035		0.015	
Variables	**β**	**P**	**β**	**P**	**β**	**P**
T7–Fz coherence	0.041	0.844	—	—	—	—
MSRS	0.080	0.571	0.077	0.580	—	—
T8–Fz coherence	0.237	0.269	0.268	0.073	0.275	0.062
WM capacity	−0.403	0.009	−0.403	0.008	−0.412	0.006

Beta (β) and P-values for predictor variables, and R^2_{adj} values and P-values for each model. Abbreviations: MSRS, movement specific reinvestment scale; WM, working memory.

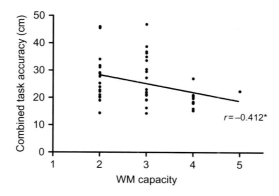

FIG. 2

The relationship of WM capacity with performance in the combined task. * indicates that the variable on the x-axis was a significant predictor of the y-variable in the regression analysis.

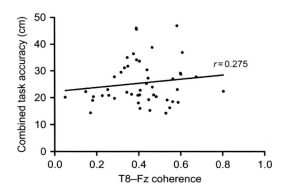

FIG. 3

The relationship of EEG T8–Fz coherence with performance in the combined task.

Table 4 Moderation Analysis for Interaction of WM Capacity and T8–Fz Coherence in Predicting Combined Task Performance

	B	SE B	β	R^2 Change
Step 1				
WM capacity	−4.079	1.425	−0.412*	
T8–Fz coherence	15.208	7.945	0.275	0.170
Step 2				
WM capacity x T8–Fz coherence	−0.516	0.897	−0.080	0.006

Note: *$P < 0.05$.

4 DISCUSSION

We explored mental processes associated with performance of a complex far-aiming task by novices, using measures of working memory (WM) capacity, corticocortical communication (EEG coherence), and the propensity for conscious control of movement (MSRS).

WM capacity was positively associated with T8–Fz coherence. Our measure of WM capacity uses an adapted version of the Rotation Span task, which involves processing of visuospatial information. People with high visuospatial information processing capacity may be more likely to process their movements visuospatially, culminating in high T8–Fz coherence (i.e., coactivation between visuospatial and motor planning regions of the brain). Buszard et al. (2016), however, reported a negative association between visuospatial WM capacity and T8–F4 coherence (F4 borders the Fz region). Further work is clearly required to disentangle the functional differences at a neural level.

WM capacity was also positively correlated with accuracy in both the single and combined tasks. For single task accuracy, backward regression analysis revealed a final model that included only WM capacity. For combined task accuracy, backward regression analysis revealed a final model that included WM capacity and T8–Fz, although T8–Fz coherence contributed in the model nonsignificantly. Moderation analysis showed that WM capacity and T8–Fz coherence did not interact to predict performance in the combined task. While WM capacity was positively associated with performance, T8–Fz displayed a trend toward a negative association with performance.

Our data imply that visuospatial WM capacity plays a substantial role in predicting early motor performance. Individuals with larger WM capacity may possess an advantage when solving motor tasks, but typically this has been attributed to verbal rather than visuospatial WM capacity. Maxwell et al. (2003), for example, argued that verbal WM is used to process movement specific information and to correct previous errors in performance. In support of this claim, Lam et al. (2010) found that people responded more slowly to an audible tone (the probe reaction time paradigm) if they previously had made an error when golf putting. Errors during movement typically are resolved by constructing and testing hypotheses about the most effective way in which to move, which is likely to load verbal rather than visuospatial WM.

However, error correction in far-aiming tasks, such as a hockey push-pass, may be more visual than verbal. During a push-pass, the performer attempts to strike a ball toward an often moving target, which can be many meters away. Error correction in this type of task requires the integration of visuospatial information with effector movements (Vickers, 1996), which is likely to make demands on visuospatial WM capacity. Other evidence supports the role of visuospatial WM capacity in early learning. For example, Anguera et al. (2010) found that rate of learning at early stages of motor adaptation depended on visuospatial WM ability and Bo and Seidler (2009) showed that greater visuospatial WM capacity was associated with more rapid learning of a motor sequence.

The role of T8–Fz coherence in predicting early motor performance is less clear in this study. The contribution of T8–Fz coherence to the final model predicting combined task performance leads us to speculate that people who display low T8–Fz coherence when first performing a single task may have a visuospatial processing advantage that allows them to cope better when decision-making during movement. Low T8–Fz coherence during single task performance might, for example, reflect more efficient visuospatial processing of movement. General brain activation patterns, such as higher alpha power (Cooke et al., 2014; Crews and Landers, 1993; Haufler et al., 2000; Hillman et al., 2000), have been linked to efficient movement processing (i.e., psychomotor efficiency), as have measures of corticocortical coherence (Deeny et al., 2009; Hatfield and Hillman, 2001). Hatfield et al. (2004) suggested that neural efficiency is represented by an appropriate "fit" of neural resources to specific task demands and a consequent reduction in irrelevant processing, which explains why experts (who rely on less verbal-analytic processing for motor control) show reduced T7–Fz (i.e., verbal-motor) coherence compared to novices. Zhu et al. (2011) also argued that low coherence between the verbal processing and the motor planning regions of the brain (T7–Fz) during performance of a surgical task represented neural efficiency. In far-aiming tasks, however, activity in visuospatial brain areas may be more indicative of error correction through visuomotor mapping, so lower coherence between the visuospatial and the motor planning areas (T8–Fz) in some novices may reflect relative neural efficiency in this task.

It is of interest that movement specific reinvestment did not appear to play a role in early motor performance in this study. Score on the Movement Specific Reinvestment Scale (MSRS) is considered to be a measure of conscious, verbal engagement in the process of moving, and previous work has revealed an association between score on the MSRS and verbal WM capacity (Buszard et al., 2013). However, given that visuospatial processing may be particularly important in the push-pass, score on the MSRS may be an inappropriate measure of the mental processes engaged during push-pass performance.

5 LIMITATIONS

A difference between the methodology of this study and Buszard et al. (2016) lies in the electrode locations used for coherence analysis. A change in electrode location by a few centimeters can have a significant influence on the outcome, and therefore interpretation of measurements (Jasper, 1958). Buszard et al.'s study used a site slightly temporal to the premotor region (F3 and F4, respectively), while we used the primary motor cortex (Fz) for both coherence calculations. An effect of hemispheric asymmetry would influence the results of both studies differentially. Intercorrelations between T7–Fz and T8–Fz coherence in this study ($r=0.734$) show how closely related these variables are, while no strong correlation between the two measurements was evident in the Buszard et al. study ($r=-0.03$).

6 CONCLUSION

Far-aiming tasks require visuomotor mapping. Capacity to process visuospatial information, rather than verbal information, may therefore be an important contributor to the ability to perform novel motor tasks that involve far aiming. When discerning measures of conscious brain processes associated with superior early motor performance, it appears that, in some cases at least, capacity trumps coactivation and character.

REFERENCES

American Electroencephalographic Society, 1994. Guideline thirteen: guidelines for standard electrode position nomenclature. J. Clin. Neurophysiol. 11, 111–113.

Anguera, J.A., Reuter-Lorenz, P.A., Willingham, D.T., Seidler, R.D., 2010. Contributions of spatial working memory to visuomotor learning. J. Cogn. Neurosci. 22 (9), 1917–1930.

Bijleveld, E., Veling, H., 2014. Separating chokers from nonchokers: predicting real-life tennis performance under pressure from behavioral tasks that tap into working memory functioning. J. Sport Exerc. Psychol. 36, 347–356.

Bo, J., Seidler, R.D., 2009. Visuospatial working memory capacity predicts the organization of acquired explicit motor sequences. J. Neurophysiol. 101 (6), 3116–3125.

Bo, J., Jennett, S., Seidler, R.D., 2011. Working memory capacity correlates with implicit serial reaction time task performance. Exp. Brain Res. 214 (1), 73–81.

Bo, J., Jennett, S., Seidler, R.D., 2012. Differential working memory correlates for implicit sequence performance in young and older adults. Exp. Brain Res. 221 (4), 467–477.

Boudet, S., et al., 2006. A global approach for automatic artifact removal for standard EEG record. Conf. Proc. IEEE Eng. Med. Biol. Soc. 1, 5719–5722.

Buszard, T., et al., 2013. Examining movement specific reinvestment and working memory capacity in adults and children. Int. J. Sport Psychol. 44 (4), 351–366.

Buszard, T., et al., 2016. The relationship between working memory capacity and cortical activity during performance of a novel motor task. Psychol. Sport Exerc. 22, 247–254.

Chatrian, G.E., Lettich, E., Nelson, P.L., 1985. Ten percent electrode system for topographic studies of spontaneous and evoked EEG activities. Am. J. EEG Technol. 25 (2), 83–92.

Chen, J., et al., 2005. Effects of anxiety on EEG coherence during dart throw, 11th World Congress of Sport Psychology, pp. 2–5.

Cooke, A., 2013. Readying the head and steadying the heart: a review of cortical and cardiac studies of preparation for action in sport. Int. Rev. Sport Exerc. Psychol. 6 (1), 122–138.

Cooke, A., et al., 2014. Preparation for action: psychophysiological activity preceding a motor skill as a function of expertise, performance outcome, and psychological pressure. Psychophysiology 51 (4), 374–384.

Crews, D.J., Landers, D.M., 1993. Electroencephalographic measures of attentional patterns prior to the golf putt. Med. Sci. Sports Exerc. 25 (1), 116–126.

Deeny, S.P., et al., 2003. Cortico-cortical communication and superior performance in skilled marksmen: an EEG coherence analysis. J. Sport Exerc. Psychol. 25, 188–204.

Deeny, S.P., et al., 2009. Electroencephalographic coherence during visuomotor performance: a comparison of cortico-cortical communication in experts and novices. J. Mot. Behav. 41 (2), 106–116.

Delorme, A., Makeig, S., 2004. EEGLAB: an open source toolbox for analysis of single-trial EEG dynamics including independent component analysis. J. Neurosci. Methods 134 (1), 9–21.

Feldman, J., Kerr, B., Streissguth, A.P., 1995. Correlational analyses of procedural and declarative learning performance. Intelligence 20 (1), 87–114.

Fitts, P.M., Posner, M., 1967. Human Performance. Brooks-Cole, Belmont, CA.

Hatfield, B.D., Haufler, A.J., Hung, T.-M., Spalding, T.W., 2004. Electroencephalographic studies of skilled psychomotor performance. J. Clin. Neurophysiol. 21 (3), 144–156.

Hatfield, B.D., Hillman, C.H., 2001. The psychophysiology of sport: a mechanistic understanding of the psychology of superior performance. In: Singer, R.N., Hausenblas, H.A., Janelle, C.M. (Eds.), Handbook of Sport Psychology. Wiley & Sons, New York, pp. 243–259.

Haufler, A.J., et al., 2000. Neuro-cognitive activity during a self-paced visuospatial task: comparative EEG profiles in marksmen and novice shooters. Biol. Psychol. 53, 131–160.

Hillman, C.H., et al., 2000. An electrocortical comparison of executed and rejected shots in skilled marksmen. Biol. Psychol. 52 (1), 71–83.

Janacsek, K., Nemeth, D., 2013. Implicit sequence learning and working memory: correlated or complicated? Cortex 49 (8), 2001–2006.

Jasper, H.H., 1958. The ten-twenty electrode system of the International Federation. Electroencephalogr. Clin. Neurophysiol. 10, 371–375.

Jurcak, V., Tsuzuki, D., Dan, I., 2007. 10/20, 10/10, and 10/5 systems revisited: their validity as relative head-surface-based positioning systems. Neuroimage 34 (4), 1600–1611.

Kaufer, D.I., Lewis, D.A., 1999. Frontal lobe anatomy and cortical connectivity. In: Miller, B.L., Cummings, J.L. (Eds.), The Human Frontal Lobes: Functions and Disorders. Guilford Press, New York, pp. 27–44.

Kaufman, S.B., et al., 2010. Implicit learning as an ability. Cognition 116 (3), 321–340.

Kerick, S.E., et al., 2001. The role of the left temporal region under the cognitive motor demands of shooting in skilled marksmen. Biol. Psychol. 58, 263–277.

Klem, G.H., et al., 1999. The ten twenty system of the international federation. In: Deuschl, G., Eisen, A. (Eds.), Recommendations for the Practice of Clinical Neurophysiology: Guidelines of the International Federation of Clinical Physiology (EEG Suppl. 52). International Federation of Clinical Neurophysiology, Cleveland, pp. 371–375.

Klimesch, W., 1999. EEG alpha and theta oscillations reflect cognitive and memory performance: a review and analysis. Brain Res. Rev. 29 (2–3), 169–195.

Laborde, S., Furley, P., Schempp, C., 2015. The relationship between working memory, reinvestment, and heart rate variability. Physiol. Behav. 139, 430–436.

Lam, W.K., Masters, R.S.W., Maxwell, J.P., 2010. Cognitive demands of error processing associated with preparation and execution of a motor skill. Conscious. Cogn. 19 (4), 1058–1061.

MacMahon, K.M.A., Masters, R.S.W., 2002. The effects of secondary tasks on implicit motor skill performance. Int. J. Sports Psychol. 33, 307–324.

Masters, R.S.W., 1992. Knowledge, knerves and know-how: the role of explicit versus implicit knowledge in the breakdown of a complex motor skill under pressure. Br. J. Psychol. 3 (83), 343–358.

Masters, R.S.W., Maxwell, J.P., 2008. The theory of reinvestment. Int. Rev. Sport Exerc. Psychol. 1 (2), 160–183.

Masters, R.S.W., Polman, R.C.J., Hammond, N.V., 1993. 'Reinvestment': a dimension of personality implicated in skill breakdown under pressure. Pers. Individ. Dif. 14, 655–666.

Masters, R.S.W., Eves, F.F., Maxwell, J.P., 2005. In: Terry, P. et al., (Ed.), Development of a movement specific reinvestment scale. ISSP 11th World Congress of Sport Psychology.

Maxwell, J.P., Masters, R.S.W., Eves, F.F., 2003. The role of working memory in motor learning and performance. Conscious. Cogn. 12 (3), 376–402.

Oswald, F.L., et al., 2015. The development of a short domain-general measure of working memory capacity. Behav. Res. Methods 47, 1343–1355.

Shah, P., Miyake, A., 1996. The separability of working memory resources for spatial thinking and language processing: an individual differences approach. J. Exp. Psychol. Gen. 125, 4–27.

Smith, M.E., McEvoy, L.K., Gevins, A., 1999. Neurophysiological indices of strategy development and skill acquisition. Cogn. Brain Res. 7, 389–404.

Stone, J.M., Towse, J.N., 2015. A working memory test battery: Java-based collection of seven working memory tasks. J. Open Res. Softw. 3, e5.

Vickers, J.N., 1996. Visual control when aiming at a far target. J. Exp. Psychol. Hum. Percept. Perform. 22 (2), 342–354.

Wood, G., Vine, S.J., Wilson, M.R., 2016. Working memory capacity, controlled attention and aiming performance under pressure. Psychol. Res. 80 (4), 510–517.

Zhu, F.F., et al., 2010. EEG activity during the verbal-cognitive stage of motor skill acquisition. Biol. Psychol. 84 (2), 221–227.

Zhu, F.F., et al., 2011. Neural co-activation as a yardstick of implicit motor learning and the propensity for conscious control of movement. Biol. Psychol. 87 (1), 66–73.

CHAPTER

Action-skilled observation: Issues for the study of sport expertise and the brain

14

April Karlinsky*, Karen Zentgraf†, Nicola J. Hodges*,1

*University of British Columbia, Vancouver, BC, Canada
†Goethe University, Frankfurt, Germany
1Corresponding author: Tel.: +1-604-822-5895; Fax: +1-604-822-6842,
e-mail address: nicola.hodges@ubc.ca

Abstract

With a growing body of research devoted to uncovering regions of the brain implicated in action observation following various action-related experiences, including sport, we ask what we know from this research, and what we still need to know, as it pertains to sport and the brain. To do this, we review and integrate knowledge garnered from developmental work, short-term motor learning studies, and most significantly sport athletes across varying skill levels. We consider various neurophysiological methods, including TMS, fMRI, and EEG, which have been used to help uncover brain regions involved in action-skilled observation. We are particularly interested in how these processes are related to action prediction and the detection of deceptive actions among athlete groups. This research is considered within broad theoretical frameworks related to action-simulation and prediction, although our main focus is on the brain regions that have been implicated in skilled action observation and the implications of this research for knowledge and further study of sport expertise.

Keywords

Motor performance, Motor learning, Perceptual processes, Expert performance, Action observation, Motor simulation, Brain processes, Perceptual–cognitive skills, Anticipation, Prediction, Neurophysiology methods, Motor development

Sports provide an interesting arena to study action experience effects on behavioral outcomes as well as mechanisms and processes that mediate or might explain these outcomes. In many sports, visual–motor experiences are closely tied to performance, and in some instances, these experiences can be separated through comparisons of novices, experts, fans, or coaches. Action experiences cover a whole range of aspects such as body-related adaptations, visual familiarity, generation of sensorimotor associations, and statistical knowledge about the occurrence of events. Of interest in the sports' expertise domain has been the cortical (and subcortical) activations that accompany or change with sport training and expertise and the relevance of these for current and future performance, decision making, training, and transfer. Here, we review how action experiences modulate observation, with reference to these neurophysiological changes. We refer to this phenomenon as "action-skilled observation," that is, observation characterized by neural processes associated with motor capabilities. Our aim is to bring together disparate lines of research that speak to skilled perception and inform our understanding of processes underpinning action-skilled observation in sport. We first present an overview of the presumed cortical processes involved in action observation, specifically those implicated in the action observation network (AON). We then provide a selective review of relevant research pertaining to the emergence of action-skilled observation through normal motor development and short-term motor learning. The main focus of our review is on expert athletes in sports and research showing brain regions that distinguish across skill. We consider the functional relevance of these differences with respect to anticipatory decisions and detection of fakes/deceptive moves and throughout we reflect on the meaning of the research for sports and sport-related research, considering potential research directions and applications.

In writing this review and selecting research to include, it is important to acknowledge that action experiences integrate many components, and in much of the research, it is hard to disentangle motor from visual experience. Further, we can only speculate on the bidirectional influences of action on perception and perception on action. Action experiences affect what you see, but what you see will also affect what you do. In addition, these experiences will change sensitivity for context and objects as well as rule-based knowledge structures.

1 THE AON AND "MIRROR PROPERTIES"

The AON is a neural umbrella network that is activated when humans visually perceive and observe other humans involved in actions. This network comprises all brain areas that are activated by the mere observation of actions (Cross et al., 2009). Many of the studied actions have not been embedded in a sports context. Based on a meta-analysis of 104 studies with predominantly hand actions, this network was reported to involve: the inferior frontal gyrus (IFG, BA44/45), the dorsal and ventral premotor cortex (dPMC, vPMC), the inferior parietal lobule (IPL), the superior parietal lobule (SPL), the inferior parietal sulcus (IPS), some sections of the primary somatosensory cortex (S1) as well as the primary motor cortex (M1), the posterior medial temporal gyrus (pMTG), the fusiform face/body area (FFA/FBA), and visual area V5 (Caspers et al., 2010). We have included a figure showing primary areas of the brain (see Fig. 1) which we use to help categorize these various areas implicated in the AON in the subsequent table. This table details key brain areas mentioned throughout this review and associated abbreviations (Table 1). It should be noted, from an analytical level, that many of the areas in this network

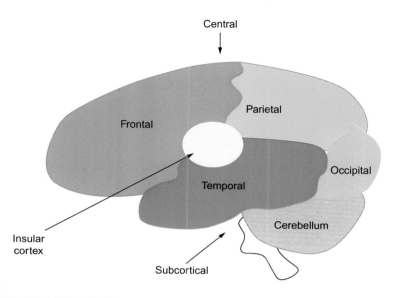

FIG. 1

Schematized brain diagram of the cerebrum and its major brain areas and the cerebellum. Please note that the insular cortex and the limbic system lie beneath the cerebrum (i.e., subcortical).

are relevant for many different processes and functions, beyond those implicated in action observation.

One body of research that underpinned the AON and subsequent research is the mirror neuron domain. Over 20 years ago, using single-cell recordings, these neurons were uncovered in area F5 of the monkey premotor cortex (comparable to the human vPMC). These were so called as they discharge both when a particular action is executed and when a similar action is observed (e.g., di Pellegrino et al., 1992; Gallese et al., 1996; for a review, see Rizzolatti and Craighero, 2004). In further experiments, Rizzolatti and other groups found neurons with similar properties also in the IPL as well as the IFG (adjacent to the primary motor cortex, e.g., Buccino et al., 2004a). In the decades since, extensive research has been conducted to determine the presence and properties of an analogous mirror neuron system (MNS) in humans. Despite such efforts, the likely neural substrates of these mirror regions remain controversial and there is still a conceptual and methodological controversy about "mirroring" per se. The vast majority of investigations involving humans have been performed using noninvasive, neuroimaging, and neurophysiological methods, providing only indirect evidence of the existence of the MNS (for reviews, see, e.g., Cattaneo and Rizzolatti, 2009; Rizzolatti and Craighero, 2004; Rizzolatti et al., 2001). However, isolated studies of single-neuron recordings in the human brain have corroborated evidence of action observation–execution matching neurons (Mukamel et al., 2010).

Table 1 Primary Areas (Cited in This Review) Implicated in the Action Observation Network (AON) and Related to Action Prediction and Identification of Familiar or Deceptive Actions

General Area	Specific Area	Acronym	Primary Study/Review Article
Frontal/central	Inferior frontal gyrus	IFG	Caspers et al. (2010)
			Wright and Jackson (2007)
	Bilateral inferior frontal gyrus (pars orbitalis)		Abreu et al. (2012)
	Dorsal and ventral premotor cortex	dPMC/vPMC	Caspers et al. (2010)
			Cross et al. (2006)
			Cross et al. (2009)
			Wright et al. (2010)
	Primary motor cortex	M1	Cisek and Kalaska (2010)
	Dorsolateral prefrontal cortex	DLPFC	Cisek and Kalaska (2010)
			Marshall et al. (2009)
			Wright et al. (2010)
	Dorsomedial prefrontal cortex	DMPFC	Kim et al. (2011)
			Wright et al. (2010)
			Wright et al. (2013)
	Ventrolateral frontal cortex		Wright et al. (2010)
	Supplementary motor area	SMA	Balser et al. (2014a)
			Cross et al. (2006)
Parietal	Inferior parietal cortex/lobe	IPL	Caspers et al. (2010)
			Cross et al. (2009)
			Wright and Jackson (2007)
	Superior parietal cortex/lobe	SPL	Balser et al. (2014a)
			Caspers et al. (2010)
			Wright and Jackson (2007)
			Wright et al. (2013)
	Inferior parietal sulcus	IPS	Caspers et al. (2010)
			Cross et al. (2006)
	Primary somatosensory cortex	S1	Caspers et al. (2010)
	Precuneus		Cisek and Kalaska (2010)
Temporal	Posterior medial temporal gyrus	pMTG	Caspers et al. (2010)
	Superior temporal gyrus	STG	Bishop et al. (2013)
	Fusiform face/body area	FFA/FBA	Caspers et al. (2010)
	Fusiform gyrus		Pilgramm et al. (2010)
	Parahippocampal gyrus		Kim et al. (2011)
	Superior temporal sulcus	STS	Cross et al. (2006)
			Wright and Jackson (2007)

Table 1 Primary Areas (Cited in This Review) Implicated in the Action Observation Network (AON) and Related to Action Prediction and Identification of Familiar or Deceptive Actions—Cont'd

General Area	Specific Area	Acronym	Primary Study/Review Article
Occipital	Visual area V5	V5	Caspers et al. (2010)
	Primary visual cortex	V1	Molenberghs et al. (2012)
	Middle temporal visual area		Wright and Jackson (2007)
	Extrastriate body area	EBA	Abreu et al. (2012)
	Middle occipital cortex		Pilgramm et al. (2010)
	Occipital gyrus		Pilgramm et al. (2010)
Subcortical	Right anterior insular cortex		Abreu et al. (2012)
			Wright et al. (2013)
	Limbic system		Molenberghs et al. (2012)
	(Anterior) cingulate cortex	ACC	Bishop et al. (2013)
			Kim et al. (2011)
			Wright et al. (2013)
	Retrosplenial cortex		Kim et al. (2011)
Cerebellum			Balser et al. (2014a)
			Calvo-Merino et al. (2006)
			Molenberghs et al. (2012)

Activation likelihood estimation meta-analysis, performed on 125 studies, has provided a quantitative index of the consistency of patterns of fMRI activity measured in human studies of action observation and action execution (Molenberghs et al., 2012). Surprisingly, the authors reported a much broader network than expected from the existing data, with 14 separate clusters identified as having "mirror properties." New clusters included the primary visual cortex (V1), the cerebellum, and parts of the limbic system. The authors argued for the existence of a brain network engaged in action observation and action execution that is complemented with areas relevant for nonmotor functions (auditory, somatosensory, affective, etc.). Similar ideas regarding a widely distributed multisensory network in the brain involved in sensorimotor matching between visual and motor aspects (including the SPL and the precuneus) have also been suggested by others (Cisek and Kalaska, 2010; Filimon et al., 2015; Ghazanfar and Schroeder, 2006).

Despite questions regarding the straightforward claim that the observation and execution of the same action solicit identical neural structures in the brain, logically, this could only happen when the observer is able to execute the observed action. Based on that, Rizzolatti et al. (2001) proposed that observation of actions within the viewer's motor repertoire could be mapped onto the viewer's own motor representations of the action. For humans, it was assumed that this common coding would result in observers covertly simulating the actions they perceive (Jeannerod, 2001).

Actions outside of the viewer's motor repertoire should then not elicit the motor system and instead should be processed more visually (e.g., Buccino et al., 2004a; Stevens et al., 2000). In this literature, motor system activation during action observation is thought of as a low-level simulation of the action (also referred to as "motor resonance," e.g., Rizzolatti and Craighero, 2004). This resonance or simulation is proposed to contribute to a number of cognitive functions, including action understanding (e.g., Rizzolatti and Craighero, 2004) and anticipation of action outcomes (for overviews, see Wilson and Knoblich, 2005; Zentgraf et al., 2011; for empirical studies, see Balser et al., 2014a,b; Bischoff et al., 2012, 2014; Mulligan et al., 2016a, b; Munzert et al., 2008).

When action and sport experiences are gathered, actors do not only build up procedural memories about how to act, but they also build up episodic memories which allow processing of contingencies between bodily movements and their sensory and environmental consequences (e.g., Wolpert and Flanagan, 2001). For example, after dribbling toward the goal, a shot, a pass, and a turnover are likely action consequences for goalkeepers anticipating actions of outfield players in soccer. Concomitantly, skilled actors or sports performers observe actions with experience-based expectations of the upcoming actions and they might also be able to prepare certain actions based on their observations or normative biases (e.g., Bar-Eli et al., 2007; Leuthold et al., 2004). There is evidence that the brain spontaneously exploits the frequency-based structure in observed action sequences (e.g., Ahlheim et al., 2014) and that this sensitivity shows up in MNS areas, as well as in the IPS and prefrontal cortex.

Conceptually, motor contributions to observation can be widespread, going beyond activation of actions to estimation of time, physical events, and mental rotation (e.g., Coull et al., 2008; Press and Cook, 2015; Schubotz, 2007; Wohlschläger, 2001). Researchers have shown that the AON is sensitive to unexpected actions (e.g., Schiffer et al., 2013) and there is evidence that vPMC activation increases with the available number of object-related actions, presumably related to some competition process in action selection (Schubotz et al., 2014). Which specific aspect of one's own action experiences causally impacts on action observation is still under debate, but it should be acknowledged that the motor system is involved in the observation of events per se, not only in dynamic human action observation.

2 NEUROPHYSIOLOGICAL METHODS

The common methods used in investigations presented in this review are briefly detailed later. These include electroencephalography (EEG), transcranial magnetic stimulation (TMS) coupled with electromyography (EMG), and functional magnetic resonance imaging (fMRI). It must be acknowledged that these are not direct measures of neuronal activity such as single-cell recordings, which are typically reserved for animal studies (cf., Mukamel et al., 2010).

EEG provides a means to capture summed neural activities at a very high temporal resolution (Nakata et al., 2010) and there are many EEG parameters available. It has been shown that the mu rhythm recorded over the motor cortex and alpha and beta band oscillations (captured over sensorimotor cortical areas within the frequency band of 8–13 Hz) decrease in power during both the execution and observation of action (e.g., Babiloni et al., 2002; Cochin et al., 1999; Muthukumaraswamy et al., 2004; Orgs et al., 2008; Pfurtscheller et al., 1997). This activity is referred to as event-related desynchronization (ERD) and is thought to reflect activity of the MNS (e.g., Lepage and Théoret, 2006; Rizzolatti and Craighero, 2004; for a review, see Pineda, 2005). The veracity of mu rhythm suppression as a marker of MNS activity has been supported through studies where both fMRI and EEG have been collected simultaneously (Arnstein et al., 2011). A meta-analytic review of 80 studies showed that mu suppression was sensitive to action observation, yielding a moderate effect size of $d=0.31$, and execution, $d=0.46$ (Fox et al., 2016).

The use of TMS in investigating processes during action observation follows the rationale that the perception of bodily motion (and skilled motion in particular) changes neural excitability in the primary motor cortex (M1). By applying TMS to the primary motor cortex during the viewing of actions, enhanced corticospinal activity can be captured using EMG of the motor-evoked potentials (MEPs—measures of corticospinal excitability) in the effector(s) involved in the viewed action (typically hand, arm, or foot muscles) (e.g., Fadiga et al., 1995). Using this technique, it has been shown that the motor neuron response elicited by the observation of an action is sensitive to the observed task requirements (e.g., weight lifted; Behrendt et al., 2016), and temporally coupled to the actual execution of that same action (Gangitano et al., 2001).

fMRI is probably the most commonly used technique to show how action observation involves the motor system and depends upon the skillset of the observer. This method of neuroimaging maps brain activity by measuring the blood-oxygen-level-dependent (BOLD) signal. While this tool obviously constrains the tasks that can be examined (movement is prevented or significantly restricted inside the scanner), it has been useful in capturing differences in response to a variety of visual stimuli as a function of the observers' individual motor repertoires. However, labeling an elevated BOLD signal in an fMRI study as signaling pure MNS activation must be regarded with caution, as it is known from direct recordings that only some neurons in this area have mirror properties. Although fMRI has its strengths in providing a valid spatial marker of changed neural processing, traditional approaches on the basis of subtraction logic and univariate analyses bear issues that constrain clear-cut conclusions. More advanced approaches (e.g., multivoxel pattern analysis and diffusion tensor imaging) have helped to broaden understanding.

Over the last decade, the notion of experience-dependent motor simulation has gained substantial interest from neuroscience. In what follows, we review current literature that speaks to how the recruitment of the simulation circuit during action observation is modulated by a broad range of processes related to the viewer's own action experiences. We briefly consider developmental work on this topic and studies related to short-term manipulations and then expertise-related effects of relevance to sports.

3 ACTION OBSERVATION AND RECRUITMENT OF THE SIMULATION CIRCUIT: A REVIEW OF THE EVIDENCE

3.1 EARLY ACTION EXPERIENCES

In order to help understand the potential significance of the AON system in adults that is responsive to motor experience, it is worthwhile briefly considering its sensitivity to motor development. Neonatal, behavioral imitation research has been interpreted by some to speak to the innate presence of an MNS (for a review, see Lepage and Théoret, 2007; for relevant critical commentaries, see also Bertenthal and Longo, 2007; Kilner and Blakemore, 2007). There is also some neurophysiological-based research supporting the existence of an MNS within the infant brain (e.g., Cochin et al., 2001; Lepage and Théoret, 2006; Nyström, 2008), as well as its modulated recruitment during action observation as a function of action experience (Southgate et al., 2009) and growing motor capabilities (van Elk et al., 2008).

Mu- and beta-desynchronization in 14–16 months infants were greatest when infants observed crawling vs walking videos, which was the motor primitive with which all infants had more experience (van Elk et al., 2008). The size of this effect was positively correlated with experience crawling. Similar results were shown in a study of 18–30 months infants watching goal-directed and imitated (nongoal-directed) actions (Warreyn et al., 2013). Mu suppression (recorded from central electrodes), was correlated with the quality of a child's imitation, pointing to its functional significance and coupling to the motor capabilities of the child. However, the relative mu suppression seen during action observation in young infants (14 months) compared to mu suppression during action execution is less than that typically shown in adults, reflecting the sensitivity of this system to development (Marshall et al., 2011).

Researchers have also shown how the action observation response is sensitive to the development of anticipatory skills in young infants. A unilateral (left hemisphere—alpha band) desynchronization was seen during observation, but it was only shown once infants could anticipate the forthcoming reach (Southgate et al., 2009). This suggests that it indexed a predictive response to action observation (bilateral desynchronization was seen during actual reaching).

The extant research in infants provides evidence that even early in life, action experiences and abilities modulate the perception of action (e.g., van Elk et al., 2008) and that the changed motor processing elicited during observation may subserve action anticipation (Southgate et al., 2009). These points will be further elucidated in the following sections, as we consider the emergence and effects of motor-experience-dependent observation in adult populations.

3.2 PHYSICAL AND IMITATIVE SHORT-TERM PRACTICE

One of the advantages of short-term motor learning studies is the potential for control of action and visual experiences. As some of the best examples of this type of research, two dance-training studies were conducted involving both behavioral and brain activity measures. In the first study, expert dancers physically practiced dance sequences for 5 weeks and underwent fMRI scanning while watching these as well as control (novel) sequences (Cross et al., 2006). As the dancers accrued more physical experience, they exhibited increased activity in the AON during observation and imagery of the rehearsed compared to the untrained sequences (differentiated by greater activity within the STS, vPMC, IPS, and SMA). Activation within the IPL and vPMC was positively correlated with the dancers' judgments of how well they could perform the movements (Cross et al., 2006; see also Gardner et al., 2015). These results provided evidence that the simulation circuit is tuned to the motor repertoire of the observer, even for short-term practice in experienced dancers (Cross et al., 2006).

In a follow-up study of novice dancers, playing a dance-based video game for 5 days, significant training-related changes in the AON were shown (Cross et al., 2009). Although there were differences in activation as a function of physical or observational experiences, there was also overlap, particularly in the left-IPL and the right-PMC regions, as a result of either experience, compared to watching novel sequences. However, it was only physical experience that produced a significant correlation between behavioral measures of performance and activity in the PMC. Competency, rather than improvement across practice, was most strongly related to AON activation.

Despite increases in the activation of the AON as a function of action and visual experiences, these experiences have also led to decreased activation in AON areas. For example, observation of unpracticed chords on a guitar evoked stronger activation in the AON compared to chords that had been practiced (physically imitated) (Vogt et al., 2007). The authors suggested that recruitment of the AON might be sensitive to the intention behind action observation, where intention to imitate decreases AON activation for practiced items. Across two studies, there was increased activation within the frontoparietal circuit (vPMC and IPL) during action observation and subsequent imitation of the unpracticed guitar chords compared to the practiced chords, which was proposed to reflect the breaking down of unfamiliar actions into familiar elements in order to imitate (Higuchi et al., 2012; Vogt et al., 2007). In other work, observation without intention to imitate enhanced motor resonance for practiced items (e.g., judgments about ability to perform or the aesthetics of a movement; Calvo-Merino et al., 2005, 2006; Cross et al., 2006; see also Buccino et al., 2004b; Higuchi et al., 2012). A decrease in activation with practice might also signal greater efficiency in action preparation, or potentially a familiarity effect associated with the lack of surprise in the stimuli.

Insight into short-term, experience-dependent action observation has also been gained through the use of EEG. Brief action experience with unfamiliar drawing actions was shown to bring about changes in the neural activity of later observations, specifically with respect to mu rhythm desynchronization at central sites

(Marshall et al., 2009). The fidelity with which the imitated action was executed was positively correlated with mu band desynchronization during observation (see also Casile and Giese, 2006). The imitation condition also resulted in greater EEG desynchronization at sites overlying the dorsolateral prefrontal cortex (DLPFC) compared to a control condition (performing a familiar writing movement). When the length of the training session was increased, no differences were observed between an observation-only group and a combined physical practice and observation group (Quandt et al., 2011). Although both showed more desynchronization than seen for novel actions, mu suppression differences as a function of practice were only noted in frontal electrode sites.

One of the advantages of EEG in such short-term studies of action experiences is that EEG can be collected in close proximity to the actual physical (or observational) practice trial, such that changes in EEG during practice can be correlated to changes at a subsequent observation period as well as with behavioral measures. This concurrent charting will allow a better understanding of the relationship between these processes and the timescale of respective changes. Because action experiences modulate the action observation–execution matching system within mere hours, this raises concerns about the discriminant validity of such cortical markers of skill. Evidently, there is still much unknown regarding the timescale of the development of skilled observation, with the most intensive training manipulation to date comprising just 5 weeks (Cross et al., 2006). Does the neurophysiological development of this system follow a power function, analogous to the Law of Practice (e.g., Crossman, 1959; Lee and Wishart, 2005), such that it might undergo dramatic changes early in practice, with incrementally smaller changes over time? It is possible, of course, that after a certain refinement of an action, further fine-tuning ceases to significantly influence perceptual–cognitive and/or neural processes, or at least those captured by current measures. Among recreational golfers, those individuals who had the more "intense" training (i.e., required fewer days to complete 40 h of training) showed more gray matter changes in the parietooccipital junction and vPMC, although these individuals were also of a higher proficiency at the start of the study (Bezzola et al., 2011).

There also needs to be a better discrimination of action and visual experiences to determine their differential impact on brain neurophysiology and what this means with respect to efficiencies and effectiveness of training. Are combined experiences (visual–motor) better than either in isolation, especially when given in an alternating schedule whereby small blocks of observation are followed by small blocks of physical practice? This schedule is most reminiscent of practice in the field where demonstrations typically precede practice attempts and then are interspersed throughout later attempts (see also Ong and Hodges, 2012; Zentgraf et al., 2011). Related to this issue, studies where individuals learn without vision (e.g., Casile and Giese, 2006; Mulligan and Hodges, 2014) could elucidate how each type of experience uniquely contributes to changes in brain neurophysiology.

The fact that visual familiarity has been shown to increase cortical activity during action observation compared to the viewing of novel stimuli (e.g., Cross et al., 2009) shows that the AON is not only sensitive to physically acquired (motor) experiences. Brief physical practice and distinct observational practice have been shown to result in similar motor resonance during subsequent action observation, although only in the frontalparietal mirror circuit and not in lower-level motor-related circuits associated with the motor cingulate–basal ganglia (Higuchi et al., 2012). The potential benefits of observational practice when physical training is not a viable option remain of interest (e.g., during rest intervals in athletic training, injury recovery periods, rehabilitation). There is some suggestion that observational practice during the retention interval of a physically practiced skill aids later retention and activation of mirror-related circuits (Zhang et al., 2011). There has also been evidence that combining observational practice with peripheral nerve stimulation (i.e., activation of the afferent system) leads to changes in the brain that are enhanced compared to observational practice alone (Bisio et al., 2015).

An alternative form of practice that has not yet been studied with respect to its potential influence on the AON is passive physical practice, where the body is guided through the motions in the absence of the individual's actively generated motor commands. Though such forms of passive physical practice do not typically support learning compared to active physical practice (for a review, see Hodges and Campagnaro, 2012), whether such action experiences are sufficient to promote the strengthening and/or building of cortical connections remains to be seen, and may be of particular relevance with respect to special populations or in situations where safety is a concern. Physical guidance is still popular in the teaching of gymnastic skills, such that cortical processes engaged following such guidance training may alert to the potential efficacy of this training as well as to the types of experiences, which engender change in the AON.

In summary, these studies give evidence for the involvement of the AON following short-term experiences. They show that the extent to which the AON is activated during action observation is related to the similarity between the perceived action event and the viewer's relevant motor representation. In these final sections, we review studies that have mostly involved cross-sectional comparisons of elite athletes in sports, in order to determine what neurological processes (and experiences) might be related to the expert advantage in perceptual–cognitive skills.

3.3 SPORT EXPERTISE AND LONG-TERM VISUAL-MOTOR EXPERIENCES

The previous section highlighted research that provides insight into how brief action experiences modulate action observation in the early phases of motor learning. At the other end of the experience spectrum, some 10,000 h of practice later are the (visual)-motor experts (Ericsson et al., 1993). These immensely dedicated individuals have reached the highest levels of motor performance and acquired the most specialized motor repertoires. Athletes in highly dynamic, open sport environments have also acquired perceptual–cognitive skills that appear to underpin their success in the field (for a meta-analysis, see Mann et al., 2007). As a result, these individuals have developed truly "skilled" observation. It is this subset of the general population that has inspired and enabled the research presented later, where we specifically focus on experts in sport. We review three broad areas of research relevant to skilled observation in sport and the brain: (1) observation as a function of action familiarity, (2) observation during predictive decisions, and (3) observation processes engaged during judgments about deceptive actions.

3.3.1 Action Observation as a Function of Familiarity

The research program of Calvo-Merino and colleagues (2005, 2006) provides an illustrative example of the modulating effects of motor expertise on the perception of action. These authors used fMRI to study the brain activity of expert ballet dancers (males and females) and capoeira martial artists during the observation of actions specific to each specialty or gender. Even though the dancers were asked to report on the aesthetics of the observed actions, increased neural activity was shown in the premotor and parietal brain cortices (areas implicated in the MNS) when movements were observed that were within the dancers' own specialized motor repertoires (e.g., ballet for ballerinas). Because expert male and female ballet dancers who trained together also showed greater activation in motor regions of the cortex and cerebellum when watching videos of their own gender-specific movements, the authors concluded that people simulate observed actions in terms of their own motor representations of the actions, not only in terms of shared visual experiences (Calvo-Merino et al., 2006). The fact that neural activity is differentiated within a subset of elite performers, who are only differentiated with respect to sex or type of dance, is important, as it shows the sensitivity of this network to specific action experiences.

Since these studies of dancers, various replications have been conducted across a number of different sports, including studies of archers, performers in racquet sports (tennis and badminton), as well as volleyball and soccer experts (e.g., Balser et al., 2014a,b; Bishop et al., 2013; Kim et al., 2011; Wright et al., 2010, 2011, 2013). For example, in an fMRI study of archers, in addition to greater activation of the AON in expert archers compared to nonarchers, the archers also exhibited greater recruitment of the dorsomedial prefrontal cortex (DMPFC), cingulate cortex, retrosplenial cortex, and parahippocampal gyrus compared to their untrained counterparts (Kim et al., 2011). These areas have been implicated in episodic memory retrieval, and as such, the authors attributed activity in these areas to the archers' enhanced familiarity with and meaningfulness of these stimuli (i.e., top-down influences, which may be related to Ericsson and Kintsch's long-term working memory ideas provided in Ericsson and Kintsch, 1995).

Complementary to the above neuroimaging findings, ERD in the alpha and beta bands was responsive to the viewer's motor repertoire in a comparison of professional dancers and nondancers viewing videos of dance and everyday movements (Orgs et al., 2008; see also Pineda, 2005). However, results contrasting to these have also been shown (rhythmic gymnastics, Babiloni et al., 2009; karate, Babiloni et al., 2010). Alpha ERD was lower in the elite compared to the less-skilled groups during action observation, with the "somewhat skilled," amateur athletes, showing intermediate cortical activity (Babiloni et al., 2010). Alpha ERD was also lower in skilled observers when they provided relatively accurate judgments of the observed actions, compared to inaccurate judgments (Babiloni et al., 2009). The authors interpreted the less pronounced alpha ERD as an index of "neural efficiency," akin to that captured in skilled athletes during action execution (e.g., Baumeister et al., 2008; Milton et al., 2007).

Despite the differences in findings, reminiscent of discrepancies in fMRI studies following short-term practice, these data support the notion that action experiences modulate the neural networks underpinning action observation. Although EEG may not be as sensitive as fMRI to experience-related changes in brain function, in addition to being a relatively inexpensive technique, EEG has the advantage of high temporal resolution compared to fMRI. The ability to capture the temporal dynamics of action observation provides opportunities to probe the predictive vs reactive nature of specific neural responses to observed action events (e.g., Southgate et al., 2009), although to date, rarely has it been used in this way in a sports-related context (cf., Amoruso et al., 2014; Jin et al., 2011).

Cross-sectional comparisons of amateur, intermediate, and more elite groups alert to the sensitivity of these neural markers or probes to characterize motor experience (e.g., Babiloni et al., 2010), in addition to the somewhat more common comparisons between strictly novice and expert groups (e.g., Kim et al., 2011; Orgs et al., 2008). However, the question remains as to how sensitive the MNS or AON is to the amount of physical training (e.g., Marshall et al., 2009). Among an fMRI study of skilled basketball players, there were no behavioral (anticipation accuracy) differences as a function of visual-motor experience (and presumably no experience-based scaling of cortical activations) despite experience ranging from 468 to 6552h of practice (Abreu et al., 2012). Further research is required to assess these potential skill-based interactions and the degree of match between motor skill/experiences, perceptual–cognitive performance, and neural substrates and circuitry associated with action observation. Neural activation during perceptual–cognitive activities is likely not simply contingent upon the number but also upon the quality of practice hours accrued (Bishop et al., 2013). Research with elite, retired athletes has the potential to inform about the permanency of these experience-related changes, especially when capabilities change.

As briefly noted, current neurophysiological measurement techniques greatly limit the movements that can be carried out by participants. While some sport researchers have attempted to require more realistic responses to visual stimuli in behavior-based studies (e.g., Farrow and Abernethy, 2003; Mann et al., 2010a,b), neurophysiology-based investigations overwhelmingly feature unnatural perception–action responses. For example, a keypress prediction of "in" or "out" is clearly not a basketball player's normal reaction to a free throw (e.g., Abreu et al., 2012; Aglioti et al., 2008), and this uncoupling may misrepresent experts' and novices' true anticipatory skill (Farrow and Abernethy, 2003) as well as the neural processes recruited during more realistic action observation scenarios (Mann et al., 2010a).

As the intention with which action observation is undertaken has been shown to modulate the underlying neural processes (e.g., Buccino et al., 2004b), it is also of interest whether an intent to respond differentially activates the MNS, perhaps also as a function of the instructional context (e.g., whether the response should comprise a complementary or emulative action; see Newman-Norlund et al., 2007).

Modulation of action observation "on the fly," so to speak, also poses interesting questions for the study and understanding of sport expertise and behavioral changes that accompany skilled observation. How might observation and perceptual–cognitive performance be affected were they executed during the course of action production? Behaviorally, there is mixed evidence regarding how the congruency of concurrently performed and observed actions impede or facilitate, for example, action–perception (e.g., Hamilton et al., 2004; Miall et al., 2006; Müsseler and Hommel, 1997; for a review, see Schütz-Bosbach and Prinz, 2007) and action anticipation (Mulligan et al., 2016a,b). Whether a neurophysiological basis for such effects might be discernable is obviously again challenged by the movement constraints of neuroimaging and electrophysiological measurement techniques. However, it is feasible that some small-scale actions would be possible even within a scanner environment, such as holding differently weighted boxes during perceptual judgment tasks (e.g., Hamilton et al., 2004), or pressing against a force plate during anticipation tasks (e.g., Mulligan et al., 2016a,b). Moreover, alternative solutions to this limitation may not be too far off. A neurophysiological technique that has gained in popularity is tDCS (transcranial direct current stimulation). Like methods involving repetitious firing of TMS (rTMS), it is possible to stimulate a specific area of the brain with effects lasting after the stimulation, such that movements are no longer constrained. tDCS works by passing a low-voltage current through an area of interest, essentially changing neural excitability. It is an inexpensive method, although there is a need to know where to apply the stimulation, which might require whole-brain fMRI before application. To date, tDCS has not been used to study the cortical simulation circuit, yet see Reis et al. (2009) and Banissy and Muggleton (2013) for a consideration of its potential value in motor learning and sports training, respectively.

3.3.2 Action Anticipation

While it is clear from the literature that motorically skilled individuals are processing familiar actions differently than their unskilled counterparts, the question remains as to the functional value of brain-related differences in observation. Recent sport expertise research sheds some light on this question, as investigators have sought to determine the neural basis of elite athletes' superior perceptual–cognitive skills, with particular focus on the anticipation of action outcomes and detection of deceptive actions (see also Smith, 2016; Yarrow et al., 2009).

The predictive brain has become a bit of a buzzword in cognitive neuroscience (Friston, 2011). When talking about action anticipation in sports, it is mostly studied as a change prediction referring to an observable event. Dynamic predictions are differentiated from probabilistic, semantic, and episodic predictions (for further details, see Schubotz, 2015). Dynamic forward predictions are related to information about events unfolding in time, such as judging and/or intercepting a moving ball or anticipating the next move or action effect from a person preparing to serve or dribble a ball.

In much of the literature in the anticipation domain, the motor system's involvement in action–perception is suggested to contribute to the prediction of observed action effects, potentially via forward models (e.g., Eskenazi et al., 2009; Gorman et al., 2013; Wilson and Knoblich, 2005; Yarrow et al., 2009). In computational models of motor control, forward models are the sensory consequences of self-generated action that are predicted based on an (efference) copy of the motor command (Miall and Wolpert, 1996). It has been proposed that when this system is activated via action observation, there is not only a simulation of the action in terms of the viewer's own motor representation but also as a corollary effect, a forward model is generated (although perhaps at a more general level of specificity), such that others' actions also elicit forward predictions of anticipated effects (Miall, 2003; Wolpert et al., 2003). The resultant sensory predictions are thought to be available to and influence cognitive processes, including the perception and anticipation of action events in others (e.g., Bischoff et al., 2012, 2014, Bischoff et al., 2015; Eskenazi et al., 2009; Frith et al., 2000; Kilner et al., 2007; Miall et al., 2006; Schubotz, 2007; Schubotz and von Cramon, 2003; Wilson and Knoblich, 2005; Wolpert and Flanagan, 2001; Zentgraf et al., 2011).

As the first functional imaging study devoted to anticipatory skill in sport, Wright and Jackson (2007) identified the neural correlates separately implicated in the viewing of sport-related motion and action, and the anticipation and judgment of action outcomes. Novice tennis players were shown videos of serves, nonserve actions (ball bouncing), and static control sequences. Using temporal occlusion (whereby videos are stopped at key points as the action and outcome unfolds), the serve sequences were edited pre- or postball–racquet contact and predicted directions were made by pressing a button while in an fMRI scanner. Compared to ball bouncing, serve sequences demanding an anticipatory judgment elicited increased activity in MNS brain areas, specifically regions in the parietal lobule (bilateral IPL, right SPL) and in the right frontal cortex (dorsal and ventral regions of the IFG). This pattern of activation was separate from the responses in areas of the brain associated with the general viewing of motion and body actions (middle temporal visual area, STS). Considering that only novice participants were included, the results suggest that the MNS can be recruited for action prediction even when observing relatively unpracticed actions.

In a cross-sectional comparison of expert, intermediate, and novice badminton players' predictions of an opposing player's shot direction, early occlusion led to increased activity in the PMC and the medial frontal cortex compared to late occlusion (Wright et al., 2010). Experts showed increased activation compared to intermediates and novices in the dPMC, ventrolateral frontal, and medial frontal cortices (areas implicated in the observation, understanding, and preparation of action), particularly when relying solely on early movement cues. In a tennis and a volleyball directional anticipation task involving comparisons of experts in these sports, in both expert groups, enhanced signals were specifically noted in the SPL, the supplementary motor area (SMA), and the cerebellum only when observing stimuli of their own sports domain (Balser et al., 2014a). We have included a figure detailing some of the results from this study, showing contrast estimates from the SPL (right), preSMA (left), and cerebellum (left) for the volleyball and tennis players in response to volleyball and tennis predictions (see Fig. 2A). Alongside this, we have also provided an illustration of brain activation differences in the preSMA area (when comparing domain expert vs domain novice differences in anticipation) (Fig. 2B). The authors suggested that the cerebellum involvement might reflect the usage of a predictive internal model (Balser et al., 2014a). In general, this research is congruent with the idea that experience performing an action prompts a low-level movement preparation which aids the expert in making anticipatory decisions (see also Balser et al., 2014b; Cacioppo et al., 2014; Wimshurst et al., 2016 for related delineations based on tennis, volleyball, and field hockey).

Frontoparietal components of the AON were similarly activated during expert and novice basketball players' prediction of the fate of a basketball free throw shot

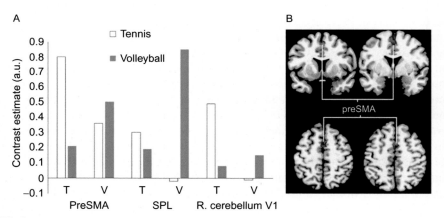

FIG. 2

(A) Contrast estimates (from fMRI) in the *left* preSMA, *right* SPL and *left* cerebellum for the volleyball and tennis players in response to predictions based on volleyball and tennis stimuli (prediction activation was calculated based on a comparison to an observe-only condition). In (B) an illustration of brain activation differences in the preSMA area when comparing domain expert vs domain novice differences in anticipation.

Data adapted from Balser, N., Lorey, B., Pilgramm, S., Naumann, T., Kindermann, S., Stark, R., Zentgraf, K., Williams, A.M., Munzert, J. (2014). The influence of expertise on brain activation of the action observation network during anticipation of tennis and volleyball serves. Front. Hum. Neurosci. 8, 1–13.

(i.e., in or out, Abreu et al., 2012). Experts showed increased activation in the extrastriate body area (EBA) during the prediction task compared to a nonprediction control task, which the authors suggested was an effect of the athletes' reliance on and interpretation of body kinematics in predicting the outcomes of others' actions (see also Abernethy and Zawi, 2007; Aglioti et al., 2008; Wright et al., 2011). When watching errors, experts also showed increased activation in the bilateral IFG (specifically pars orbitalis) and right anterior insular cortex, areas associated with error monitoring and awareness (Abreu et al., 2012; see also Aglioti et al., 2008, for TMS-based evidence of elite basketball players' superior ability to predict action outcomes based on early body kinematics, and their enhanced sensitivity to errors).

Although not based on neurophysiological methods, motor-based simulation mechanisms have also been implicated in short- and longer-term practice experiences in predictions of dart outcomes. In a series of studies, an action-incongruent secondary task (i.e., pushing lightly against a force gauge with the throwing arm) interfered with prediction accuracy among physically (not perceptually) trained individuals (Mulligan et al., 2016a,b). What is important about this work is that prediction accuracy was linked to motor-based interference. Although no measures of brain activity were taken during these anticipatory decisions, there was evidence that the areas or processes responsible for action execution needed to be "available" in order to aid action prediction accuracy. Using TMS, temporary disruptions to the motor cortex have been shown to impair recognition and anticipation of others' actions (Michael et al., 2014; Stadler et al., 2012). It appears that the action-interference effect is not only specific to action experiences, but it can also be erased by perceptual training (Mulligan and Hodges, 2016). This suggests that motor-related brain mechanisms that operate during action prediction are either not automatic, or that (assuming time allows) they can be overridden by more strategic, perceptually based mechanisms (Mulligan and Hodges, 2016).

3.3.3 Detection of Deceptive Actions
It might be expected that motor resonance among skilled athletes would be sensitive to detection of errors and deceptive actions given a discrepancy between an athlete's expectations and actual outcomes. However, if an athlete is likely to activate their own motor program for action when viewing an action that has a surprising ending, it may be that they will more likely be deceived. Indeed, the research seems to highlight some potential discrepancies, depending on the methods used to study deception.

Observers ranging in soccer experience were classified as low-, intermediate-, or high-skill anticipators, based on their accuracy in predicting the direction change of an oncoming soccer player (Bishop et al., 2013). When viewing deceptive actions, high-skill anticipators were distinguished from the less-skilled groups by activation within the right anterior cingulate cortex, which the authors suggested reflected the suppression of (deceived) responses to the deceptive action and the monitoring of incorrect decisions (Bishop et al., 2013). In a related study involving different skill groups, soccer players were required to determine (from a point-light model) whether (i) a move by an attacker was "normal" or "deceptive" and (ii) the resulting direction of the ball (Wright et al., 2013). Both conditions resulted in activation in MNS areas, and patterns of activation in these areas were sensitive to skill, with highest activations among the most skilled, male players. Detection of deceptive actions was most strongly delineated with respect to higher-level cortical processing areas (i.e., prefrontal cortex, medial frontal cortex, anterior insula, cingulate gyrus) in comparison to ball location anticipation. The authors suggested that these differences may reflect the novelty of judging deceptive actions and the cognitive effort this entails, in comparison to anticipating outcomes, which are more of an automatic consequence of observation.

TMS has also been used to show how interactions between the perceptual and motor systems affect expert athletes' sensitivity to deceptive movements. Expert soccer kickers were more likely to be fooled by fake actions (where an observed kicker's body kinematics and the ball trajectory did not match) than novices and expert goalkeepers (Tomeo et al., 2013). Based on MEPs in response to TMS, "fooling" actions elicited similar lower-limb motor facilitation as that seen for real actions, only in the expert kickers. Differences between goalkeepers and expert kickers suggest that anticipatory decisions are a result of different mechanisms in players who have primarily motor experience, in comparison to goalkeepers who have acquired more visual experience predicting kick direction. There have, however, been criticisms of these methods and conclusions (Mann et al., 2013). Some of this criticism is based on the stimuli used to show "deceptive" actions (where there is no intentional modification of body kinematics by the actor, which would be better representative of "fooling"). The ability to not be fooled or identify deception is likely to be a skill that is both a product and a curse of motor expertise. Although with playing experience players should be better able to detect valid from invalid cues (as confirmed across many studies, e.g., Dicks et al., 2011; Jackson et al., 2006; Mori and Shimada, 2013), the potential to misinterpret cues due to action-simulation processes, particularly when the actions change in an unexpected way, is not surprising. What remains to be established is how training might moderate these effects and potentially lead to different types of prediction processes. Although the data are only behavioral, there is evidence that motor simulation processes that are engaged to aid anticipation can be trumped by acquisition of perceptually based strategies (Mulligan and Hodges, 2016).

In summary, the studies cited in these last three sections highlight the role of the (cortical) motor system in detecting deceptive actions and generating predictions about what might happen (and ultimately what to do) in sport-specific scenarios. There are of course limits to some of these approaches with respect to the types of tasks that can be studied and the general difficulty in assessing performance when people are realistically responding. Moreover, just because a particular area of the brain is activated during these perceptual–cognitive tasks, this still does not allow us to distinguish what participants are actually doing when they make decisions, that is, whether they access perceptual or motor images, whether they generate actions and then suppress them, or whether they rely on some generative action processes in addition to recognition of past events. These techniques and ideas should, however, prompt reflection as to what it means to argue for motor-driven skilled observation and guide future inquiries into how observation and associated perceptual–cognitive activities are bound within the action capabilities of the observer.

While researchers have attended to the action experiences and performance of their observer groups, it is important for such design vigilance to extend to the actors featured in experimental stimuli. There is behavioral evidence showing that the closer the match between an actor and observer's motor skills, the better the viewer's action recognition (Loula et al., 2005) and action prediction (Knoblich and Flach, 2001; Knoblich et al., 2002; Mulligan et al., 2016a). Therefore, consideration of how the similarity between the actor and observer groups' motor capabilities factors into the observers' processing of action events, perceptual–cognitive performance, and the resulting inferences is needed. To date, there have been no attempts to compare perception and anticipation differences or motor resonance when a novice performer has provided the visual stimuli in addition to the typical skilled model (although Ikegami and Ganesh, 2014, have shown that motor experts can change their predictions as a result of observing and receiving feedback about prediction accuracy when watching low-skilled models).

There has been little attempt to examine behavioral differences in perception or anticipation of self-generated (vs other generated) sporting actions (cf., Bischoff et al., 2012; Jackson et al., 2008; Knoblich and Flach, 2001; Mulligan et al., 2016a) and no attempts to study neurophysiological differences underpinning the perception of these actions. This may have implications for effective training methods (based on self vs other observation; for a related review, see Hodges and Ste-Marie, 2013). The inclusion of self- and novice action stimuli can provide insight into the sensitivity of action-simulation systems for the anticipation of action outcomes and potentially alert as to how expert athletes' perception may be worse when action consequences are less predictable (e.g., when associated with novices' less consistent motor performance), in comparison to that of more novice or intermediate athletes. Moreover, it may be the case that novices exhibit "skilled" levels of observation and perceptual–cognitive abilities when given the opportunity to view motor performances that more closely match their own action representations. This similarity-driven process would also raise some issues about coaching and refereeing and what type of person (in terms of past or current motor skills or experiences) would make the best perceptual–cognitive decisions (for an illustration of such interactions based on motor experience among referees, see Dosseville et al., 2011; Pizzera, 2012).

4 SUMMARY AND CONCLUSIONS

With the advancement and wider availability of neurophysiological measurement techniques, there has been a growing interest in and understanding of the neural processes subserving action observation in recent years. This has been particularly evidenced by research into sport expertise and the human brain. There is now a compelling body of research demonstrating both the involvement of the viewer's motor system during action observation, as well as the importance of the viewer's individual motor repertoire toward a multifaceted understanding of observed action events. Converging evidence that action experiences modulate the neural structures and circuitry recruited when watching motor performances and promote "skilled" observation of familiar movements has been provided through the study of motor experts and their novice counterparts, training manipulations, and natural motor development.

Even though substantial progress has been made over the last decade toward elucidating the complexities associated with experience-dependent motor resonance, there is still much to learn. The susceptibility of the AON to different forms and durations of training remains to be determined, as does the importance of the similarity between an actor and observer toward observation-induced simulation. What is important from a sports and the brain perspective is to know the sensitivity of these networks to current and future performance and whether neural signatures during action observation give insight into subsequent trainability or performance on the field. Also, it would be important to better determine how sensitive observation-induced activations in the brain are to specific skills of the performer, including such things as physical and tactical ability, creativity in decision processes, and behaviors typically demonstrated (or observed) by players in the field. The issue of transfer across domains also has the potential to be enhanced by the study of brain activations during observation to elucidate shared processes that may be involved in tactical processes (such as decision-making in soccer and hockey), motor performance (e.g., jumping in hurdles vs jumping in long jump), as well as observation for aesthetic judgments (like those required in gymnastics or ice-dancing judges) as a function of different experiences as a performer or coach.

REFERENCES

Abernethy, B., Zawi, K., 2007. Pickup of essential kinematics underpins expert perception of movement patterns. J. Mot. Behav. 39 (5), 353–367.

Abreu, A.M., Macaluso, E., Azevedo, R.T., Cesari, P., Urgesi, C., Aglioti, S.M., 2012. Action anticipation beyond the action observation network: a functional magnetic resonance imaging study in expert basketball players. Eur. J. Neurosci. 35 (10), 1646–1654.

Aglioti, S.M., Cesari, P., Romani, M., Urgesi, C., 2008. Action anticipation and motor resonance in elite basketball players. Nat. Neurosci. 11, 1109–1116.

Ahlheim, C., Stadler, W., Schubotz, R.I., 2014. Dissociating dynamic probability and predictability in observed actions—an fMRI study. Front. Hum. Neurosci. 8, 1–13.

Amoruso, L., Sedeño, L., Huepe, D., Tomio, A., Kamienkowski, J., Hurtado, E., Cardona, J.F., Álvarez González, M.Á., Rieznik, A., Sigman, M., Manes, F., Ibáñez, A., 2014. Time to tango: expertise and contextual anticipation during action observation. Neuroimage 98, 366–385.

Arnstein, D., Cui, F., Keysers, C., Maurits, N.M., Gazzola, V., 2011. mu-suppression during action observation and execution correlates with BOLD in dorsal premotor, inferior parietal, and SI cortices. J. Neurosci. 31 (40), 14243–14249.

Babiloni, C., Babiloni, F., Carducci, F., Cincotti, F., Cocozza, G., Del Percio, C., Moretti, D.V., Rossini, P.M., 2002. Human cortical electroencephalography (EEG) rhythms during the observation of simple aimless movements: a high-resolution EEG study. Neuroimage 17, 559–572.

Babiloni, C., Del Percio, C., Rossini, P.M., Marzano, N., Iacoboni, M., Infarinato, F., Lizio, R., Piazza, M., Pirritano, M., Berlutti, G., Cibelli, G., Eusebi, F., 2009. Judgment of actions in experts: a high-resolution EEG study in elite athletes. Neuroimage 45 (2), 512–521.

Babiloni, C., Marzano, N., Infarinato, F., Iacoboni, M., Rizza, G., Aschieri, P., Cibelli, G., Soricelli, A., Eusebi, F., Del Percio, C., 2010. "Neural efficiency" of experts' brain during judgment of actions: a high-resolution EEG study in elite and amateur karate athletes. Behav. Brain Res. 207 (2), 466–475.

Balser, N., Lorey, B., Pilgramm, S., Naumann, T., Kindermann, S., Stark, R., Zentgraf, K., Williams, A.M., Munzert, J., 2014a. The influence of expertise on brain activation of the action observation network during anticipation of tennis and volleyball serves. Front. Hum. Neurosci. 8, 1–13.

Balser, N., Lorey, B., Pilgramm, S., Stark, R., Bischoff, M., Zentgraf, K., Williams, A.M., Munzert, J., 2014b. Prediction of human actions: expertise and task-related effects on neural activation of the action observation network. Hum. Brain Mapp. 35 (8), 4016–4034.

Banissy, M.J., Muggleton, N.G., 2013. Transcranial direct current stimulation in sports training: potential approaches. Front. Hum. Neurosci. 7, 1–3.

Bar-Eli, M., Azar, O.H., Ritov, I., Keidar-Levin, Y., Schein, G., 2007. Action bias among elite soccer goalkeepers: the case of penalty kicks. J. Econ. Psychol. 28 (5), 606–621.

Baumeister, J., Reinecke, K., Liesen, H., Weiss, M., 2008. Cortical activity of skilled performance in a complex sports related motor task. Eur. J. Appl. Physiol. 104 (4), 625–631.

Behrendt, F., de Lussanet, M.H.E., Zentgraf, K., Zschorlich, V.R., 2016. Motor-evoked potentials in the lower back are modulated by visual perception of lifted weight. PLoS One 11 (6), 1–13.

Bertenthal, B.I., Longo, M.R., 2007. Is there evidence of a mirror system from birth? Dev. Sci. 10 (5), 526–529.

Bezzola, L., Mérillat, S., Gaser, C., Jäncke, L., 2011. Training-induced neural plasticity in golf novices. J. Neurosci. 31 (35), 12444–12448.

Bischoff, M., Zentgraf, K., Lorey, B., Pilgramm, S., Balser, N., Baumgartner, E., Hohmann, T., Stark, R., Vaitl, D., Munzert, J., 2012. Motor familiarity: brain activation when watching kinematic displays of one's own movements. Neuropsychologia 50 (8), 2085–2092.

Bischoff, M., Zentgraf, K., Pilgramm, S., Stark, R., Lorey, B., Munzert, J., 2014. Anticipating action effects recruits audiovisual movement representations in the ventral premotor cortex. Brain Cogn. 92, 39–47.

Bischoff, M., Zentgraf, K., Pilgramm, S., Krueger, B., Balser, N., et al., 2015. Anticipating action effects with different attention foci is reflected in brain activation. Percept. Mot. Skills 120 (1), 39–56.

Bishop, D.T., Wright, M.J., Jackson, R.C., Abernethy, B., 2013. Neural bases for anticipation skill in soccer: an fMRI study. J. Sport Exerc. Psychol. 35 (1), 98–109.

Bisio, A., Avanzino, L., Gueugneau, N., Pozzo, T., Ruggeri, P., Bove, M., 2015. Observing and perceiving: a combined approach to induce plasticity in human motor cortex. Clin. Neurophysiol. 126 (6), 1212–1220.

Buccino, G., Vogt, S., Ritzi, A., Fink, G.R., Zilles, K., Freund, H.J., Rizzolatti, G., 2004b. Neural circuits underlying imitation learning of hand actions: an event-related fMRI study. Neuron 42 (2), 323–334.

Buccino, G., Lui, F., Canessa, N., Patteri, I., Lagravinese, G., Benuzzi, F., Porro, C.A., Rizzolatti, G., 2004a. Neural circuits involved in the recognition of actions performed by non-conspecifics: an fMRI study. J. Cogn. Neurosci. 16 (1), 114–126.

Cacioppo, S., Fontang, F., Patel, N., Decety, J., Monteleone, G., Cacioppo, J.T., 2014. Intention understanding over T: a neuroimaging study on shared representations and tennis return predictions. Front. Hum. Neurosci. 8, 1–18.

Calvo-Merino, B., Glaser, D.E., Grèzes, J., Passingham, R.E., Haggard, P., 2005. Action observation and acquired motor skills: an fMRI study with expert dancers. Cereb. Cortex 15 (8), 1243–1249.

Calvo-Merino, B., Grèzes, J., Glaser, D.E., Passingham, R.E., Haggard, P., 2006. Seeing or doing? Influence of visual and motor familiarity in action observation. Curr. Biol. 16 (19), 1905–1910.

Casile, A., Giese, M.A., 2006. Nonvisual motor training influences biological motion perception. Curr. Biol. 16 (1), 69–74.

Caspers, S., Zilles, K., Laird, A.R., Eickhoff, S.B., 2010. ALE meta-analysis of action observation and imitation in the human brain. Neuroimage 50, 1148–1167.

Cattaneo, L., Rizzolatti, G., 2009. The mirror neuron system. Arch. Neurol. 66 (5), 557–560.

Cisek, P., Kalaska, J.F., 2010. Neural mechanisms for interacting with a world full of action choices. Annu. Rev. Neurosci. 33, 269–298.

Cochin, S., Barthelemy, C., Roux, S., Martineau, J., 1999. Observation and execution of movement: similarities demonstrated by quantified electroencephalography. Eur. J. Neurosci. 11 (5), 1839–1842.

Cochin, S., Barthelemy, C., Roux, S., Martineau, J., 2001. Electroencephalographic activity during perception of motion in childhood. Eur. J. Neurosci. 13 (9), 1791–1796.

Coull, J.T., Nazarian, B., Vidal, F., 2008. Timing, storage, and comparison of stimulus duration engage discrete anatomical components of a perceptual timing network. J. Cogn. Neurosci. 20 (12), 2185–2197.

Cross, E.S., Hamilton, A.F., Grafton, S.T., 2006. Building a motor simulation de novo: observation of dance by dancers. Neuroimage 31 (3), 1257–1267.

Cross, E.S., Kraemer, D.J., Hamilton, A.F., Kelley, W.M., Grafton, S.T., 2009. Sensitivity of the action observation network to physical and observational learning. Cereb. Cortex 19 (2), 315–326.

Crossman, E.R.F.W., 1959. A theory of the acquisition of speed-skill. Ergonomics 2 (2), 153–166.

Dicks, M., Uehara, L., Lima, C., 2011. Deception, individual differences and penalty kicks: implications for goalkeeping in association football. Int. J. Sports Sci. Coach. 6 (4), 515–521.

di Pellegrino, G., Fadiga, L., Fogassi, L., Gallese, V., Rizzolatti, G., 1992. Understanding motor events: a neurophysiological study. Exp. Brain Res. 91 (1), 176–180.

Dosseville, F., Laborde, S., Raab, M., 2011. Contextual and personal motor experience effects in judo referees' decisions. Sport Psychol. 25 (1), 67–81.

Ericsson, K.A., Kintsch, W., 1995. Long-term working memory. Psychol. Rev. 102 (2), 211–245.

Ericsson, K.A., Krampe, R.T., Tesch-Romer, C., 1993. The role of deliberate practice in the acquisition of expert performance. Psychol. Rev. 100 (3), 363–406.

Eskenazi, T., Grosjean, M., Humphreys, G.W., Knoblich, G., 2009. The role of motor simulation in action perception: a neuropsychological case study. Psychol. Res. 73 (4), 477–485.

Fadiga, L., Fogassi, L., Pavesi, G., Rizzolatti, G., 1995. Motor facilitation during action observation: a magnetic stimulation study. J. Neurophysiol. 73 (6), 2608–2611.

Farrow, D., Abernethy, B., 2003. Do expertise and the degree of perception-action coupling affect natural anticipatory performance? Perception 32 (9), 1127–1139.

Filimon, F., Rieth, C.A., Sereno, M.I., Cottrell, G.W., 2015. Observed, executed, and imagined action representations can be decoded from ventral and dorsal areas. Cereb. Cortex 25 (9), 3144–3158.

Fox, N.A., Bakermans-Kranenburg, M.J., Yoo, K.H., Bowman, L.C., Cannon, E.N., Vanderwert, R.E., Ferrari, P.F., van IJzendoorn, M.H., 2016. Assessing human mirror activity with EEG mu rhythm: a meta-analysis. Psychol. Bull. 142 (3), 291–313.

Friston, K., 2011. What is optimal about motor control? Neuron 72, 488–498.

Frith, C.D., Blakemore, S.J., Wolpert, D.M., 2000. Abnormalities in the awareness and control of action. Philos. Trans. R. Soc. Lond. B Biol. Sci. 355 (1404), 1771–1788.

Gallese, V., Fadiga, L., Fogassi, L., Rizzolatti, G., 1996. Action recognition in the premotor cortex. Brain 119, 593–609.

Gangitano, M., Mottaghy, F.M., Pascual-Leone, A., 2001. Phase specific modulation of cortical motor output during movement observation. Neuroreport 12 (7), 1489–1492.

Gardner, T., Goulden, N., Cross, E.S., 2015. Dynamic modulation of the action observation network by movement familiarity. J. Neurosci. 35 (4), 1561–1572.

Ghazanfar, A.A., Schroeder, C.E., 2006. Is neocortex essentially multisensory? Trends Cogn. Sci. 10 (6), 278–285.

Gorman, A.D., Abernethy, B., Farrow, D., 2013. Is the relationship between pattern recall and decision-making influenced by anticipatory recall? Q. J. Exp. Psychol. 66 (11), 2219–2236.

Hamilton, A., Wolpert, D., Frith, U., 2004. Your own action influences how you perceive another person's action. Curr. Biol. 14 (6), 493–498.

Higuchi, S., Holle, H., Roberts, N., Eickhoff, S.B., Vogt, S., 2012. Imitation and observational learning of hand actions: prefrontal involvement and connectivity. Neuroimage 59 (2), 1668–1683.

Hodges, N.J., Campagnaro, P., 2012. Physical guidance research: assisting principles and supporting evidence. In: Hodges, N.J., Williams, A.M. (Eds.), Skill Acquisition in Sport: Research, Theory and Practice, second ed. Routledge, New York, NY, pp. 150–169.

Hodges, N.J., Ste-Marie, D., 2013. Observation as an instructional method. In: Farrow, D., Baker, J., MacMahon, C. (Eds.), Developing Sport Expertise: Researchers and Coaches Put Theory Into Practice, second ed. Routledge, New York, NY, pp. 115–128.

Ikegami, T., Ganesh, G., 2014. Watching novice action degrades expert motor performance: causation between action production and outcome prediction of observed actions by humans. Sci. Rep. 4, 1–7.

Jackson, R.C., Warren, S., Abernethy, B., 2006. Anticipation skill and susceptibility to deceptive movement. Acta Psychol. (Amst.) 123 (3), 355–371.

Jackson, R.C., van der Kamp, J., Abernethy, B., 2008. Experts do, experts see? Common coding versus perceptual experience in anticipation skill. J. Sport Exerc. Psychol. 30 (S1), S95.

Jeannerod, M., 2001. Neural simulation of action: a unifying mechanism for motor cognition. Neuroimage 14 (1), S103–S109.

Jin, H., Xu, G., Zhang, J.X., Gao, H., Ye, Z., Wang, P., Lin, H., Mo, L., Lin, C.D., 2011. Event-related potential effects of superior action anticipation in professional badminton players. Neurosci. Lett. 492 (3), 139–144.

Kilner, J.M., Blakemore, S.J., 2007. How does the mirror neuron system change during development? Dev. Sci. 10 (5), 524–526.

Kilner, J.M., Friston, K.J., Frith, C.D., 2007. Predictive coding: an account of the mirror neuron system. Cogn. Process. 8 (3), 159–166.

Kim, Y.-T., Seo, J.-H., Song, H.-J., Yoo, D.-S., Lee, H.J., Lee, J., Lee, G., Kwon, E., Kim, J.G., Chang, Y., 2011. Neural correlates related to action observation in expert archers. Behav. Brain Res. 223 (2), 342–347.

Knoblich, G., Flach, R., 2001. Predicting the effects of actions: interactions of perception and action. Psychol. Sci. 12 (6), 467–472.

Knoblich, G., Seigerschmidt, E., Flach, R., Prinz, W., 2002. Authorship effects in the prediction of handwriting strokes. Q. J. Exp. Psychol. 55A, 1027–1046.

Lee, T.D., Wishart, L.R., 2005. Motor learning conundrums (and possible solutions). Quest 57 (1), 67–78.

Lepage, J.-F., Théoret, H., 2006. EEG evidence for the presence of an action observation-execution matching system in children. Eur. J. Neurosci. 23 (9), 2505–2510.

Lepage, J.-F., Théoret, H., 2007. The mirror neuron system: grasping others' actions from birth? Dev. Sci. 10 (5), 513–523.

Leuthold, H., Sommer, W., Ulrich, R., 2004. Preparing for action: inferences from CNV and LRP. J. Psychophysiol. 18 (2–3), 77–88.

Loula, F., Prasad, S., Harber, K., Shiffrar, M., 2005. Recognizing people from their movement. J. Exp. Psychol. Hum. Percept. Perform. 31 (1), 210–220.

Mann, D.T., Williams, A.M., Ward, P., Janelle, C.M., 2007. Perceptual-cognitive expertise in sport: a meta-analysis. J. Sport Exerc. Psychol. 29 (4), 457–478.

Mann, D.L., Abernethy, B., Farrow, D., Davis, M., Spratford, W., 2010b. An event-related visual occlusion method for examining anticipatory skill in natural interceptive tasks. Behav. Res. Methods 42 (2), 556–562.

Mann, D.L., Abernethy, B., Farrow, D., 2010a. Action specificity increases anticipatory performance and the expert advantage in natural interceptive tasks. Acta Psychol. (Amst) 135 (1), 17–23.

Mann, D., Dicks, M., Cañal-Bruland, R., van der Kamp, J., 2013. Neurophysiological studies may provide a misleading picture of how perceptual-motor interactions are coordinated. I-Perception 4 (1), 78–80.

Marshall, P.J., Bouquet, C.A., Shipley, T.F., Young, T., 2009. Effects of brief imitative experience on EEG desynchronization during action observation. Neuropsychologia 47 (1), 2100–2106.

Marshall, P.J., Young, T., Meltzoff, A.N., 2011. Neural correlates of action observation and execution in 14-month-old infants: an event-related EEG desynchronization study. Dev. Sci. 14 (3), 474–480.

Miall, R.C., 2003. Connecting mirror neurons and forward models. Neuroreport 14 (17), 2135–2137.

Miall, R.C., Wolpert, D.M., 1996. Forward models for physiological motor control. Neural Netw. 9 (8), 1265–1279.

Miall, R.C., Stanley, J., Todhunter, S., Levick, C., Lindo, S., Miall, J.D., 2006. Performing hand actions assists the visual discrimination of similar hand postures. Neuropsychologia 44 (6), 966–976.

Michael, J., Sandberg, K., Skewes, J., Wolf, T., Blicher, J., Overgaard, M., Frith, C.D., 2014. Continuous theta-burst stimulation demonstrates a causal role of premotor homunculus in action understanding. Psychol. Sci. 25 (4), 963–972.

Milton, J., Solodkin, A., Hlustik, P., Small, S.L., 2007. The mind of expert motor performance is cool and focused. Neuroimage 35 (2), 804–813.

Molenberghs, P., Cunnington, R., Mattingley, J.B., 2012. Brain regions with mirror properties: a meta-analysis of 125 human fMRI studies. Neurosci. Biobehav. Rev. 36 (1), 341–349.

Mori, S., Shimada, T., 2013. Expert anticipation from deceptive action. Atten. Percept. Psychophys. 75 (4), 751–770.

Mukamel, R., Ekstrom, A.D., Kaplan, J., Iacoboni, M., Fried, I., 2010. Single-neuron responses in humans during execution and observation of actions. Curr. Biol. 20 (8), 750–756.

Mulligan, D., Hodges, N.J., 2014. Throwing in the dark: improved prediction of action outcomes following motor training without vision of the action. Psychol. Res. 78 (5), 692–704.

Mulligan, D., Hodges, N.J., 2016. Evidence against the automaticity of motor simulation in action prediction: separately acquired visual-motor and visual representations can be used flexibly to aid in prediction accuracy. J. Exerc. Mov. Sport 48 (1).

Mulligan, D., Lohse, K.R., Hodges, N.J., 2016a. An action-incongruent secondary task modulates prediction accuracy in experienced performers: evidence for motor simulation. Psychol. Res. 80 (4), 496–509.

Mulligan, D., Lohse, K.R., Hodges, N.J., 2016b. Evidence for dual mechanisms of action prediction dependent on acquired visual-motor experiences. J. Exp. Psychol. Hum. Percept. Perform. 42 (10), 1615–1626.

Munzert, J., Zentgraf, K., Stark, R., Vaitl, D., 2008. Neural activation in cognitive motor processes: comparing motor imagery and observation of gymnastic movements. Exp. Brain Res. 188 (3), 437–444.

Müsseler, J., Hommel, B., 1997. Blindness to response-compatible stimuli. J. Exp. Psychol. Hum. Percept. Perform. 23 (3), 861–872.

Muthukumaraswamy, S.D., Johnson, B.W., McNair, N.A., 2004. Mu rhythm modulation during observation of an object-directed grasp. Cogn. Brain Res. 19 (2), 195–201.

Nakata, H., Yoshie, M., Miura, A., Kudo, K., 2010. Characteristics of the athletes' brain: evidence from neurophysiology and neuroimaging. Brain Res. Rev. 62 (2), 197–211.

Newman-Norlund, R.D., van Schie, H.T., van Zuijlen, A.M.J., Bekkering, H., 2007. The mirror neuron system is more active during complementary compared with imitative action. Nat. Neurosci. 10 (7), 817–818.

Nyström, P., 2008. The infant mirror neuron system studied with high density EEG. Soc. Neurosci. 3 (3–4), 334–347.

Ong, N.T., Hodges, N.J., 2012. Mixing it up a little: how to schedule observational practice. In: Hodges, N.J., Williams, A.M. (Eds.), Skill Acquisition in Sport: Research, Theory and Practice, second ed. Routledge, New York, NY, pp. 22–39.

Orgs, G., Dombrowski, J.-H., Heil, M., Jansen-Osmann, P., 2008. Expertise in dance modulates alpha/beta event-related desynchronization during action observation. Eur. J. Neurosci. 27 (12), 3380–3384.

Pfurtscheller, G., Neuper, C., Andrew, C., Edlinger, G., 1997. Foot and hand area mu rhythms. Int. J. Psychophysiol. 26 (1–3), 121–135.

Pilgramm, S., Lorey, B., Stark, R., Munzert, J., Vaitl, D., Zentgraf, K., 2010. Differential activation of the lateral premotor cortex during action observation. BMC Neurosci. 11 (89), 1–7.

Pineda, J.A., 2005. The functional significance of mu rhythms: translating "seeing" and "hearing" into "doing". Brain Res. Rev. 50 (1), 57–68.

Pizzera, A., 2012. Gymnastic judges benefit from their own motor experience as gymnasts. Res. Q. Exerc. Sport 83 (4), 603–607.

Press, C., Cook, R., 2015. Beyond simulation: domain-general motor contributions to perception. Trends Cogn. Sci. 19 (4), 176–178.

Quandt, L.C., Marshall, P.J., Bouquet, C.A., Young, T., Shipley, T.F., 2011. Experience with novel actions modulates frontal alpha EEG desynchronization. Neurosci. Lett. 499 (1), 37–41.

Reis, J., Schambra, H.M., Cohen, L.G., Buch, E.R., Fritsch, B., Zarahn, E., Celnik, P.A., Krakauer, J.W., 2009. Noninvasive cortical stimulation enhances motor skill acquisition over multiple days through an effect on consolidation. Proc. Natl. Acad. Sci. U.S.A. 106 (5), 1590–1595.

Rizzolatti, G., Craighero, L., 2004. The mirror-neuron system. Annu. Rev. Neurosci 27, 169–192.

Rizzolatti, G., Fogassi, L., Gallese, V., 2001. Neurophysiological mechanisms underlying the understanding and imitation of action. Nat. Rev. Neurosci. 2 (9), 661–670.

Schiffer, A.-M., Ahlheim, C., Ulrichs, K., Schubotz, R.I., 2013. Neural changes when actions change: adaptation of strong and weak expectations. Hum. Brain Mapp. 34, 1713–1727.

Schubotz, R.I., 2007. Prediction of external events with our motor system: towards a new framework. Trends Cogn. Sci. 11 (5), 211–218.

Schubotz, R.I., 2015. Prediction and expectation. In: Toga, A.W. (Ed.), Brain Mapping: An Encyclopedic Reference, vol. 3, pp. 295–302.

Schubotz, R.I., von Cramon, D.Y., 2003. Functional–anatomical concepts of human premotor cortex: evidence from fMRI and PET studies. Neuroimage 20 (S1), S120–S131.

Schubotz, R.I., Wurm, M.F., Wittmann, M., von Cramon, D.Y., 2014. Objects tell us what action we can expect: dissociating brain areas for retrieval and exploitation of action knowledge during action observation in fMRI. Front. Psychol. 5, 1–15.

Schütz-Bosbach, S., Prinz, W., 2007. Perceptual resonance: action-induced modulation of perception. Trends Cogn. Sci. 11 (8), 349–355.

Smith, D.M., 2016. Neurophysiology of action anticipation in athletes: a systematic review. Neurosci. Biobehav. Rev. 60, 115–120.

Southgate, V., Johnson, M.H., Osborne, T., Csibra, G., 2009. Predictive motor activation during action observation in human infants. Biol. Lett. 5 (6), 769–772.

Stadler, W., Ott, D.V.M., Springer, A., Schubotz, R.I., Schütz-Bosbach, S., Prinz, W., 2012. Repetitive TMS suggests a role of the human dorsal premotor cortex in action prediction. Front. Hum. Neurosci. 6, 1–11.

Stevens, J.A., Fonlupt, P., Shiffrar, M., Decety, J., 2000. New aspects of motion perception: selective neural encoding of apparent human movements. Neuroreport 11 (1), 109–115.

Tomeo, E., Cesari, P., Aglioti, S.M., Urgesi, C., 2013. Fooling the kickers but not the goalkeepers: behavioral and neurophysiological correlates of fake action detection in soccer. Cereb. Cortex 23 (11), 2765–2778.

van Elk, M., van Schie, H.T., Hunnius, S., Vesper, C., Bekkering, H., 2008. You'll never crawl alone: neurophysiological evidence for experience-dependent motor resonance in infancy. Neuroimage 43 (4), 808–814.

Vogt, S., Buccino, G., Wohlschläger, A.M., Canessa, N., Shah, N.J., Zilles, K., Eickhoff, S.B., Freund, H.J., Rizzolatti, G., Fink, G.R., 2007. Prefrontal involvement in imitation learning of hand actions: effects of practice and expertise. Neuroimage 37 (4), 1371–1383.

Warreyn, P., Ruysschaert, L., Wiersema, J.R., Handl, A., Pattyn, G., Roeyers, H., 2013. Infants' mu suppression during the observation of real and mimicked goal-directed actions. Dev. Sci. 16 (2), 173–185.

Wilson, M., Knoblich, G., 2005. The case for motor involvement in perceiving conspecifics. Psychol. Bull. 131 (3), 460–473.

Wimshurst, Z.L., Sowden, P.T., Wright, M., 2016. Expert-novice differences in brain function of field hockey players. Neuroscience 315, 31–44.

Wohlschläger, A., 2001. Mental object rotation and the planning of hand movements. Percept. Psychophys. 63 (4), 709–718.

Wolpert, D.M., Flanagan, J.R., 2001. Motor prediction. Curr. Biol. 11 (18), R729–R732.

Wolpert, D.M., Doya, K., Kawato, M., 2003. A unifying computational framework for motor control and social interaction. Philos. Trans. R. Soc. Lond. B Biol. Sci. 358 (1431), 593–602.

Wright, M.J., Jackson, R.C., 2007. Brain regions concerned with perceptual skills in tennis: an fMRI study. Int. J. Psychophysiol. 63 (2), 214–220.

Wright, M.J., Bishop, D.T., Jackson, R.C., Abernethy, B., 2010. Functional MRI reveals expert-novice differences during sport-related anticipation. Neuroreport 21 (2), 94–98.

Wright, M.J., Bishop, D.T., Jackson, R.C., Abernethy, B., 2011. Cortical fMRI activation to opponents' body kinematics in sport-related anticipation: expert-novice differences with normal and point-light video. Neurosci. Lett. 500 (3), 216–221.

Wright, M.J., Bishop, D.T., Jackson, R.C., Abernethy, B., 2013. Brain regions concerned with the identification of deceptive soccer moves by higher-skilled and lower-skilled players. Front. Hum. Neurosci. 7, 1–15.

Yarrow, K., Brown, P., Krakauer, J.W., 2009. Inside the brain of an elite athlete: the neural processes that support high achievement in sports. Nat. Rev. Neurosci. 10, 585–596.

Zentgraf, K., Munzert, J., Bischoff, M., Newman-Norlund, R.D., 2011. Simulation during observation of human actions: theories, empirical studies, applications. Vision Res. 51 (8), 827–835.

Zhang, X., de Beukelaar, T.T., Possel, J., Olaerts, M., Swinnen, S.P., Woolley, D.G., Wenderoth, N., 2011. Movement observation improves early consolidation of motor memory. J. Neurosci. 31 (32), 11515–11520.

CHAPTER

Gunslingers, poker players, and chickens 1: Decision making under physical performance pressure in elite athletes

15

Beth L. Parkin*,†,1, Katie Warriner‡, Vincent Walsh†

Department of Psychology, University of Westminster, London, United Kingdom
†ICN, UCL, London, United Kingdom
‡Rugby Football Union, Twickenham, United Kingdom
1Corresponding author: Tel.: +44-20-7911-5000, e-mail address: b.parkin@westminster.ac.uk

Abstract

Background: The cognitive skills required during sport are highly demanding; accurate decisions based on the processing of dynamic environments are made in a fraction of a second (Walsh, 2014). Optimal decision-making abilities are crucial for success in sporting competition (Bar-Eli et al., 2011; Kaya, 2014). Moreover, for the elite athlete, decision making is required under conditions of intense mental and physical pressure (Anshel and Wells, 2000), yet much of the work in this area has largely ignored the highly stressful context in which athletes operate. A number of studies have shown that conditions of elevated pressure influence athletes' decision quality (Kinrade et al., 2015; Smith et al., 2016), response times (Hepler, 2015; Smith et al., 2016) and risk taking (Pighin et al., 2015). However, almost all of this work has been undertaken in nonelite athletes and participants who do not routinely operate under conditions of high stress. Thus, there is very little known about the influence of pressure on decision making in elite athletes.

Objective: This study investigated the influence of physical performance pressure on decision making in a sample of world-class elite athletes. This allowed an examination of whether findings from the previous work in nonelite athletes extend to those who routinely operate under conditions of high stress. How this work could be applied to improve insight and understanding of decision making among sport professionals is examined. We sought to introduce a categorization of decision making useful to practitioners in sport: gunslingers, poker players, and chickens.

Methods: Twenty-three elite athletes who compete and have frequent success at an international level (including six Olympic medal winners) performed tasks relating to three categories of decision making under conditions of low and high physical pressure. Decision making under risk was measured with performance on the Cambridge Gambling Task

(CGT; Rogers et al., 1999), decision making under uncertainty with the Balloon Analogue Risk Task (BART; Lejuez et al., 2002), and fast reactive responses and interference with the Stroop Task (Stroop, 1935). Performance pressures of physical exhaustion was induced via an exercise protocol consisting of intervals of maximal exertion undertaken on a watt bike.

Results: At a group level, under physical pressure elite athletes were faster to respond to control trials on the Stroop Task and to simple probabilistic choices on the CGT. Physical pressure was also found to increase risk taking for decisions where probability outcomes were explicit (on the CGT), but did not affect risk taking when probability outcomes were unknown (on the BART). There were no significant correlations in the degree to which individuals' responses changed under pressure across the three tasks, suggesting that elite athletes did not show consistent responses to physical pressure across measures of decision making. When assessing the applicability of results based on group averages to individual athletes, none of the sample showed an "average" response (within 1 SD of the mean) to pressure across all three decision-making tasks.

Conclusion: There are three points of conclusion. First, an immediate scientific point that highlights a failure of transfer of work reported from nonelite athletes to elite athletes in the area of decision making under pressure. Second, a practical conclusion with respect to the application of this work to the elite sporting environment, which highlights the limitations of statistical approaches based on group averages and thus the beneficial use of individualized profiling in feedback sessions. Third, the application of this work in a sports setting is described, in particular the development and implementation of a decision-making taxonomy as a framework to conceptualize and communicate psychological skills among elite sporting professionals.

Keywords
Decision making, Physical pressure, Elite athletes, Exercise, Exhaustion, Risk taking

1 INTRODUCTION
Optimal decision making is a crucial component of successful performance in sporting competition (Bar-Eli et al., 2011; Kaya, 2014). A significant factor often overlooked in previous work is the context in which decisions are made (Hepler, 2015). In particular, elite athletes routinely operate under a diverse array of mental and physical pressures. Commonly cited sources of acute stress include physical exhaustion, injury, the psychological impact of errors, negative feedback from the crowd, coaches and teammates, interpersonal conflict, rivalry, and the pressure to obtain highly valued awards (Anshel and Wells, 2000; Walsh, 2014). Given such a demanding environment, it is not uncommon for athletes to perform significantly below expectation despite high levels of motivation, a phenomenon termed "choking" (Jackson et al., 2013). Indeed, reduced performance when pressure is at its greatest has been shown even at an elite level. For instance, analysis of European Championship football penalty shootouts (from 1976 to 2004) revealed significant differences in performance under high vs low pressure. Players perform worse for penalty shots which would cost the team winning the game (high pressure, 62% success rate) when compared to penalty shots which would secure the team the win (low pressure, 92%

success rate) (Jordet and Hartman, 2008). Similarly, analysis of Professional Golfing Association tournament scores (from 1983 to 2010) showed that golfers played worse on the final round, when pressure is at its highest, compared to the penultimate round of a tournament (Wells and Skowronski, 2012).

There are a handful of studies that have explored decision making in the sporting domain under conditions of high pressure. These studies tend to examine nonelite athlete's decision quality and reaction times to sport-specific decision-making tasks (rather than the cognitive processes underlying decision making such as risk taking). In these studies, performance pressure is operationalized in a variety of ways that fall into broad categories of psychological and physical. Psychological performance pressure has been shown to impair task performance in nonelite athletes. For example, conditions of elevated social evaluation were reported to negatively impact the decision quality of novice basketball players but only in highly complex game scenarios (Kinrade et al., 2015). Moreover, mental exhaustion (induced by performance of the Stroop Task for 30min) was shown to impair decision accuracy and response times of nonelite athletes on a football specific decision-making task (Smith et al., 2016). Hepler (2015) compared mental and physical pressure in nonelite athletes and found that the time taken to generate decision outcomes was longer under conditions of mental stress (performance on a dual subtraction task), while conditions of physical exertion had no effect.

Other studies have focused on the physical performance pressure of intense physical exertion. A recent review paper examined the influence of physical load on perceptual and cognitive tasks in athletes of differing levels of expertise. This revealed that the inverted-U relationship between physical exertion and reaction times established in nonathletic samples, with moderate exertion inducing a facilitatory effect while high-intensity exertion inducing a detrimental effect on reaction times, did not extend to expert athletes. Expert athletes were found to show a general facilitation in response time measures under conditions of both moderate and intense physical pressure and were more positively affected than novice athletes. (Schapschröer et al., 2016). While the review had a broad inclusion criterion of an expert athlete, including all those competing at a national level, it importantly highlighted athletic expertise as a factor in determining the influence of physical pressure on indicators of performance.

There have been two studies to date that examined the influence of physical pressure on risk taking both undertaken in nonelite athletes. One found physical exertion to increase risk taking on decision making under uncertainty in a sample of male adolescent athletes (Black et al., 2013). The other reported physical exertion to induce an increase in risk taking in male and a decrease in risk taking in female athletes (Pighin et al., 2015).

These results are similar to that found in laboratory studies that have examined the precise influence that stress (operationalized as elevated cortisol levels) has on decision making (for a review see Starcke and Brand, 2012). The majority of these studies report increased risk taking under conditions of elevated stress for both decisions made when probability outcomes were explicit (Pabst et al., 2013; Starcke et al., 2008) and when probability outcomes were unknown (Lighthall et al., 2009; Preston et al., 2007; van den Bos et al., 2009). In line with Pighin et al. (2015) a number of other studies have reported an interaction with gender, where males show increased risk taking, while females show decreased risk taking, under conditions of acute stress (Lighthall et al., 2009, 2012; Preston et al., 2007).

Due to the difficulties of access to elite populations, most of the research examining decision making under pressure in sport has been undertaken with undergraduate students or nonelite athletes. By the very nature of training and selection of world-class abilities, elite athletes may perform differently. Unsurprisingly, expertise has been shown to affect decision making, with elite athletes making faster, more accurate decisions (Vaeyens et al., 2007). Furthermore, skilled athletes show quicker responses on simple choice reaction time tasks following acute physical exertion (Schapschröer et al., 2016). Therefore, elite athletes may show more resilience to the effects of performance pressure than is evident in the literature, as they are well equipped and practiced at operating under conditions of limited resources. On the other hand, elite athletes may also be subject to the detrimental effects of performance pressure, along with their more inexperienced counterparts. Indeed the presence of "choking" is well documented in elite athletic performances (Jordet and Hartman, 2008; Wells and Skowronski, 2012).

In light of previous work which (a) explored decision making in sport and has largely ignored the context in which decisions occur, and (b) has been mostly studied in nonelite participants, the aim of this study is to investigate how decision making is influenced by performance pressure in a sample of elite athletes. A key motivation of this research is to examine how it can provide useful insights for those working in elite sport.

In this study, world-class elite athletes undertook tasks assessing three categories of decision making under low and high physical performance pressure. We chose the tasks to reflect three kinds of decision: Fast, in the moment decisions which we call *Gunslinging*; probabilistic decisions which we call *Playing Poker*; and nonprobabilistic decisions which we call *Playing Chicken*. Probabilistic decision making was examined via performance of the CGT (Rogers et al., 1999), decision making under uncertainty via the BART (Lejuez et al., 2002), and fast reactive responses and interference via the Stroop Task (Stroop, 1935). Performance pressure was induced by a physical exhaustion protocol consisting of intervals of maximal exertion exercise on a watt bike. The results of the study will shed light on whether these different categories of decision making are influenced by physical pressure in elite athletes and therefore examine whether findings in the current literature transfer from nonelite to elite athletes. Additionally, the study will discuss the application of this work in a sports setting.

2 METHOD
2.1 PARTICIPANTS
The sample consisted of 23 world-class elite athletes (Swann et al., 2015) (12 males) aged 23–36 years (mean age: 28.25). All athletes belonged to a national training program for competition in the upcoming Olympic Games (Rio 2016). They all fell

within the 'world-class' and 'successful-elite' expertise categories defined by Swann et al (2015). To emphasize the caliber of the athletes, all had represented the United Kingdom at international championships and six were medal winners at London 2012 or Beijing 2008. The approximate average age of entry to this sport was 8 years; thus these elite athletes had approximately 20 years' experience in their given domain.

Recruitment occurred with the assistance of Team GB Sport Psychologists and Coaches during training camps. The squad was initially informed of the aims and procedures of the study during a group meeting. Upon expressing interest, a testing session was scheduled, during which the aims and procedures were reexplained and written informed consent obtained. While there was no financial compensation for participation, a cash prize was awarded to the top three performers to add a competitive element into task performance. Upon completion, in conjunction with Team GB Sport Psychologists, each athlete received a detailed debrief. The study and consent procedures were approved by the UCL ethics board in compliance with the principles of the Declaration of Helsinki.

2.2 PROTOCOL

A within-subject design was used, whereby the decision making of elite athletes was assessed under conditions of low and high physical pressure. Each participant was tested within a single session. Performance on three tasks was recorded; the BART (Lejuez et al., 2002), the CGT (Rogers et al., 1999), and the Stroop Task (Stroop, 1935). Initially participants were presented with instructions and undertook a short practice of each task. Participants then performed the decision-making tasks "at rest," i.e., in the absence of any additional performance pressure. Under conditions of physical performance pressure, participants performed these tasks immediately following a protocol of intense physical exertion on a watt bike (see Section 2.3). Performance of CGT always followed one of the physical exertion sessions, and performance of the BART and Stroop proceeded the other session (the latter two tasks being paired together as they were of shorter duration); the order in which the decision-making tasks were undertaken was counterbalanced across participants. After completion of the experimental procedures, participants were debriefed during a one-to-one session with the Sports Psychologist, Coach, and member of the research team. Here psychological concepts relating to decision making were discussed and applied to the sporting context, and athletes were also provided with a profile of their individual performance.

2.3 PHYSICAL PRESSURE INDUCTION

Physical pressure was induced via a protocol of maximal intensity exertion on an ergometer. Athletes completed two sessions of eight repetitions of 30s at maximal exertion (i.e., an all-out sprint), followed by 30s of recovery. Maximal exertion was derived from self-report, and athletes were given the following instruction: "You are required to cycle for 8 sequences of 30 second on and 30 second rest. We ask that you give maximum effort for each of the 30 second periods and recover afterwards." During these intense sprints, verbal encouragement was provided. During the protocol, participants pedaled a watt bike, with the resistance set by the individual athlete to a level that allowed maximal exertion. At the end of each session, participants immediately undertook the decision-making tasks. Two separate exercise sessions were undertaken to ensure that physical exhaustion was maintained across all tasks.

As a warm-up, athletes completed 5 min of low resistance exercise, which included at least two maximal intensity exertion sprints, to minimize the risk of injury. The 30 s duration of maximal intensity exertion was based on the Wingate test (Vandewalle et al., 1987), a protocol devised to measure anaerobic capacity. In developing this test, it was noted that participants were reluctant to endure protocols longer than 30 s at a time and when required to do so would initially reduce their maximal effort to save energy (Vandewalle et al., 1987). The number of repetitions was decided upon after discussion with Team GB Coaches and Sports Physiologists. The intent was to induce physical exhaustion in line with the type of physical exertion that the athlete endures during competition.

2.4 DECISION-MAKING TASKS

All decision-making tasks were delivered via a laptop computer, with a 17-in. display screen, and run via Inquisit software version 4.0.7.0 (Millisecond Software, Seattle, WA). Responses were made through mouse click or button press, and the software automatically recorded choice outcomes and response times for subsequent analysis.

2.4.1 Balloon analogue risk task

The BART (Lejuez et al., 2002; Fig. 1) is a standardized measure of risk taking under uncertainty. The task requires participants to inflate a series of computerized balloons for which the participant accrues money (5p per pump). However, the winnings from each balloon can only be kept if they are "banked" before the balloon bursts. Participants are faced with the decision, in light of not knowing the bursting point, when the optimal point is to stop inflating the balloon and transfer the winnings into a safe wallet. While participants did not receive cash equating to the final sum accumulated in the safe wallet, the top three highest scorers on the task (average of low- and high-pressure performance) received a cash prize. The participants' objective was therefore to obtain the largest amount of money on the task in order to win the cash prize. This method of reimbursement has been used in previous studies (Fecteau et al., 2007; Sela et al., 2012).

Here we used a shortened version of the task with 20 balloons. This version has been employed by a number of studies to date (e.g., Cheng and Lee, 2016; Derefinko et al., 2014; Ryan et al., 2013; Vaca et al., 2013) following the observation that there is no overall change in the measurements acquired.

The *average adjusted number of pumps* is the standard measure of risky decision making on the BART, which was used for the analysis in this study. This is the number of pumps for the balloons that did not burst and thus removes the variation that

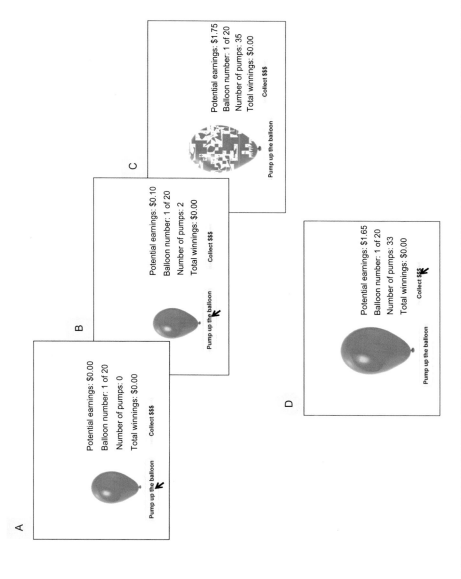

FIG. 1

The Balloon Analogue Risk Task. (A) Participants are required to inflate a balloon presented on screen by selecting the "pump up the balloon" button with the mouse. (B) With each pump 5p is earned, which is added to the potential earnings. (C) If the balloon bursts the potential earnings collected are lost and the next balloon appears. (D) Participants have to decide the optimal point at which to bank the potential earnings and move them to their total winnings, they do so by clicking the "collect $$$" button.

occurs as a result of the computer generated random explosion point. In the previous work, this variable was shown to be a strong predictor of real-world risk taking behaviors (Aklin et al., 2005; Crowley et al., 2006; Lejuez et al., 2003a,b, 2005).

2.4.2 Cambridge gambling task

The CGT (Rogers et al., 1999; Fig. 2) was used to assess decision making under risk, where information relating to the probability of different outcomes is explicit. The task displays a simple probabilistic decision where the participant is required to guess the location of a yellow token hidden in 1 of 10 boxes presented on screen. The boxes are colored either blue or red, and in any given trial the proportions of these vary across ratios of 9:1, 8:2, 7:2, and 6:4. Participants are required to choose where they think the token is hidden, in the red or the blue boxes, by clicking on rectangles at the bottom of the screen labeled red or blue. The participant is then given the opportunity

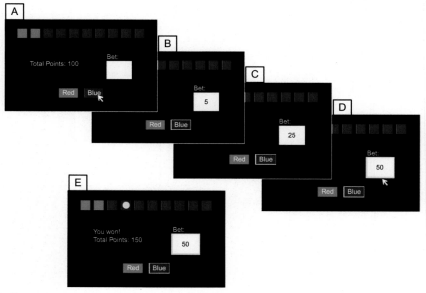

FIG. 2

The Cambridge Gambling Task. (A) The task displays a simple probabilistic decision whereby participants choose whether to look in the *red* or *blue box* for the *yellow token*. The proportions of *red* or *blue boxes* vary across ratios of 9:1, 8:2, 7:2, and 6:4. The participants clicks on the *red* or *blue icon* to indicate their response. (B) Following this a number appears in the *white box* which represents the amount possible to gamble on their decision. (C) In the ascending trial this amount increases every 3s. (D) When the *number* in the *box* represents the value the participants would like to gamble, they are required to click on it. (E) The position of the *yellow token* is then revealed to the participants: if correct the value is added to the total points, if incorrect the value is subtracted.

to select the number of points they wish to gamble. The amounts that can be bet appear as a number on screen, and they are always a proportion of the participant's total points and are presented in a sequence. There are two types of bet presentation, the ascending version, whereby the amount of points one can gamble starts small and increases in magnitude (from 5%, 25%, 75% to 95% of total points). Conversely, in the descending version the bets start large and decrease in magnitude (from 95%, 75%, 25% to 5%). There was an interval between the presentations of each point score of 3 s. The participant is required to select the number on screen when it represents the amount of points they wish to gamble on each trial. Following this, the location of the token is then displayed to the participant; if the participant is correct in their choice, the text "You Won" appears and the amount of points are added to their total. If the participant is incorrect, the text "You Lost" appears and the amount of points gambled are deducted from their total. Sound effects from the task were delivered via headphones, including beeps for each bet presentation, with a high pitch beep for an increasing bet, and lower pitch sound for a decreasing bet.

The task consisted of a total of 48 trials, whereby the 4 different probability types (1:9, 2:8, 3:7, 4:6) were presented 12 times in a predefined pseudo-randomized order, half the time there were more red, and half the time more blue boxes. The trials were presented in blocks, with eight blocks in total each containing six trials. In each block, participants began with 100 points, and the points accumulated were reset at the end of each block. The location of the token was pseudo-randomly determined, whereby one in every six trials the better choice (the color with the highest probability) led to a loss. There were four blocks with ascending points (a total of 24 trials, with each trial type being presented 6 times) presented consecutively and three blocks with descending points. The order of ascending or descending blocks was counterbalanced across participants, but kept the same within participants across repetition of the task. Due to time constraints when testing with a specialized sample, the task used in this instance consisted of a fewer number of trials than presented in the standard version of the task, which has a total of 72 trials (Rogers et al., 1999).

The task is administered via standardized instructions (outlined in Manes et al., 2002; Rogers et al., 1999). In short, the task procedures were outlined to the participant, along with the instruction to collect as many points as possible. There were four practice trials for participants to familiarize themselves with the task, two of these consisted of the decision phase only, and the other two mimicked the task.

There are four indices of task performance that were used in the analysis. In the initial decision-making phase the *response time* is measured as the duration between when the trial is presented and when the participant indicate their choice to look in the red or blue box via a mouse click. This measure is expected to interact with the probability ratio of the trial; in particular, deliberation time is likely to be less in trials with 9:1 ratios, in comparison to 6:4 trials. The *error rates* represent the quality of the decision, measured by the proportion of trials where participants chose the most likely box color. A winning choice of blue is only counted as a correct decision if blue represents the most number of boxes in the trial.

In the second phase, the performance measure the *mean percentage points* gambled represents the degree of risk taking on the task, with high-risk takers gambling a larger number of points. The task allows the disassociation of risk taking and motor impulsivity. Motor impulsivity can be derived by examining the difference between the amount bet on ascending trials and that bet on descending trials. In ascending trials, participants have to patiently wait for the appearance of a more risky bet, whereas in the descending trials participants can make risky bets immediately. Therefore, participants high in impulsivity will bet an amount that occurs early in the sequence: for ascending trials bets will be small and for descending trials bets will be large. There was a decreased duration in point presentation in this version of the task (3 s) compared to that developed by Rogers et al. (1999) (5 s).

Lastly, a measure of *risk adjustment* can be deduced from the amount gambled across different trial ratios and quantifies participants' ability to vary their risk taking in response to task contingencies. Optimal behavior on the task is where larger bets are made on trials where there is a higher likelihood of winning (i.e., those with the odds ratio 9:1), in comparison to trials where the likelihood of winning is lower (i.e., those with the odds ratio 6:4). Risk adjustment is calculated as the degree to which risk taking differed between ratios and is calculated in a manner designed to be independent from the total amount gambled (Rogers et al., 1999). It is calculated using the following equation: $((2 * \% \text{ bet at } 9{:}1) + (\% \text{ bet at } 8{:}2) - (\% \text{ bet at } 7{:}3) - (2 * \% \text{ bet at } 6{:}4))/\text{average } \% \text{ bet}$. A score of 0 represents no risk adjustment, whereby participants do not adjust their bets according to the different betting ratios. This is thought to indicate a failure to use information relating to the decision (Manes et al., 2002; Rogers et al., 1999).

2.4.3 Stroop task

The Stroop Task (first presented by Stroop, 1935) (Fig. 3) is a widely researched measure of interference and processing speed (for a review, see MacLeod, 1991). In this task, participants were instructed to name the color of items presented on screen by pressing a corresponding key on the keyboard (d for red, f for green, j for blue, and k for black).

The stimuli presented consisted of color words (red, blue, green, and black) presented in red, blue, green, and black ink, as well as solid rectangle blocks in these same colors on a white background. There were three trial types: congruent trials, where the word and the ink color are the same (i.e., the word "red" written in red ink); incongruent trials, where the word and ink color are different (i.e., the word "red" written in blue ink); and control trials, which simply measure reaction times to identify a solid block of color. The Stroop interference effect refers to the increased amount of time it takes to name the color of a word when the ink color and the word are incongruent, compared to when the ink color and the word are

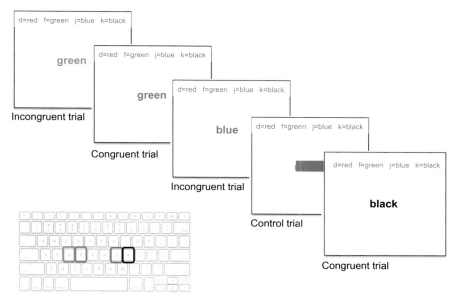

FIG. 3

The Stroop Task. Participants are required to identify the color of the word presented to them by a button press: d key for red, f key for green, j key for blue, and the k key for black (highlighted on the keyboard). A congruent trial is when the ink color is the same as the word (i.e., green written in green ink) and an incongruent trial is when the ink color is different to the word meaning (i.e., green written in red ink). A control trial presents a solid block of color.

congruent. This is thought to result from the automatic access of word naming being overridden in incongruent trials (MacLeod, 1991).

The task consisted of a total of 84 trials, the order of which was randomized. There were 28 congruent trials and 28 incongruent trials, with each of the 4 color words being presented 7 times. For the incongruent trials, each color word was presented in the three different colors twice (i.e., red presented in blue ink, green ink, and black ink), and one of the color–word pairing (randomly selected) was presented an additional time. There were also 28 control trials, which presented solid blocks of color, and again each of the 4 colors was presented 7 times.

The task was self-paced, whereby the stimuli remained on screen until participants made a response. Participants were instructed to make their responses as quickly and accurately as possible. There were two measures used in the analysis of this task: reaction times to control trials and the interference effect, measured as the reaction time to incongruent minus the reaction time to congruent stimuli. Before starting, participants undertook a short practice consisting of 12 trials (4 of each trial type) to familiarize themselves with the task.

2.5 ANALYSIS

To explore whether decision making on the three tasks is influenced by performance pressure, ANOVA and paired t-tests (two tailed) were undertaken to compare task performance under conditions of low and high physical pressure. Where relevant, Mauchly's test of sphericity was performed and Greenhouse–Geisser correction applied. Bonferroni correction was used to correct for multiple comparisons. For each task, the dependent variables used in this analysis are laid out below.

2.5.1 Balloon analogue risk task

The mean adjusted number of pumps was analyzed as a measure of risk taking under uncertainty. A three-way mixed ANOVA was undertaken to compare performance of the task under conditions of low and high physical pressure (within-subject factor of pressure: low pressure, high pressure) broken down by time (within-subject factor of balloon numbers: 1–10, 11–20). A between-subject factor of gender was also included (male, female).

2.5.2 Cambridge gambling task

The response times (the duration from the trial appearing on screen and the participant choosing to look in a red or blue box) and the error rates (the percentage of trials in which the participant chose the most likely box color) were analyzed using a repeated measures ANOVA. This was undertaken to compare the dependent variable (response time or error rates) under conditions of low and high physical pressure (within-subject factor of pressure: low pressure, high pressure) with the dependent variables broken down by the odd ratios presented in the trial (within-subject factor of odds ratio: 1:9, 2:8, 3:7, 4:6). This part of the analysis was collapsed across the ascending and descending trials as the presentation and selection of bets occurred after these variables were recorded.

In order to explore gambling behavior in probabilistic decision making the mean percentage number of points bet on the task was analyzed. This analysis included trials in which participants chose the most likely outcome in order to not confuse betting behavior and decision making. A three-way mixed ANOVA was used to compare the amount of points bet for ascending and descending trials (within-subject factor point presentation: ascending, descending) under conditions of high and low physical performance pressure (within-subject factor of pressure: low-performance pressure, high-performance pressure). A between-subject factor of gender was included (male, female).

Lastly, a measure of risk adjustment was derived from the data and a paired t-test was performed to compare the degree of risk adjustment under conditions of low and high physical pressure.

2.5.3 Stroop task
One participant was not included in the analysis due to experimenter error during data collection. Therefore, the analysis consisted of 22 participants. Paired *t*-tests were undertaken to compare reaction times on control trials and the Stroop interference effect under conditions of low and high physical pressure. Reaction times of correct responses only were included in the analysis.

2.5.4 Correlation analysis assessing response to pressure across tasks
It was also of interest to examine whether an individual's responses to pressure were consistent across the three decision-making tasks. The difference under pressure score was calculated for key indicators of task performance, by subtracting the score under high pressure from baseline. The variables chosen for indicators of task performance are the average adjusted number of pumps on the BART (a positive score represents better performance under pressure), the reaction time to control trials on the Stroop Task (a negative score represents better performance under pressure), and the risk adjustment score on the CGT (a positive score represents better performance under pressure). The difference under pressure scores for each task was compared, using the Pearson correlation coefficient, in order to examine whether participants showed a consistent response to pressure across tasks. The participant who did not complete the Stroop Task was excluded due to incomplete data; therefore, 22 participants were included in this analysis.

2.5.5 Applicability of group data to individuals
To explore the degree to which group data applies to individuals, the number of athletes whose change under pressure showed an "average" response across the three tasks were assessed. In accordance with the previous work an average response was calculated as a score that fell within 1 SD of the mean (0.5 SD above and below) (Daniels, 1952; Rose, 2016) for the average adjusted number of pumps on the BART, the reaction time to control trials on the Stroop Task, and the risk adjustment score on the CGT. Again, the participants who did not complete the Stroop Task were excluded due to incomplete data.

3 RESULTS
3.1 DOES PERFORMANCE ON THE BART CHANGE UNDER PRESSURE?
To assess whether decision making changes under conditions of low and high physical pressure performance on the BART, CGT and Stroop Tasks were examined. For the BART the mean adjusted number of pumps was analyzed as a measure of risk taking under uncertainty (Fig. 4). The ANOVA revealed no effect of gender ($F_{(1,21)}=1.33$, $P=0.26$) or no effect of performance pressure ($F_{(1,21)}=3.65$,

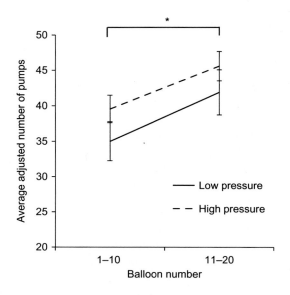

FIG. 4

BART results. In elite athletes, there was no significant effect of physical pressure on the mean adjusted number of pumps, a measure of risk taking under uncertainty. There was a significant effect of balloon number. The mean adjusted number of pumps was higher for the later part of the task, indicative of usual learning effects. *Statistical significance at $P<0.05$. Error bars represent the standard error of the mean.

$P=0.07$), while there was a significant effect of balloon number ($F_{(1,21)}=43.90$, $P>0.01$). In addition, there were no significant interactions between performance pressure and balloon number ($F_{(1,21)}=0.13$, $P=0.71$), performance pressure and gender ($F_{(1,21)}=2.17$, $P=0.15$), or balloon number and gender ($F_{(1,21)}=0.22$, $P=0.64$). Finally, there was no three-way interaction between balloon number, gender, and performance pressure ($F_{(1,21)}=0.01$, $P=0.91$). Hence, in this sample of elite athletes, physical exhaustion did not induce any significant changes in the degree of risk taking. However, it was found that the average adjusted number of pumps was higher for the later part of the task (mean: balloon numbers 1–10, 37.38 pumps; balloon numbers 11–20, 43.87 pumps); this reflects the usual learning effects found on this task.

3.2 DOES PERFORMANCE ON THE CGT CHANGE UNDER PRESSURE?
3.2.1 Response times
For the CGT a number of different performance measures were examined. First, a repeated measures ANOVA performed on the response times (Fig. 5A) revealed a significant effect of physical pressure ($F_{(1,22)}=5.63$, $P=0.03$), with the mean

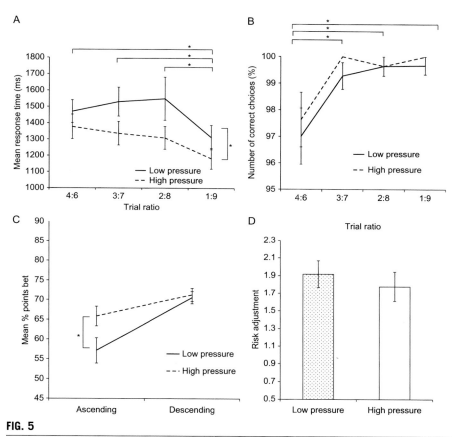

FIG. 5

CGT results. (A) The elite athletes showed significantly faster response times under physical pressure. (B) The elite athletes were less accurate on trials with odds ratios 4:6 compared to 3:7, 2:8, and 1:9. (C) Under physical pressure the elite athletes showed an increased amount of points gambled on ascending trials, indicating increased risk taking which is unlikely to be a result of increased impulsivity. (D) The degree of risk adjustment was unaffected by physical pressure. *Statistical significance at $P<0.05$. Error bars represent standard error of the mean.

deliberation time being less under conditions of high physical pressure (mean: low pressure, 1470.71 ms; high pressure, 1377.45 ms). In addition, we found a significant effect of trial ratio ($F_{(2.12,46.70)} = 8.66, P < 0.01$). Pairwise comparisons revealed that the response time to trials with 1:9 ratios was significantly quicker than that with 4:6 ($P < 0.01$), 3:7 ($P < 0.01$), and 2:8 ratios ($P < 0.01$). There was no significant interaction of trial ratio and physical pressure ($F_{(1.91,42.02)} = 1.34, P = 0.27$). This indicates that elite athletes tend to respond faster under physical pressure, and when the trial odds ratios were higher.

3.2.2 Error rates
A repeated measures ANOVA was performed on the error rates (Fig. 5B), and this revealed a nonsignificant effect of physical pressure ($F_{(1,22)}=0.95$, $P=0.34$) and a significant effect of trial ratio ($F_{(1.30,28.53)}=8.56$, $P<0.01$). There was a no significant interaction of ratio and physical pressure ($F_{(1.54,33.80)}=0.18$, $P=0.77$). Pairwise comparisons revealed that participants made more errors on trials with 4:6 ratios compared to those with 3:7 ($P<0.05$), 2:8 ($P<0.05$), and 1:9 ratios ($P<0.05$). Elite athletes were less accurate and opted for the most likely box color on fewer occasions, when the odds ratios were lower.

3.2.3 Number of points gambled
A mixed ANOVA on the mean percentage points gambled was analyzed as the main measure of risk taking (Fig. 5C). This revealed an effect of physical pressure ($F_{(1,21)}=9.08$, $P<0.01$), as well as an effect of point presentation (ascending or descending $F_{(1,21)}=8.98$, $P<0.01$), but no effect of gender ($F_{(1,21)}=0.89$, $P=0.36$). In addition, there was a significant interaction between physical pressure and point presentation ($F_{(1,21)}=9.45$, $P<0.01$), but neither between physical pressure and gender ($F_{(1,21)}=0.15$, $P=0.70$) nor between gender and point presentation ($F_{(1,21)}=4.02$, $P=0.53$). The three-way interaction between pressure, point presentation, and gender was also nonsignificant ($F_{(1,21)}=0.03$, $P=0.87$). Post hoc t-tests revealed that for ascending trials, there were a significantly higher number of points gambled under conditions of high physical pressure compared to low physical pressure ($t_{(22)}=-3.69$, $P<0.01$; mean: low pressure, 57.13%; high pressure, 65.86%). On descending trials, there were no significant differences in the number of points gambled under conditions of high and low physical pressure ($t_{(22)}=-0.43$, $P=0.67$; mean: low pressure, 70.62%; high pressure, 71.32%). Hence, athletes made significantly higher bets on ascending trials under physical pressure, indicating a significant increase in risk taking under physical pressure. On ascending trials the participant has to wait patiently for the number of points to increase; therefore, this increase in risk taking is unlikely to be due to increased motor impulsivity. Male and female elite athletes did not differ in terms of the effect that physical pressure has on the number of points bet on the CGT.

3.2.4 Risk Adjustment
A paired t-test revealed that there were no significant differences in the measure of risk adjustment between conditions of low and high physical pressure ($t_{(22)}=1.18$, $P=0.25$) (Fig. 5D); therefore the tendency for elite athletes to modify the amount bet according to the different reward and loss contingencies was not influenced by conditions of high physical pressure.

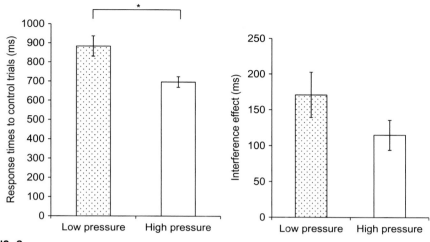

FIG. 6

Stroop Task results. The elite athlete was significantly faster responding to control trials under physical performance pressure. There were no significant differences under conditions of low and high pressure in the Stroop interference effect, i.e., the additional time taken to respond to incongruent trials compared to congruent trials. *Statistical significance at $P<0.05$. Error bars represent standard error of the mean.

3.3 DOES PERFORMANCE ON THE STROOP TASK CHANGE UNDER PRESSURE?

There were significant differences in the reaction times to identify the color of control trials under conditions of low and high physical pressure ($t_{(21)}=5.85, P<0.01$) (Fig. 6). The elite athletes performed significantly faster under conditions of physical pressure (mean: high physical pressure, 696.88 ms; low physical pressure, 883.62 ms). There was no significant difference in the Stroop interference effect under conditions of low- and high-performance pressures ($t_{(21)}=2.01, P=0.06$). Therefore, increased pressure did not influence the additional time taken to identify incongruent trials in comparison to congruent trials in elite athletes. The Stroop Task was performed with a high degree of accuracy under conditions of both high (97.17% accuracy) and low physical pressure (97.03% accuracy).

3.4 WERE INDIVIDUAL PERFORMANCE CHANGES UNDER PRESSURE CONSISTENT ACROSS TASKS?

To determine whether changes under pressure on key indicators of performance were consistent across the three decision-making tasks, correlation analyses were performed. However, no significant correlations between the degree of change under pressure on performance of the BART and CGT ($r_{(22)} = -0.12, P = 0.59$), BART and Stroop Task ($r_{(22)} = -0.11, P = 0.63$), or CGT and Stroop Task ($r_{(22)} = -0.06, P = 0.80$) were found. Therefore, individual participants' responses to pressure were not consistent across key indices of decision making over the three tasks. This highlights the importance of individualized profiling of results especially in a setting where the understanding the performance of the individual athlete is most

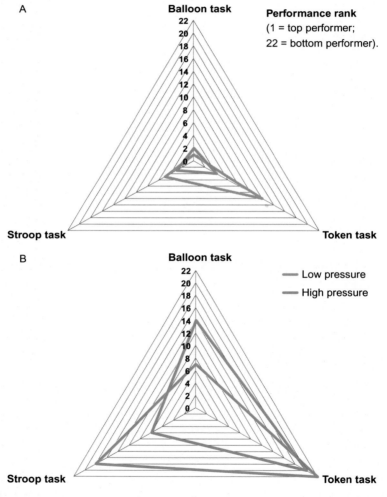

FIG. 7

Two examples of athletes' individual performance feedback. (A) Example athlete 1; (B) Example athlete 2. The radar graphs display individual performance on each task under low (blue line) and high (red line) pressure. 1 = top ranking performer; 22 = bottom ranking performer within the sample.

important. Fig. 7 shows example results of two athletes and example feedback. It is evident that within this group, there are individuals who showed very different patterns of responses to physical pressure. For example, Athlete 1 ranked highly within the cohort on the BART and Stroop and made small improvements in his/her rankings under pressure. On the CGT the athlete improved his/her rank from 12th to 4th place under pressure. Athlete example 2, however, showed a much more variable response to physical pressure; on the BART he/she moved down on the ranking under pressure from 6th to 14th place; on the Stroop Task he/she improved his/her ranking from 16th to 7th in the squad and the ranking on the CGT remained stable.

3.5 APPLICABILITY OF GROUP DATA TO INDIVIDUALS

In order to explore the degree to which group data apply to individual athletes, an additional analysis was undertaken whereby the number of athletes who showed an average response (defined as falling within 0.5 SD above and below the mean) to pressure across the three tasks was assessed. In line with the previous work (Daniels, 1952; Rose, 2016), this categorization of an "average" responder meant that at least the middle 35% of the group was categorized as falling within the "average" on each variable. The results of this analysis revealed that there was not one participant who showed mean responses to pressure across all three indices of decision making which included the average adjusted number of pumps on the BART, the reaction time to control trials on the Stroop Task, and the risk adjustment score on the CGT. As the group average scores were not representative of the behavior of a single individual athlete across the three measures, this highlights the importance of feeding back individual results in an elite setting where providing insight at an individual level is paramount (Fig. 7).

4 DISCUSSION

The aim of this study was to further our understanding about decision making in elite sport, by investigating the influence of physical pressure on decision making in a sample of world-class elite athletes. We also sought to introduce different categories of decision making to show that one's decision-making strengths and weaknesses at baseline may not, and indeed didn't, predict performance under physical pressure. The results revealed that under increased physical pressure, elite athletes showed faster response times. They also displayed an increase in risk taking for decisions where probability outcomes were explicit, while there were no significant differences in risk taking when probability outcomes were unknown.

In particular, reaction times were significantly faster following physical pressure on the Stroop Task and CGT. This observation coincides with the findings of a recent systematic review outlining the effects of physical exertion in athletes. Here, athletes at an expert level were reported to show an increased facilitation in response time measures on perceptual-cognitive tasks under conditions of moderate and intense physical activity, in comparison to novices (Schapschröer et al., 2016). In this review, expert athletes were broadly defined as those competing at a national level. It was proposed that the reaction time facilitation may be due to increases in exercise-induced arousal. Schapschröer et al. (2016) highlight that sporting expertise is a significant factor in how one responds to physical performance pressure, the findings of the current study suggests that this may extend to those competing at the highest elite standard. However, this conclusion is made in light of possible learning effects due to the experimental design in this study.

The decision making of elite athletes showed an increase in risk taking under physical pressure in response to decisions where the reward and loss outcomes were explicit. In particular, there was a significant increase in the amount bet on the CGT under physical pressure on trials with ascending presentation of points. This indicates that the increased propensity for risk was unlikely to be due to increased motor impulsivity as elite athletes were able to wait longer durations for the bet to increase. In line with this, risk adjustment did not change under conditions of increased physical pressure. Thus despite an increase in risk taking, elite athletes retained the ability to bet appropriately according to outcome probabilities and therefore use information relating to the decision scenario under physical pressure. For decisions made under uncertainty, there was no significant effect of physical pressure on risk taking.

Together, these findings partially support work that has shown risky decision making to be modulated under conditions of elevated stress as the previous literature has shown this to affect both decision making under uncertainty (Lighthall et al., 2009; Preston et al., 2007; van den Bos et al., 2009) and risk. Studies that examined the effect of physical exertion revealed that nonelite athletes showed increased risk taking on the BART in males (Black et al., 2013; Pighin et al., 2015) but decreased risk taking on the BART in females (Pighin et al., 2015). In this study, there were no effects of gender on measures of risk taking under pressure on either the BART or the CGT. In the case of elevated risk taking on the CGT, this was found to be similar for males and females, with the shift to more cautious responding under stress in females not evident in this sample of elite female athletes. Again, these discrepancies may allude to further differences between nonelite and elite athletes in terms of decision making under pressure.

On the one hand, the finding that elite athletes take more risks when the contingencies were known, but not unknown, could reflect differences in expertise in the two types of decision making. Decision making under uncertainty scenarios are much more prevalent in everyday life, and the dynamic sporting environment is no exception (Schonberg et al., 2011). As such, elite athletes are likely to have more experience in the specific coupling of decision making under uncertainty and acute physical pressure which may account for the robustness in performance seen in this study. An alternate explanation for the increase in risk taking specifically in situations where outcomes are explicit is one of calculable risk taking. Under pressure the elite athletes may be more willing to take risks, but only when they can deduce favorable chances of that risk paying off. This is in line with increased risk taking on ascending trials, and a resilient ability of risk adjustment, under physical pressure.

Together the findings highlight two important points. First, that physical exhaustion has an influence on the performance of elite athletes and therefore it is important to consider the context when studying the decision making of athletes. While on the whole elite performers' decision making was resilient to the effects of pressure, there were a speeding of reaction times and an increased propensity toward risk on probabilistic decision-making tasks. Second, that previous work in decision making and sport may not extend to elite athletes or those with a high level of expertise in operating under conditions of high pressure. For example, the results of this study support the observation that quicker reaction times were observed in elite athletes under intense physical exertion, yet this was not the case in studies undertaken in healthy volunteers (Schapschröer et al., 2016). The previous work in nonelite athletes has also shown an effect of gender on risk taking under physical pressure (Pighin et al., 2015), whereas we found no differences in performance between males and females. While direct comparisons are difficult due to divergent methodologies, together these findings raise concerns over the application of decision-making research undertaken in nonelite athletes to elite athletes.

As this work was undertaken outside of a laboratory context in a specialized sample group, it has high ecological validity. However, with this there was less control over the experiment and a number of compromises were made in terms of scientific rigor. One limitation is that the order of low and high physical pressure conditions was not counterbalanced. This meant that all elite athletes performed the tasks under conditions of low physical pressure first, followed by the tasks under condition of increased physical pressure. When performing tasks under physical pressure, participants were also undertaking the tasks for the second time, and thus the changes observed may reflect learning effects. Due to the time constraints of working with an elite population, it was not possible to design the experiment with separate testing sessions. In order to reduce possible learning effects in this study, attempts were made to choose appropriate tasks. For example, the BART has been shown to have high test–retest reliability, with small increases in risk taking (two average adjusted number of pumps) when repeated testing occurred within a single day (White et al., 2008). Moreover the CGT was designed to assess decision making in the absence of other learning processes, with trials reported to be relatively independent of one another (Rogers et al., 1999). These limits were considered a price worth paying precisely because athletes will be required to perform under pressure, skills they have practiced and learned previously.

A further interesting result of this study is that the degree to which participants changed under pressure did not correlate across key indicators of decision making over the three tasks. This indicates that the effects of physical pressure are not uniform across indices of decision making in individuals; for example, an athlete who improved on one task under pressure did not necessarily show similar improvements under pressure on a different task. This highlights that responses to physical stress are specific to the type of decision making, and also specific to the individual. This finding emphasizes a key issue when examining how one assesses and applies psychological data in an elite sporting environment—the central aim of this work—relating to the application of results based on group averages in a setting where understanding the behavior of the individual athlete is most important.

While statistical approaches based on group means used in this chapter are prevalent in almost all of psychology, and informative regarding the average response of a group of individuals, they may not be informative regarding any one individual (Rose, 2016). One example of this is a study by Daniels (1952) who studied 4063 pilots across 10 dimensions of size in order to design the optimal cockpit. He reported that within this sample, no one individual was average on all 10 dimensions (average defined as scores which fell within the middle 30% of the range of values) and only 3.5% of the sample showed average characteristics on three dimensions of size. Thus, if a cockpit were designed to fit the average individual, it would be a poor fit for any individual pilot. This has special relevance to elite sports psychology where understanding the behavior of the individual athlete is paramount. Indeed in this study, there was no one athlete that presented a mean response to pressure across all three indices of decision making. This indicates that applying statistical approaches based on group averages to the individual athlete would be incorrect in the majority of cases. Additionally these approaches disguises the unique profile of strength and weaknesses of an individual (Rose, 2016). In the case of the elite athletes, it may be that this individuality provides them with their competitive edge. Along these lines, in the current work feedback sessions were undertaken, whereby data relating to individual performance as well as group averages were provided (see Fig. 7).

Furthermore, to facilitate insight into decision making in elite sporting professionals a requirement was to increase understanding of key psychological principles and examine how to apply these in a sporting context. In order for these psychological concepts to have meaning in a sporting context, in debriefing discussions with athletes and coaches, examples of these types of decision making within sport were developed. In the application of this framework, a key challenge was in how best to communicate the different decision-making concepts to an audience with no prior knowledge of psychology; this centered on the use of analogies to represent concepts relevant to the type of decision making, which provided accessible, easily memorable terminology. Decision making under risk on the CGT was described as "poker playing," to represent a scenario in which one takes risks considering information about the reward and loss probabilities—i.e., the cards in their hand. This type of decision making was identified as most applicable to tactical decisions in sport. Regarding decision making under uncertainty on the BART, this was described as "playing chicken," with reference to a well-known game devised to test the nerve of each contender whereby two people drive straight toward each other; the first car to swerve (and, thus, miss a head on collision) is named the "chicken." As neither driver knows the behavior of the other, this type of risk taking was thought a relevant metaphor for decision making under uncertainty. In sport, this type of decision making was common to dynamic play, where athletes make decisions—i.e., decide who to pass to—with incomplete knowledge relating to opponents' positions and intentions. Fast reactive decision making was described as "gun slinging"; a shooting match whereby a person must shoot a target as quickly as possible but with the requirement to not hit innocent bystanders. These terms were key in in facilitating understanding that decision making is not a singular concept and of the different types of decision-making scenarios. It also provided an accessible shared language by which sport professionals could use to understand their own, and others' decision making, and communicate these constructs with one another.

5 CONCLUSION

The findings of this study show that the decision making of elite athletes was influenced by physical exhaustion. In particular under increased physical pressure elite athletes showed faster response times, increased risk taking for decisions where probability outcomes were explicit, but no significant change in risk taking when probability outcomes were unknown. These responses were different to those reported in the literature in nonelite athletes and healthy controls. In addition, individual changes in performance under pressure did not correlate across tasks, highlighting that response to physical pressure across tasks was specific to the individual athlete. The limitation of statistical approaches based on group means are highlighted as a key issue when examining how one applies psychological data in elite sport psychology. The development and implementation of a decision-making taxonomy provided a useful framework to conceptualize and communicate psychological skills among elite sporting professionals.

REFERENCES

Aklin, W.M., Lejuez, C.W., Zvolensky, M.J., Kahler, C.W., Gwadz, M., 2005. Evaluation of behavioral measures of risk taking propensity with inner city adolescents. Behav. Res. Ther. 43 (2), 215–228. http://dx.doi.org/10.1016/j.brat.2003.12.007.

Anshel, M., Wells, B., 2000. Sources of acute stress and coping styles in competitive sport. Anxiety Stress Coping. Retrieved from: http://www.tandfonline.com/doi/abs/10.1080/10615800008248331.

Bar-Eli, M., Plessner, H., Raab, M., 2011. Judgment, decision-making and success in sport. Retrieved from: https://books.google.co.uk/books?hl=en&lr=&id=IpsL6MKd5h0C&oi=fnd&pg=PT4&dq=Judgment,+Decision-Making+and+Success+in+Sport+&ots=EOnHiD-vX-&sig=9zvr9xpqxQrgP_Kiyaq6O4rW4Fg.

Black, A.C., Hochman, E., Rosen, M.I., 2013. Acute effects of competitive exercise on risk-taking in a sample of adolescent male athletes. J. Appl. Sport Psychol. 25 (2), 175–179.

Cheng, G.L.F., Lee, T.M.C., 2016. Altering risky decision-making: influence of impulsivity on the neuromodulation of prefrontal cortex. Soc. Neurosci. 11 (4), 353–364. http://dx.doi.org/10.1080/17470919.2015.1085895.

Crowley, T.J., Raymond, K.M., Mikulich-Gilbertson, S.K., Thompson, L.L., Lejuez, C.W., 2006. A risk-taking "set" in a novel task among adolescents with serious conduct and substance problems. J. Am. Acad. Child Adolesc. Psychiatry 45 (2), 175–183. http://dx.doi.org/10.1097/01.chi.0000188893.60551.31.

Daniels, G.S., 1952. The "Average Man"? (No. TN-WCRD-53-7). Air Force Aerospace Medical Research Lab. Wright-Patterson AFB, OH.

Derefinko, K.J., Peters, J.R., Eisenlohr-Moul, T.A., Walsh, E.C., Adams, Z.W., Lynam, D.R., 2014. Relations between trait impulsivity, behavioral impulsivity, physiological arousal, and risky sexual behavior among young men. Arch. Sex. Behav. 43 (6), 1149–1158. http://dx.doi.org/10.1007/s10508-014-0327-x.

Fecteau, S., Pascual-Leone, A., Zald, D.H., Liguori, P., Théoret, H., Boggio, P.S., Fregni, F., 2007. Activation of prefrontal cortex by transcranial direct current stimulation reduces appetite for risk during ambiguous decision making. J. Neurosci. 27 (23), 6212–6218. http://dx.doi.org/10.1523/JNEUROSCI.0314-07.2007.

Hepler, T., 2015. Decision-making in sport under mental and physical stress. International Journal of Kinesiology & Sports. Retrieved from: http://search.proquest.com/openview/db0918af03476fd663ac34020c77146c/1?pq-origsite=gscholar&cbl=2041008.

Jackson, R., Beilock, S.L., Kinrade, N.P., 2013. "Choking" in sport. In: Farrow, D., Baker, J., MacMahon, C. (Eds.), Developing Sport Expertise: Researchers and Coaches Put Theory Into Practice, 177–190. Routledge.

Jordet, G., Hartman, E., 2008. Avoidance motivation and choking under pressure in soccer penalty shootouts. J. Sport Exerc. Psychol. 30, 450–457.

Kaya, A., 2014. ScienceDirect decision making by coaches and athletes in sport. Procedia. Soc. Behav. Sci. 152, 333–338. http://dx.doi.org/10.1016/j.sbspro.2014.09.205.

Kinrade, N., Jackson, R., Ashford, K., 2015. Reinvestment, task complexity and decision making under pressure in basketball. Psychol. Sport Exerc. Retrieved from: http://www.sciencedirect.com/science/article/pii/S1469029215000308.

Lejuez, C.W., Read, J.P., Kahler, C.W., Richards, J.B., Ramsey, S.E., Stuart, G.L., et al., 2002. Evaluation of a behavioral measure of risk taking: the balloon analogue risk task (BART). J. Exp. Psychol. Appl. http://dx.doi.org/10.1037//1076-898X.8.2.75.

Lejuez, C.W., Aklin, W., Bornovalova, M., Moolchan, E.T., 2005. Differences in risk-taking propensity across inner-city adolescent ever- and never-smokers. Nicotine Tob. Res. 7 (1), 71–79. http://dx.doi.org/10.1080/14622200412331328484.

Lejuez, C.W., Aklin, W.M., Jones, H.A., Richards, J.B., Strong, D.R., Kahler, C.W., Read, J.P., 2003a. The balloon analogue risk task (BART) differentiates smokers and nonsmokers. Exp. Clin. Psychopharmacol. 11 (1), 26–33. http://dx.doi.org/10.1037/1064-1297.11.1.26.

Lejuez, C.W., Aklin, W.M., Zvolensky, M.J., Pedulla, C.M., 2003b. Evaluation of the Balloon Analogue Risk Task (BART) as a predictor of adolescent real-world risk-taking behaviours. J. Adolesc. 26 (4), 475–479. http://dx.doi.org/10.1016/S0140-1971(03)00036-8.

Lighthall, N.R., Mather, M., Gorlick, M.A., Jones, L., Yoon, T., Kim, J., et al., 2009. Acute stress increases sex differences in risk seeking in the balloon analogue risk task. PLoS One 4 (7), e6002. http://dx.doi.org/10.1371/journal.pone.0006002.

Lighthall, N.R., Sakaki, M., Vasunilashorn, S., Nga, L., Somayajula, S., Chen, E.Y., et al., 2012. Gender differences in reward-related decision processing under stress. Soc. Cogn. Affect. Neurosci. 7 (4), 476–484. http://dx.doi.org/10.1093/scan/nsr026.

MacLeod, C.M., 1991. Half a century of research on the Stroop effect: an integrative review. Psychol. Bull. 109 (2), 163. Retrieved from: http://psycnet.apa.org/journals/bul/109/2/163/.

Manes, F., Sahakian, B., Clark, L., Rogers, R., Antoun, N., Aitken, M., Robbins, T., 2002. Decision-making processes following damage to the prefrontal cortex. Brain 125 (Pt 3), 624–639. Retrieved from: http://www.ncbi.nlm.nih.gov/pubmed/11872618.

Mellalieu, S.D., Neil, R., Hanton, S., Fletcher, D., 2009. Competition stress in sport performers: stressors experienced in the competition environment. J. Sports Sci. 27 (7), 729–744. http://dx.doi.org/10.1080/02640410902889834.

Pabst, S., Schoofs, D., Pawlikowski, M., Brand, M., Wolf, O.T., 2013. Paradoxical effects of stress and an executive task on decisions under risk. Behav. Neurosci. 127 (3), 369–379. http://dx.doi.org/10.1037/a0032334.

Pighin, S., Savadori, L., Bonini, N., Andreozzi, L., Savoldelli, A., Schena, F., 2015. Acute exercise increases sex differences in amateur athletes' risk taking. Int. J. Sports Med. 94 (10), 858–863. Retrieved from: https://www.thieme-connect.com/products/ejournals/html/10.1055/s-0034-1398677.

Preston, S.D., Buchanan, T.W., Stansfield, R.B., Bechara, A., 2007. Effects of anticipatory stress on decision making in a gambling task. Behav. Neurosci. 121 (2), 257. http://dx.doi.org/10.1037/0735-7044.121.2.257.

Rogers, R.D., Owen, A.M., Middleton, H.C., Williams, E.J., Pickard, J.D., Sahakian, B.J., Robbins, T.W., 1999. Choosing between small, likely rewards and large, unlikely rewards activates inferior and orbital prefrontal cortex. J. Neurosci. 19 (20), 9029–9038. http://www.ncbi.nlm.nih.gov/pubmed/10516320.

Rose, T., 2016. The End of Average: How to Succeed in a World That Values Sameness. Penguin, UK.

Ryan, K.K., MacKillop, J., Carpenter, M.J., 2013. The relationship between impulsivity, risk-taking propensity and nicotine dependence among older adolescent smokers. Addict. Behav. 38 (1), 1431–1434.

Schapschröer, M., Lemez, S., Baker, J., Schorer, J., 2016. Physical load affects perceptual-cognitive performance of skilled athletes: a systematic review. Sports Med. 2 (1), 37. http://dx.doi.org/10.1186/s40798-016-0061-0.

Schonberg, T., Fox, C.R., Poldrack, R.A., 2011. Mind the gap: bridging economic and naturalistic risk-taking with cognitive neuroscience. Trends Cogn. Sci. 15 (1), 11–19. http://dx.doi.org/10.1016/j.tics.2010.10.002.

Sela, T., Kilim, A., Lavidor, M., 2012. Transcranial alternating current stimulation increases risk-taking behavior in the balloon analog risk task. Front. Neurosci. 6, 22. http://dx.doi.org/10.3389/fnins.2012.00022.

Smith, M.R., Zeuwts, L., Lenoir, M., Hens, N., De Jong, L.M.S., Coutts, A.J., 2016. Mental fatigue impairs soccer-specific decision-making skill. J. Sports Sci. 34 (14), 1297–1304. http://dx.doi.org/10.1080/02640414.2016.1156241.

Starcke, K., Brand, M., 2012. Decision making under stress: a selective review. Neurosci. Biobehav. Rev. 36 (4), 1228–1248. http://dx.doi.org/10.1016/j.neubiorev.2012.02.003.

Starcke, K., Wolf, O.T., Markowitsch, H.J., Brand, M., 2008. Anticipatory stress influences decision making under explicit risk conditions. Behav. Neurosci. 122 (6), 1352–1360. http://dx.doi.org/10.1037/a0013281.

Stroop, J.R., 1935. Studies of interference in serial verbal reactions. J. Exp. Psychol. 18 (6), 643–662. http://dx.doi.org/10.1037/h0054651.

Swann, C., Moran, A., Piggott, D., 2015. Defining elite athletes: issues in the study of expert performance in sport psychology. Psychol. Sport Exerc. 16, 3–14.

Vaca, F.E., Walthall, J.M., Ryan, S., Moriarty-Daley, A., Riera, A., Crowley, M.J., Mayes, L.C., 2013. Adolescent balloon analog risk task and behaviors that influence risk of motor vehicle crash injury. Ann. Adv. Automot. Med. 57, 77–88. Retrieved from: http://www.ncbi.nlm.nih.gov/pubmed/24406948.

Vaeyens, R., Lenoir, M., Williams, A.M., Mazyn, L., Philippaerts, R.M., 2007. The effects of task constraints on visual search behavior and decision-making skill in youth soccer players. J. Sport Exerc. Psychol. 29 (2), 147–169. http://dx.doi.org/10.1123/jsep.29.2.147.

van den Bos, R., Harteveld, M., Stoop, H., 2009. Stress and decision-making in humans: performance is related to cortisol reactivity, albeit differently in men and women. Psychoneuroendocrinology 34 (10), 1449–1458. http://dx.doi.org/10.1016/j.psyneuen.2009.04.016.

Vandewalle, H., Péerès, G., Monod, H., 1987. Standard anaerobic exercise tests. Sports Med. 4 (4), 268–289. Retrieved from: http://link.springer.com/article/10.2165/00007256-198704040-00004.

Walsh, V., 2014. Is sport the brain's biggest challenge? Curr. Biol. 24 (18), R859–R860. http://dx.doi.org/10.1016/j.cub.2014.08.003.

Wells, B., Skowronski, J., 2012. Evidence of choking under pressure on the PGA tour. Basic Appl. Soc. Psychol. 34 (2), 175–182. Retrieved from: http://www.tandfonline.com/doi/abs/10.1080/01973533.2012.655629.

White, T.L., Lejuez, C.W., de Wit, H., 2008. Test-retest characteristics of the Balloon Analogue Risk Task (BART). Exp. Clin. Psychopharmacol. 16 (6), 565–570. http://dx.doi.org/10.1037/a0014083.

CHAPTER

Gunslingers, poker players, and chickens 2: Decision-making under physical performance pressure in subelite athletes

16

Beth L. Parkin*,†,1, **Vincent Walsh**†

**University of Westminster, London, United Kingdom*
†*ICN, UCL, London, United Kingdom*
[1]*Corresponding author: Tel.: +44-20-7911-5000, e-mail address: b.parkin@westminster.ac.uk*

Abstract

Background: Having investigated the influence of acute physical exhaustion on decision-making in world-class elite athletes in Parkin et al. (2017), here a similar method is applied to subelite athletes. These subelite athletes were enrolled on a Team GB talent development program and were undergoing training for possible Olympic competition in 4–8 years. They differ from elite athletes examined previously according to expertise and age. While considered elite (Swann et al., 2015), the subelite athletes had approximately 8 years fewer sporting experience and were yet to obtain sustained success on the international stage. Additionally, the average age of the subelite sample is 20 years; thus, they are still undergoing the behavioral, cognitive, and neuronal changes that occur during the transition from late adolescence to young adulthood (Blakemore and Robbins, 2012). Previous work has used broad definitions of elite status in sport, and as such overlooked different categories within the spectrum of elite athletes (Swann et al., 2015). Therefore it is important to consider subelite athletes as a discrete point on the developmental trajectory of elite sporting expertise.

Objective: This work aims to investigate the influence of physical pressure on key indicators of decision-making in subelite athletes. It forms part of a wider project examining decision-making across different stages of the developmental trajectory in elite sport. In doing so, it aims to examine how to apply and develop psychological insights useful to an elite sporting environment.

Methods: 32 subelite athletes (18 males, mean age: 20 years) participated in the study. Performance across three categories of decision-making was assessed under conditions of low and high physical pressure. Decision-making under risk was measured with performance of the Cambridge Gambling Task (CGT; Rogers et al., 1999), decision-making under uncertainty with the Balloon Analogue Risk Task (BART; Lejuez et al., 2002), and fast reactive responses and inhibition via the Stop Signal Reaction Time Task (SSRT; Logan, 1994). Physical exhaustion was induced via intervals of maximal exertion exercise on a wattbike.

Results: Under pressure subelite athletes showed increased risk taking for both decisions where probability outcomes were explicit (on the CGT), and those where probability outcomes were unknown (on the BART). Despite making quicker decisions under pressure, with fewer errors, on the CGT, subelite athletes showed a reduced ability to optimally adjust betting behavior according to reward and loss contingencies. Fast reactive responses to perceptual stimuli and response inhibition did not change as a result of physical pressure. Individual responses to pressure showed a negative correlation in that a decrease in reaction times on the SSRT Task under pressure was associated with an increase in risk taking on the BART. When assessing the applicability of results based on group averages to individual athletes, 17% of the sample showed an "average" response (within 1 SD of the mean) to pressure across all three decision-making tasks.

Conclusion: Indicators of decision-making in a sample of subelite athletes are influenced by physical pressure, with a shift toward increased indiscriminate risk taking. The influence that physical pressure has on decision-making was different to that observed in world-class elite athletes; this highlights the importance of distinguishing between athletes at the elite level (Swann et al., 2015). The application of this work to a novel subgroup of elite athletes, including the implementation of a decision-making taxonomy, is discussed.

Keywords

Decision-making, Elite athletes, Physical pressure, Exercise, Exhaustion, Risk taking

1 INTRODUCTION

Having noted decision-making and performance under pressure as crucial factors to sporting success (Bar-Eli et al., 2011; Jordet and Hartman, 2008; Kaya, 2014; Wells and Skowronski, 2012), this paper forms part of a wider project examining these abilities across different developmental stages of elite sporting expertise (see Parkin and Walsh, 2017; Parkin et al., 2017). Using a similar protocol to that employed in Parkin et al. (2017) the influence of physical pressure on decision-making is examined in a sample of subelite athletes. These athletes were enrolled on a Team GB talent development program, training for competition at an Olympic level in 4–8 years. While these athletes make a living from sport and compete internationally, several hallmarks of elite status (Swann et al., 2015), they are yet to reach the highest levels of performance. Thus, it is important to consider them as a unique class of athlete to provide a more nuanced view of expertise at the elite level. Importantly, they differ from the world-class elite athletes studied previously according to two key factors, experience and age.

When it comes to the development of expertise, there is the widespread acceptance that it takes 10 years, or 10,000 h, of accumulated, deliberate practice to reach expert status within a given field (Ericsson et al., 1993). On average the subelite athletes have approximately 12 years of sporting expertise and are considered within the top 50 national players. In comparison to the world-class elite athletes, they have on average 8 years less sporting experience; a difference that is likely to impact decision-making competencies and responses to performance pressure. Unsurprisingly, previous work has shown that decision-making competencies within sport develop with expertise (Vaeyens et al., 2007). Moreover, a recent review revealed that sporting expertise modified responses to physical pressure. In particular following intense physical exercise, athletes with a higher level of expertise showed faster responding on simple choice reaction time tasks compared to novices (Schapschröer et al., 2016).

Almost all of the research examining decision-making and expertise has explored the differences between elite and nonelite athletes, rather than examining the spectrum of elite athletes. Indeed sports psychology has been criticized for its considerably broad definition of elite athlete status, ranging from Olympic champions, to those included in a regional or university sports team (Swann et al., 2015). This led Swann et al. (2015) to propose a categorization system to distinguish along the spectrum of expertise at an elite sporting level. The subelite athletes included in this sample would fall into *semi-elite* and *competitive-elite* expertise categories outlined by Swann et al. (2015), due to their inclusion of talent development programs, competitive success at a national level, and infrequent success at international competition, while the elite athletes included in Parkin et al. (2017) would fall into their *successful-elite* or *world-class elite* expertise category, due to their frequent appearance and sustained success in globally recognized competition.

In addition, the subelite athletes are distinct in terms of age. They are in their late adolescence and early 20s (average age 20), in contrast to the elite athletes who were in their late 20s (average age 28). This is relevant as the subelite athletes are undergoing cognitive changes and brain developments characteristic of adolescence, a process that does not cease until the mid-20s (Arain et al., 2013). Many of these developments relate to decision-making (for a review, see Blakemore and Robbins, 2012). In terms of behavior, this period is characterized by a tendency to engage increased risk taking in relation to adults (Defoe et al., 2015), likely to arise from a heighted responsiveness to incentives and increased influence of socioemotional factors (Blakemore and Robbins, 2012; Chein et al., 2011). In terms of anatomical brain development, regions that show late structural maturity include the DLPFC, an area involved with impulse control and weighing up the consequences of decision-making (Giedd, 2004).

Functional brain differences between early to late adulthood have also been observed. For example, Veroude et al. (2013) revealed that young adults (aged 23–25) and adolescents (aged 18–19) engage different brain regions when performing cognitive tasks. In particular when undertaking the Stroop Task, the young adults showed stronger activation in the DLPFC, left inferior frontal, left middle temporal gyrus, and middle cingulate, when compared to those in the adolescence group. Although such functional changes are yet to be fully deciphered, differences in the neural underpinnings of cognitive control are notable. These ongoing developmental changes reiterate the importance of considering subelite athletes as a discrete stage of elite sporting expertise.

In light of work that has highlighted the importance of considering the high-pressured context in which decisions in sport occur (Hepler, 2015; Kinrade et al., 2015; Pighin et al., 2015; Smith et al., 2016), the aim of the current study was to examine this topic in subelite athletes. The subelite athletes will perform tasks assessing decision-making under conditions of low and high physical performance pressure. Decision-making under risk will be assessed via performance of the CGT (Rogers et al., 1999), decision-making under uncertainty via the BART (Lejuez et al., 2003), and fast reactive responses and inhibition via the SSRT Task (Logan, 1994). Performance pressure was induced by physical exhaustion protocol consisting of intervals of maximal exertion exercise on a wattbike.

This will allow a greater insight into decision-making under pressure in subelite athletes, which may prove useful in understanding the developmental trajectory of expertise in elite sport. An additional aim is to examine how to apply this work in an elite sporting context, including the development of the decision-making taxonomy as a framework for conceptualizing decision-making in sport.

2 METHOD
2.1 PARTICIPANTS

The sample consisted of 31 subelite athletes (16 males), aged between 18 and 27 (mean age: 20.05 years). All athletes were part of a Team GB national centralized program focused on training for Olympic competition in 4–8 years, i.e., not the next Olympic games but the one after that, in this case Tokyo 2020. To emphasize the elite caliber of these athletes, they are considered as a top 50 national performer in their given sport. According to the Swann et al (2015) classification, these athletes would be defined as competitive-elite and semi-elite. All athletes included in the sample were from the same sport. The approximate age of entry for this sport is 8 years, and thus these subelite athletes had been training for approximately 12 years.

Recruitment and testing occurred during a 7-day residential training camp, with the assistance of Team GB sports psychologist and coaches. The squad was initially informed of the aims and procedures of the study via a group presentation, then a testing session was scheduled with athletes wishing to participate. While participants did not receive financial reimbursement for their participation, a cash prize was awarded to the top three performers on the BART. Upon completion a detailed debriefing session was undertaken in conjunction with Team GB coaches and a sport psychologist. The study and consent procedures were approved by the UCL ethics board, in compliance with the principles of the Declaration of Helsinki.

2.2 PROTOCOL

The protocol in this study was similar to that outlined in Parkin et al. (2017). A within-subject design was used, whereby the decision-making of subelite athletes was assessed under conditions of low and high physical pressure. Testing was undertaken within a single session, at a Team GB training facility. Performance on three decision-making tasks was examined via a laptop computer; these included the BART (Lejuez et al., 2002), the CGT (Rogers et al., 1999), and the SSRT Task (Logan, 1994). Initially participants received instructions and completed a short practice of each task. The subelite athletes then undertook these tasks at rest, i.e., in the absence of additional physical pressure. Following this, conditions of elevated physical pressure were induced via a protocol of intense physical exertion undertaken on a wattbike. Subelite athletes undertook two sessions of six repetitions of 30 s maximal exertion sprints, followed by 30 s of recovery. The CGT was performed following one of these sessions; the SSRT Task and BART followed the other. Taking this into account, the order in which the decision-making tasks were undertaken was counterbalanced across participants. After completion the subelite athletes were debriefed in a feedback session where the results of the study were discussed and applied to the sporting context. Athletes were also provided with a profile of their own performance.

2.3 PHYSICAL PRESSURE INDUCTION

A similar physical exertion protocol as described in Parkin et al. (2017) was undertaken to induce conditions of elevated physical pressure, adjusted to the fitness level of the subelite athletes. The protocol was undertaken on a wattbike. Athletes undertook two sessions of six repetitions of 30 s at maximal exertion (all out sprints), followed by 30 s of recovery. Maximal exertion was derived as a matter of self-report and athletes were instructed, "You are required to cycle for 6 sequences of 30 second on and 30 second rest. We ask that you give maximum effort for each of the 30 second periods and recover afterwards." The athlete was instructed to decide on an appropriate resistance level that reflected the usual parameters they use in training. Verbal encouragement and a 5 s countdown to each sprint were given to ensure high levels of motivation and adherence to the protocol. The aim was to induce physical exhaustion in accordance with the intermittent type of physical exertion that the athlete endures during competition. At the end of each session, participants quickly proceeded to perform the decision-making tasks. Two separate exercise sessions were undertaken to ensure that physical exhaustion was maintained across all decision-making tasks. Prior to the procedure, athletes undertook a warm up, consisting of low-resistance exercise and a number of maximal intensity exertion sprints to reduce risk of injury.

2.4 DECISION-MAKING TASKS

Tasks were delivered via a laptop computer, with a 17-in. display screen, run using Inquisit software version 4.0.7.0 (Millisecond Software, Seattle, WA); used to automatically record responses for subsequent analysis. Participants made responses via the use of a mouse or button press.

2.4.1 Balloon Analogue Risk Task

The BART (Lejuez et al., 2002) is a standardized measure of risk taking under uncertainty. The task requires participants to inflate a series of computerized balloons in order to accrue money (5p per pump). The winnings from each balloon can only be added to the total if they are "banked" before the balloon bursts. Participants do not know when the balloon will burst and must decide when to transfer winnings in order to obtain the highest amount of money. The degree to which the balloon is inflated, in particular the mean adjusted number of pumps, provides a measure of risk taking and is used for the analysis. This is the number of pumps for the balloons that did not burst, thus removing the variation that occurs as a result of the randomly generated explosion point.

Full task parameters and delivery is outlined in Parkin et al. (2017), Participants undertook a shorten version of the task with 20 balloons. This has been used by a number of studies to date (e.g., Cheng and Lee, 2016; Derefinko et al., 2014; Ryan et al., 2013; Vaca et al., 2013) and was employed to make testing as efficient as possible in an elite sample. While participants did not receive cash equating to the final sum accumulated in the safe wallet, the top three highest scorers on the task (average of low- and high-pressure performance) received a cash prize.

2.4.2 Cambridge Gambling Task

The CGT (Rogers et al., 1999) was used to assess decision-making under risk, where the probability of different outcomes is explicit. The task presents the participant with 10 boxes; these are colored red or blue according to different ratios (9:1, 8:2, 7:2, and 6:4). There is a token hidden in one of these boxes, and the participant must choose whether to look in the red or blue boxes to locate the token. The participant is then given the opportunity to select the amount of points they wish to gamble on their decision. The amounts that can be bet appear as a number on screen; they are a proportion of the participant's total points and are presented in either in an ascending (from 5%, 25%, 75% to 95% of total points) or descending (from 95%, 75%, 25% to 5%) sequence. The participant is required to click on the value when it represents the value they wish to gamble, if they are correct this value is added to the score and if they are incorrect this value is taken away from their score. The task is administered via standardized instructions which include a short practice of the task (outlined in Manes et al., 2002; Rogers et al., 1999). Further details of task parameters and delivery are described in Parkin et al. (2017).

Performance on the task is measured by a number of dependent variables which were used in the analysis of this task. *The response time* is measured as the duration from trial presentation to when the participant identifies whether they chose to look in the blue or red box. *The error rate* is measured as the proportion of trials whereby participants choose to look in the most likely box color. The *mean percentage points gambled* is used to represent the degree of risk taking; participants who show risky decision-making gamble a higher number of points. This is examined in the context of ascending or descending point presentations, to provide an indication of motor impulsivity. *Risk adjustment* was calculated from the amount gambled across different trial types, it quantifies a participants ability to vary their risk taking in response to task contingencies. Optimal behavior is where larger bets are made on trials where there is a higher likelihood of winning (i.e., those with the odds ratio 9:1), in comparison to trials where the likelihood of winning is lower (i.e., those with the odds ratio 6:4). Risk adjustment is calculated using the following equation: $((2 * \% \text{ bet at } 9:1) + (\% \text{bet at } 8:2) - (\% \text{bet at } 7:3) - (2 * \% \text{ bet at } 6:4))/\text{average } \% \text{ bet}$, in accordance with previous work (Manes et al., 2002; Rogers et al., 1999).

2.4.3 SSRT Task

The SSRT Task (Logan, 1994) was used to measure reaction times and response inhibition (Fig. 1). Response inhibition refers to the ability to suppress a response that is no longer required; it is an executive control process that allows behavior to be adapted in response to a dynamic environment (Logan, 1994; Verbruggen et al., 2008).

In this task, participants are instructed to respond as fast as possible to a "go" stimuli, in this case the appearance of an arrow. Participants are required to indicate whether the arrow is pointed to the left (by pressing the D key) or to the right (by pressing the Y key). Participants are instructed to inhibit their responses (and not make a button press) if the arrow appears alongside a "stop" signal, in this case a beep noise. The stop signal (beep) always appears after the go signal (arrow). The duration at which the beep is presented following presentation of the arrow is called the stop signal delay (SSD).

The task adjusts the SSD in a stepwise procedure, according to performance. When a participant is unsuccessful at responding to a stop signal and fails to inhibit their response, the SSD is decreased; therefore making it easier on the next trial. When the participant successfully responds to the stop signal and inhibits their button press, the SSD is increased; thus, increasing the difficulty on the next trial. The SSD

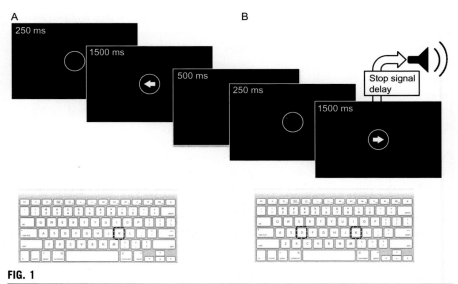

FIG. 1

The SSRT Task: (A) On *go trials*, an outline of a *circle* appears on screen (for 250 ms) to alert the participant to the presentation of the *arrow*. The *arrow* appears in the *circle*, which points either to the left or the right. If the *arrow* points to the left the participant must press the D key and if it points to the right they must press the K key. Following this a blank screen is presented for 500 ms until the next trial appears. (B) On *stop trials*, the presentation of the *arrow* is shortly followed by a beep. The beep indicates that no response to the *arrow* is required.

is initially set to a duration of 250 ms and is increased or decreased by 50 ms each time. This stepwise procedure allows the SSD duration at which the participant is able to withhold their response in half of the trials to be deduced. This is necessary to calculate the SSRT.

The SSRT refers to the time taken to inhibit the response provoked by the go signal. This is inferred indirectly and calculated as the average response time to go trials minus the SSD duration for which the participants are successful at withholding the response 50% of the time.

A standard version of the task was administered (as laid out in Verbruggen et al., 2008); this had a total number of 192 trials, of which 42 were stop trials and 150 go trials. These were presented in 3 blocks of 64 trials, each of which had 14 stop trials and 50 go trials. In half the trials the arrow pointed right and in the other half the arrow pointed left. For each block the order of the trial type (stop or go) or arrow direction (left or right) were randomized. The blocks were separated by a black screen containing a brief summary of performance; this was presented for 10 s, following which the task automatically proceeded to the next trial. The task started with a 10-trial practice to familiarize the participant. The instructions given to the participant were standardized and laid out in Verbruggen et al. (2008). Importantly, it was emphasized that the participant should not wait to respond to the go trials as the program adapts, nor should they worry if they are not successful as response inhibition was designed to be difficult. The two dependent variables on the task will be used in the analysis: the reaction time to go trials and the SSRT.

2.5 ANALYSIS

To explore whether decision-making on the three tasks is influenced by performance pressure, ANOVA or paired t-tests (two tailed) were undertaken to compare task performance under conditions of low and high physical pressure. Where relevant, Mauchly's test of sphericity was performed and Greenhouse–Geisser correction applied. Bonferroni correction was used to correct for multiple comparisons. For each task, the dependent variables used in this analysis are laid out below.

2.5.1 Balloon Analogue Risk Task
The mean adjusted number of pumps was analyzed as a measure of risk taking under uncertainty. A three-way mixed ANOVA was undertaken to compare performance of the task under conditions of low and high physical pressure (within-subjects factor of pressure: low pressure, high pressure) broken down by time (within-subject factor of balloon number: 1–10, 11–20). A between-subject factor of gender was included (male, female), due to previous work that has shown gender to influence risk taking under pressure.

2.5.2 Cambridge Gambling Task
The response times and the error rates were analyzed using a repeated measures ANOVA. This was undertaken to compare these variables under conditions of low and high pressure (within-subjects factor of physical pressure: low pressure, high pressure) broken down by the odd ratios presented in the trial (within-subject factor of odds ratio: 1:9, 2:8, 3:7, 4:6). This part of the analysis was collapsed across the ascending and descending trials as the presentation and selection of bets occurred after these variables were recorded.

In order to explore gambling behavior in decision-making under risk, the mean percentage number of points bet on the task was analyzed. This analysis included trials in which participants chose the most likely outcome in order to not confuse betting behavior and decision-making. A three-way mixed ANOVA was used to compare the amount of points bet for ascending and descending trials (within-subject factor point presentation: ascending, descending) under conditions of high and low physical pressure (within-subjects factor of pressure: low pressure, high pressure). A between-subject factor of gender was included (male, female) due to previous work that has shown gender to influence risk taking under pressure. Overall, higher gambles are indicative of increased risk taking; a large difference in the amount bet on ascending compared to descending trials is indicative of impulsivity.

Lastly, a measure of risk adjustment was derived from the data and a paired *t*-test was performed to compare the degree of risk adjustment under conditions of low and high physical pressure.

2.5.3 Stop Signal Reaction Time Task
One participant was excluded due to experimenter error (failure of audio presentation); therefore 30 participants were included in the analysis. Paired *t*-tests were undertaken to compare reaction times on go trials and the SSRT under conditions of low and high physical pressure.

2.5.4 Correlation analysis assessing response to pressure across tasks
The following analysis was undertaken with 30 participants; this number excludes the participant who has incomplete data on the SSRT Task. To examine whether an individual's responses to pressure were consistent across the three decision-making tasks, the difference under pressure score was calculated for key indicators of task performance by subtracting the score under high pressure from that under low pressure. The variables chosen as indicators of task performance are the average adjusted number of pumps on the BART (a positive score represents better performance under pressure); the reaction time to go trials on the SSRT Task (a negative score represents better (quicker) performance under pressure), and; the risk adjustment score on the CGT (a positive score represents better performance under pressure). The difference under pressure scores for each task were compared, using the Pearson Correlation Coefficient, to examine whether participants showed a consistent response to pressure across tasks.

2.5.5 Applicability of group data to individuals
To explore the degree to which group data applied to individuals, the number of athletes whose change under pressure showed an "average" response across the three tasks were assessed. In accordance with previous work, an average response was calculated as a score that fell within 1 SD of the mean (0.5 SD above and below) (Daniels, 1952; Rose, 2016) for the average adjusted number of pumps on the BART, the reaction time to go trials on the SSRT Task, and the risk adjustment score on the CGT. The participant who did not complete the SSRT Task was excluded due to incomplete data; therefore, 30 participants were included in this analysis.

3 RESULTS
3.1 DOES PERFORMANCE ON THE BART CHANGE UNDER PRESSURE?
To explore the influence that pressure has on the decision-making of subelite athletes, the performance of the BART, CGT, and SSRT Task under conditions of low and high physical pressure was assessed. For the BART, the mean adjusted number

of pumps was analyzed as a measure of risk taking under uncertainty (Fig. 2). The ANOVA revealed a significant effect of physical pressure ($F_{(1,29)} = 6.38, P = 0.02$) and a significant effect of balloon number ($F_{(1,29)} = 5.12, P = 0.03$), while there was no significant effect of gender ($F_{(1,29)} = 0.52, P = 0.47$). Pressure and balloon number ($F_{(1,29)} = 1.40, P = 0.25$), gender and pressure ($F_{(1,29)} = 0.01, P = 0.94$), or gender and balloon number ($F_{(1,29)} = 1.40, P = 0.25$). There was also no significant three-way interaction of pressure, balloon number, and gender ($F_{(1,29)} = 0.71, P = 0.41$). Hence, in this sample of subelite athletes physical pressure increased the average adjusted number of pumps; this is indicative of increased risk taking (mean: low pressure, 33.79 pumps; high pressure, 39.72 pumps). Also the average adjusted number of pumps were higher for the latter part of the task (mean: balloon number 1–10, 32.39 pumps; balloon number 11–20, 38.12 pumps); this reflects the usual learning effects found on this task. There were no differences in the risk taking of male and female subelite athletes as a result of increased physical pressure.

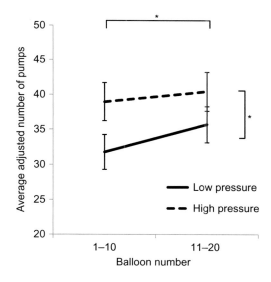

FIG. 2

BART results: in subelite athletes there was a significant increase in the average adjusted number of pumps under high physical pressure, indicative of increased risk taking. There was also a significant effect of balloon number, with the average adjusted number of pumps found to be larger for the last half of the task (indicative of the usual learning effects on this task). * denotes statistical significance at $P < 0.05$. Error bars represent standard error of the mean.

3.2 DOES PERFORMANCE ON THE CGT CHANGE UNDER PRESSURE?

3.2.1 Response times

Next, the influence of physical exhaustion on the CGT was examined. First, a repeated measures ANOVA was performed on response times (Fig. 3A). This revealed a significant main effect of physical pressure ($F_{(1,30)}=22.08$, $P<0.01$), with the mean deliberation time being less under conditions of high physical pressure (mean: low pressure, 2168.11 ms; high pressure, 1661.23 ms). In addition, there was also a significant effect of trial ratio ($F_{(3,90)}=8.67$, $P<0.01$); pairwise comparisons revealed that the response times to trials with 1:9 ratios were significantly quicker than compared to those with 4:6 ($P=0.04$) (mean: 1:9, 1665.82 ms; 4:6, 1960.50 ms) and to 2:8 ($P<0.01$) (mean: 2:8, 2170.23 ms). There was no significant interaction of trial ratio and physical pressure ($F_{(2.13,63.80)}=0.64$, $P=0.54$); this indicates that elite athletes tend to respond faster under pressure and when trial ratios were higher.

3.2.2 Error rates

A repeated measures ANOVA was performed on the error rates (Fig. 3B); this revealed a significant main effect of physical pressure ($F_{(1,30)}=4.49$, $P=0.04$) and a significant main effect of trial ratio ($F_{(2.06,61.67)}=1.33$, $P=0.27$). There was a nonsignificant interaction of ratio and physical pressure ($F_{(2.09,62.83)}=0.33$, $P=0.73$). Subelite athletes were less accurate, and opted for the most likely box color on fewer occasions, under conditions of low pressure (mean % correct: low pressure, 96.8%; high pressure, 99.1%).

3.2.3 Number of points gambled

A mixed ANOVA was performed on the mean number of points gambled (%) as a measure of risk taking (Fig. 3C). This revealed a significant effect of physical pressure ($F_{(1,29)}=39.16$, $P<0.01$) and a significant effect of point presentation (ascending or descending) ($F_{(1,29)}=47.73$, $P<0.01$). There was also a significant interaction of physical pressure and point presentation ($F_{(1,29)}=11.99$, $P<0.01$). Post hoc tests revealed that there were significantly more points gambled under high pressure on both ascending trials ($t_{(30)}=-7.09$, $P<0.01$) (mean: low pressure ascending, 50.84%; low pressure descending, 76.40%) and on descending trials ($t_{(30)}=-3.09$, $P<0.01$) (mean: high pressure ascending, 64.20%; high pressure descending, 82.20%).

Additionally, there was no significant main effect of gender on the number of points gambled ($F_{(1,29)}<0.01$, $P=0.97$) and no significant interactions of gender and other main effects; including physical pressure and gender ($F_{(1,29)}=1.02$, $P=0.32$) or points presentation and gender ($F_{(1,29)}=0.03$, $P=0.86$). There was also no significant three-way interaction of gender, point presentation, and physical pressure ($F_{(1,29)}=0.34$, $P=0.57$).

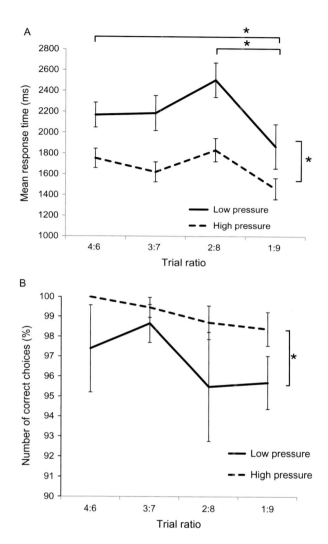

FIG. 3

Cambridge Gambling Task results: (A) the subelite athlete showed significantly faster response times under physical pressure. (B) Under physical pressure, subelite athletes made significantly fewer errors. (C) Under physical pressure the subelite athletes also showed increased amount of points gambled, indicating increased risk taking. There was a significant increase on both ascending and descending trials. (D) The degree of risk adjustment was significantly reduced under conditions of increased pressure. * denotes statistical significance at $P<0.05$. Error bars represent standard error of the mean.

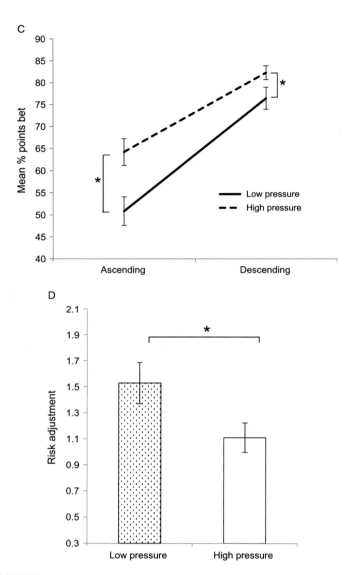

FIG. 3—CONT'D

3.2.4 Risk adjustment

A paired t-test revealed there was a significant difference in the measure of risk adjustment between conditions of low and high physical pressure ($t_{(30)} = 2.77$, $P < 0.05$) (Fig. 3D). The tendency for subelite athletes to modify the amount bet according to the different reward and loss contingencies was reduced under conditions of high physical pressure (mean risk adjustment score: low pressure, 1.52; high pressure, 1.10).

3.3 DOES PERFORMANCE ON THE SSRT TASK CHANGE UNDER PRESSURE?

Paired t-test results show there were no significant differences in the reaction time to respond to go trials under conditions of low and high physical pressure ($t_{(29)}=0.63$, $P=0.53$). There was also no significant difference in the SSRT under conditions of low and high physical pressure ($t_{(29)}=-0.06, P=0.95$) (Fig. 4). Therefore increased physical pressure did not influence reaction times to "go" trials or response inhibition processes as assessed by the SSRT.

3.4 WERE INDIVIDUAL PERFORMANCE CHANGES UNDER PRESSURE CONSISTENT ACROSS TASKS?

Pearson correlations were undertaken to examine whether individual responses to pressure were consistent across key indicators of performance of the three decision-making tasks. There was a significant correlation when comparing the difference under pressure performance on the BART and on the SSRT Task ($r_{(30)}=-0.62, P<0.01$); this showed that as reaction times decreased under pressure on the SSRT Task, the degree of risk taking under pressure on the BART increased (Fig. 5). There were no further significant correlations when comparing the degree of change under pressure on the BART and the CGT ($r_{(30)}=0.16, P=0.39$) or on the BART and SSRT Task ($r_{(30)}<0.01, P=0.99$).

3.5 APPLICABILITY OF GROUP DATA TO INDIVIDUALS

To explore the degree to which group data applied to individual athletes, the number of athletes who showed average responses (defined as 0.5 SD above and below the mean) to pressure across key performance indicators of decision-making tasks were

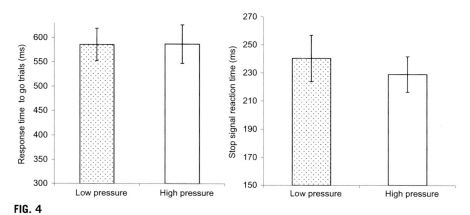

FIG. 4

SSRT Task results: the subelite athletes showed no significant differences under conditions of low and high pressure in the time taken to respond to "go" trials (*left*) or on the SSRT (*right*). *denotes statistical significance at $P<0.05$. Error bars represent standard error of the mean.

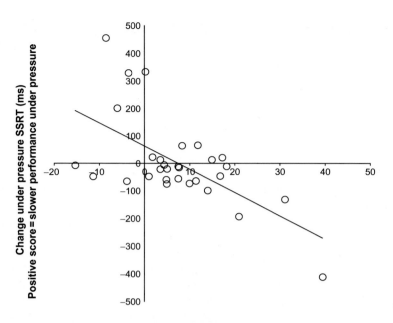

FIG. 5

There was a significant negative correlation when comparing the difference under pressure performance on the BART and the SSRT Task.

assessed. Inline with previous work (Daniels, 1952; Rose, 2016), this categorization of an "average" responder meant that at least the middle 43% of the group were categorized as falling within the "average" on each variable. The results of this analysis revealed that there was 16.67% of the sample were average responders across the average adjusted number of pumps on the BART, the reaction time to "go" trials on the SSRT Task, and the risk adjustment score on the CGT.

As the group average scores were not representative of the behavior of athletes across the three measures in the large majority of cases, this highlights the importance of feeding back individual results in an elite setting, where providing insight at an individual level is paramount.

4 DISCUSSION

The goal of the current study was to investigate the influence of physical exhaustion on key indicators of decision-making in subelite athletes. The study formed part of a wider project assessing decision-making across different developmental stages of elite sporting expertise and, in doing so, examined how to increase psychological insight in this environment. The main findings were that, under conditions of physical pressure, subelite athletes showed increased risk taking, as well as a reduced ability to modify behavior in line with explicit reward and loss contingencies. There was no change to fast reactive responses to perceptual stimuli and response inhibition under conditions of increased physical pressure.

The decision-making of subelite athletes showed an increase in risk taking under physical pressure, in response to both decision-making under risk and uncertainty. On the BART, there was an increase in the mean adjusted number of pumps under conditions of high pressure; which may be notable given the high correlation that performance on this task has with real-world risk taking behaviors (Aklin et al., 2005; Lejuez et al., 2003). On the CGT, while subelite athletes made fewer errors and were faster to respond to simple probabilistic decisions under physical pressure, they also opted to gamble a higher number of points. In this case the increase in risk taking was evident for both ascending and descending point presentation trial types. These findings support previous work that has shown modulations to risk taking following physical exertion (Black et al., 2013; Pighin et al., 2015). Inline with the notion that the physiological responses to exercise are akin to those observed under stress/arousal, the results also align with work that has shown risk taking to be affected by other sources of acute stress (for a review: Starcke and Brand, 2012).

One finding from previous work not supported by the current study is the influence that gender has on risk taking under pressure. In particular, males have been reported to show an increase, and females a decrease, in risk taking under pressure, in both nonelite athletes under physical exertion (Pighin et al., 2015) and nonathletic healthy samples following stress induction paradigms (Lighthall et al., 2009, 2012; Preston et al., 2007; van den Bos et al., 2009). In the current study, there was no effect of gender on measures of risk taking across two different decision-making tasks, and the behavior of male and female subelite athletes did not significantly differ from one another. The shift toward more cautious decision-making reported in female healthy volunteers and nonelite athletes was not evident in subelite female athletes.

On the CGT, the subelite athletes also showed a significant reduction in risk adjustment, the ability to gamble appropriately according to different probability ratios, under pressure. This measure is thought to reflect the degree to which the participant can use information relating to the decision to appropriately modify one's propensity for risk (Clark et al., 2003; Rogers et al., 1999). Together, the results of the CGT show an increase in risk taking (in the number of points bet) and a decrease in the responsiveness to optimal betting scenarios (reduced risk adjustment). Thus, such a pattern may indicate a suboptimal shift in risky decision-making under pressure in subelite athletes. In a sporting context this may be noteworthy; while reductions in risk taking have been linked to performance decrements in elite sport (Jordet and Hartman, 2008; Paserman, 2007), the relationship between risk taking and sporting performance is likely to be one of adaptability. It is clear that in some circumstances, taking risks would provide the athlete with the competitive edge; i.e., in a football match when the losing team substitutes a defender for an attacker as they enter injury time. While in others, the potential exposure to negative outcomes that risk entails means playing it safe is optimal; i.e., in a football match when the team which is in the lead decides to adopt a defensive strategy as they enter injury time.

In relation to reaction times, fast reactive responses to perceptual stimuli, i.e., the time taken to respond to go trials on the SSRT Task, did not change under increased physical pressure. There were also no differences in the time taken to inhibit responses measured by the SSRT. Considering the intense level of the physical exhaustion protocol undertaken by the subelite athletes the consistency in responding may be indicative of expertise. However, in consideration of previous research, the lack of significant improvements on these measures are noteable. In particular, a recent review in this area reported high-intensity levels of physical exertion to have facilitatory effects on response time measures on perceptual–cognitive tasks, but only in athletes with high levels of sporting expertise (Schapschröer et al., 2016). While the subelite athletes were within the top 50 national players in their chosen sport, faster responding under physical exertion were not evident. These differences may allude to the variation of abilities across elite samples which, due to the broad definition in elite status in the current review, could not be teased apart (Schapschröer et al., 2016).

At an individual level there was a consistent response to physical pressure across indices of performance on two of the decision-making tasks. There was a significant correlation of performance on the BART and the SSRT Task, with subelite athletes who show quicker reaction times to go trials under pressure also showing increases in risk taking on the BART. This indicates that subelite athletes may show a general orientation at an individual level in terms of responding to physical pressure.

Together the indiscriminate increases in risk taking and lack of facilitation in fast reactive responses to perceptual stimuli in subelite athletes may be indicative of reduced resilience to physical exhaustion. This is in comparison to world-class elite athletes who performed a similar protocol and showed faster reaction times in perceptual stimuli (although notably to a different task) and were not impaired on their abilities to adjust risk taking according to task contingencies under pressure (see Parkin et al., 2017). The differences here highlight that elite athletes are not a homogenous group and the importance of considering variation in expertise at the elite level (Swann et al., 2015). As well as expertise, the two groups of athletes were distinct in terms of age, which may also help explain the differences noted here. The subelite athletes who are in their late teens to early 20s are still undergoing the cognitive changes and brain developments characteristics of adolescence that are important in determining decision-making abilities (Blakemore and Robbins, 2012). In particular there is a heightened tendency to engage in increased risk taking during this time in relation to young adults (Defoe et al., 2015), which is inline with the findings from the current study.

A further goal of this study was to explore how psychology can be applied to improve insight into decision-making within the elite sporting environment. In accordance with Parkin et al. (2017), the results highlight the limitation of statistical approaches based on group averages. When examining the number of athletes that showed average responses to increased physical performance pressure, 16% of the sample presented a mean response to pressure across three indices of decision-making. Thus, in the large majority of cases, it would be incorrect to apply results based on group means to the individual athlete. Therefore, the feedback of individual results along with group means for comparison provided the most useful in a context where understanding the behavior of the individual is paramount.

Across feedback sessions, the taxonomy of decision-making was applied to this new sample. In line with previous work the taxonomy classification was based on the three types of decision-making assessed during the study. Accessible terminology was used to provide real-world analogies of each type of decision-making. In particular, probabilistic decision-making was described as "playing poker" and decision-making under uncertainty was described as "playing chicken" (with reference to a game devised to test the nerve of each contender, whereby two cars drive straight toward each other, and the first car to serve is named the "chicken"). Fast reactive decision-making was described as "gun slinging." These terms facilitated the development of a common language of decision-making to aid understanding and communication of decision-making among sporting professionals. The feedback sessions provided the athlete, coaches, and sport psychologist the opportunity to apply these concepts, and insights from undertaking the study, to the performance of an individual athlete. This meant that the application was mainly practitioner led, and in a number of cases decision-making interventions focused on a particular type of decision-making were developed.

5 CONCLUSION

The findings of the current study show that the decision-making of subelite athletes was influenced by physical exhaustion. Under physical pressure, subelite athletes showed increased risk taking for both decisions where probability outcomes were explicit and where outcomes were unknown, as well as a reduced ability to adjust risk taking behavior according to odds. There were no differences in fast reactive responses to perceptual stimuli under physical pressure despite previous findings of facilitatory effects on these measures. The influence of physical exhaustion on subelite athletes was different from those observed previously in elite world-class athletes, highlighting the importance of considering the differences between athletes at the elite level. The application of this work to a novel subgroup of elite athletes was examined. The importance of feedback of individual patterns of behavior using the decision-making taxonomy is highlighted.

REFERENCES

Aklin, W.M., Lejuez, C.W., Zvolensky, M.J., Kahler, C.W., Gwadz, M., 2005. Evaluation of behavioral measures of risk taking propensity with inner city adolescents. Behav. Res. Ther. 43 (2), 215–228.

Arain, M., Haque, M., Johal, L., Mathur, P., Nel, W., Rais, A., et al., 2013. Maturation of the adolescent brain. Neuropsychiatr. Dis. Treat. 9, 449–461. http://dx.doi.org/10.2147/NDT.S39776.

Bar-Eli, M., Plessner, H., Raab, M., 2011. Judgment, decision-making and success in sport. Retrieved from, https://books.google.co.uk/books?hl=en&lr=&id=IpsL6MKd5h0C&oi=fnd&pg=PT4&dq=Judgment,+Decision-Making+and+Success+in+Sport+&ots=EOnHiDvX-&sig=9zvr9xpqxQrgP_Kiyaq6O4rW4Fg.

Black, A.C., Hochman, E., Rosen, M.I., 2013. Acute effects of competitive exercise on risk-taking in a sample of adolescent male athletes. J. Appl. Sport Psychol. 25 (2), 175–179. http://dx.doi.org/10.1080/10413200.2012.704621.

Blakemore, S., Robbins, T., 2012. Decision-making in the adolescent brain. Nat. Neurosci. 15, 1184–1191. Retrieved from, http://www.nature.com/neuro/journal/v15/n9/full/nn.3177.html%3FWT.ec_id%3DNEURO-201209.

Chein, J., Albert, D., O'Brien, L., Uckert, K., Steinberg, L., 2011. Peers increase adolescent risk taking by enhancing activity in the brain's reward circuitry. Dev. Sci. 14 (2), F1–10. http://dx.doi.org/10.1111/j.1467-7687.2010.01035.x.

Cheng, G.L.F., Lee, T.M.C., 2016. Altering risky decision-making: influence of impulsivity on the neuromodulation of prefrontal cortex. Soc. Neurosci. 11 (4), 353–364. http://dx.doi.org/10.1080/17470919.2015.1085895.

Clark, L., Manes, F., Antoun, N., Sahakian, B., Robbins, T., 2003. The contributions of lesion laterality and lesion volume to decision-making impairment following frontal lobe damage. Neuropsychologia 41, 1474–1483. Retrieved from, http://www.sciencedirect.com/science/article/pii/S0028393203000812.

Daniels, G.S., 1952. The "Average Man"? (No. TN-WCRD-53-7). Air Force Aerospace Medical Research Lab, Wright-Patterson AFB, OH.

Defoe, I., Dubas, J., Figner, B., 2015. A meta-analysis on age differences in risky decision making: adolescents versus children and adults. Psychol. Bull. 141, 48–84. Retrieved from, http://psycnet.apa.org/psycarticles/2014-45086-001.

Derefinko, K.J., Peters, J.R., Eisenlohr-Moul, T.A., Walsh, E.C., Adams, Z.W., Lynam, D.R., 2014. Relations between trait impulsivity, behavioral impulsivity, physiological arousal, and risky sexual behavior among young men. Arch. Sex. Behav. 43 (6), 1149–1158. http://dx.doi.org/10.1007/s10508-014-0327-x.

Ericsson, K., Krampe, R., Tesch-Römer, C., 1993. The role of deliberate practice in the acquisition of expert performance. Psychol. Rev. 100, 363–406. Retrieved from, http://doi.apa.org/psycinfo/1993-40718-001.

Giedd, J.N., 2004. Structural magnetic resonance imaging of the adolescent brain. Ann. N. Y. Acad. Sci. 1021 (1), 77–85. http://dx.doi.org/10.1196/annals.1308.009.

Hepler, T., 2015. Decision-making in sport under mental and physical stress. Int. J. Kinesiol. Sports 3, 79–83. Retrieved from, http://search.proquest.com/openview/db0918af03476fd663ac34020c77146c/1?pq-origsite=gscholar&cbl=2041008.

Jordet, G., Hartman, E., 2008. Avoidance motivation and choking under pressure in soccer penalty shootouts. J. Sport Exerc. Psychol. 30, 450–457.

Kaya, A., 2014. ScienceDirect decision making by coaches and athletes in sport. Procedia. Soc. Behav. Sci. 152, 333–338. http://dx.doi.org/10.1016/j.sbspro.2014.09.205.

Kinrade, N., Jackson, R., Ashford, K., 2015. Reinvestment, task complexity and decision making under pressure in basketball. Psychol. Sport Exerc. 20, 11–19. Retrieved from, http://www.sciencedirect.com/science/article/pii/S1469029215000308.

Lejuez, C.W., Read, J.P., Kahler, C.W., Richards, J.B., Ramsey, S.E., Stuart, G.L., et al., 2002. Evaluation of a behavioral measure of risk taking: the Balloon Analogue Risk Task (BART). J. Exp. Psychol. Appl. 8 (2), 75–84. http://dx.doi.org/10.1037/1076-898X.8.2.75.

Lejuez, C.W., Aklin, W.M., Zvolensky, M.J., Pedulla, C.M., 2003. Evaluation of the Balloon Analogue Risk Task (BART) as a predictor of adolescent real-world risk-taking behaviours. J. Adolesc. 26 (4), 475–479.

Lighthall, N.R., Mather, M., Gorlick, M.A., Jones, L., Yoon, T., Kim, J., et al., 2009. Acute stress increases sex differences in risk seeking in the balloon analogue risk task. PLoS One 4 (7), e6002. http://dx.doi.org/10.1371/journal.pone.0006002.

Lighthall, N.R., Sakaki, M., Vasunilashorn, S., Nga, L., Somayajula, S., Chen, E.Y., et al., 2012. Gender differences in reward-related decision processing under stress. Soc. Cogn. Affect. Neurosci. 7 (4), 476–484. http://dx.doi.org/10.1093/scan/nsr026.

Logan, G., 1994. On the ability to inhibit thought and action: a users' guide to the stop signal paradigm. Retrieved from, http://psycnet.apa.org/psycinfo/1994-97487-005.

Manes, F., Sahakian, B., Clark, L., Rogers, R., Antoun, N., Aitken, M., Robbins, T., 2002. Decision-making processes following damage to the prefrontal cortex. Brain J. Neurol. 125 (Pt. 3), 624–639. Retrieved from, http://www.ncbi.nlm.nih.gov/pubmed/11872618.

Parkin, B.L., Walsh, V., 2017. Gunslingers, poker players, and chickens 3: decision making under mental pressure in elite-junior athletes. In: Wilson, M.R., Walsh, V., Parlin, B.L. (Eds.), Sport and the Brain: The Science of Preparing, Enduring and Winning, Part B, vol. 234. Elsevier, Amsterdam, The Netherlands, pp. 339–359.

Parkin, B.L., Warriner, K., Walsh, V., 2017. Gunslingers, poker players, and chickens 1: decision making under physical performance pressure in elite athletes. In: Wilson, M.R., Walsh, V., Parlin, B.L. (Eds.), Sport and the Brain: The Science of Preparing, Enduring and Winning, Part B, vol. 234. Elsevier, Amsterdam, The Netherlands, pp. 291–316.

Paserman, M.D., 2007. Gender Differences in Performance in Competitive Environments: Evidence from Professional Tennis Players. IZA Discussion Paper No. 2834. Available at SSRN. https://ssrn.com/abstract=997269.

Pighin, S., Savadori, L., Bonini, N., 2015. Acute exercise increases sex differences in amateur athletes' risk taking. Int. J. Sports Med. 36, 858–863. Retrieved from, https://www.thieme-connect.com/products/ejournals/html/10.1055/s-0034-1398677.

Preston, S.D., Buchanan, T.W., Stansfield, R.B., Bechara, A., 2007. Effects of anticipatory stress on decision making in a gambling task. Behav. Neurosci. 121 (2), 257.

Rogers, R.D., Owen, A.M., Middleton, H.C., Williams, E.J., Pickard, J.D., Sahakian, B.J., Robbins, T.W., 1999. Choosing between small, likely rewards and large, unlikely rewards activates inferior and orbital prefrontal cortex. J. Neurosci. 19 (20), 9029–9038. Retrieved from, http://www.ncbi.nlm.nih.gov/pubmed/10516320.

Rose, T., 2016. The End of Average: How to Succeed in a World That Values Sameness. Penguin, UK.

Ryan, K.K., MacKillop, J., Carpenter, M.J., 2013. The relationship between impulsivity, risk-taking propensity and nicotine dependence among older adolescent smokers. Addict. Behav. 38, 1431–1434.

Schapschröer, M., Lemez, S., Baker, J., Schorer, J., 2016. Physical load affects perceptual-cognitive performance of skilled athletes: a systematic review. Sports Med. Open 2 (1), 37. http://dx.doi.org/10.1186/s40798-016-0061-0.

Smith, M.R., Zeuwts, L., Lenoir, M., Hens, N., De Jong, L.M.S., Coutts, A.J., 2016. Mental fatigue impairs soccer-specific decision-making skill. J. Sports Sci. 34 (14), 1297–1304. http://dx.doi.org/10.1080/02640414.2016.1156241.

Starcke, K., Brand, M., 2012. Decision making under stress: a selective review. Neurosci. Biobehav. Rev. 36 (4), 1228–1248. http://dx.doi.org/10.1016/j.neubiorev.2012.02.003.

Swann, C., Moran, A., Piggott, D., 2015. Defining elite athletes: issues in the study of expert performance in sport psychology. Psychol. Sport Exerc. 16, 3–14. Retrieved from, http://www.sciencedirect.com/science/article/pii/S1469029214000995.

Vaca, F.E., Walthall, J.M., Ryan, S., Moriarty-Daley, A., Riera, A., Crowley, M.J., Mayes, L.C., 2013. Adolescent Balloon Analog Risk Task and behaviors that influence risk of motor vehicle crash injury. Ann. Adv. Automot. Med. 57, 77–88. Retrieved from, http://www.ncbi.nlm.nih.gov/pubmed/24406948.

Vaeyens, R., Lenoir, M., Williams, A.M., Mazyn, L., Philippaerts, R.M., 2007. The effects of task constraints on visual search behavior and decision-making skill in youth soccer players. J. Sport Exerc. Psychol. 29 (2), 147–169. http://dx.doi.org/10.1123/jsep.29.2.147.

van den Bos, R., Harteveld, M., Stoop, H., 2009. Stress and decision-making in humans: performance is related to cortisol reactivity, albeit differently in men and women. Psychoneuroendocrinology 34 (10), 1449–1458. http://dx.doi.org/10.1016/j.psyneuen.2009.04.016.

Verbruggen, F., Logan, G.D., Stevens, M.A., 2008. STOP-IT: Windows executable software for the stop-signal paradigm. Behav. Res. Methods 40 (2), 479–483. http://dx.doi.org/10.3758/BRM.40.2.479.

Veroude, K., Jolles, J., Croiset, G., Krabbendam, L., 2013. Changes in neural mechanisms of cognitive control during the transition from late adolescence to young adulthood. Dev. Cogn. Neurosci. 5, 63–70. http://dx.doi.org/10.1016/j.dcn.2012.12.002.

Wells, B., Skowronski, J., 2012. Evidence of choking under pressure on the PGA tour. Basic Appl. Soc. Psychol. 34, 175–182. Retrieved from, http://www.tandfonline.com/doi/abs/10.1080/01973533.2012.655629.

CHAPTER 17

Gunslingers, poker players, and chickens 3: Decision making under mental performance pressure in junior elite athletes

Beth L. Parkin*,†,1, **Vincent Walsh**†

*Department of Psychology, University of Westminster, London, United Kingdom
†ICN, UCL, London, United Kingdom
[1]Corresponding author: Tel.: +44-20-7911-5000, e-mail address: b.parkin@westminster.ac.uk

Abstract

Background: Having investigated the decision making of world class elite and subelite athletes (see Parkin and Walsh, 2017; Parkin et al., 2017), here the abilities of those at the earliest stage of entry to elite sport are examined. Junior elite athletes have undergone initial national selection and are younger than athletes examined previously (mean age 13 years). Decision making under mental pressure is explored in this sample. During performance an athlete encounters a wide array of mental pressures; these include the psychological impact of errors, negative feedback, and requirements for sustained attention in a dynamic environment (Anshel and Wells, 2000; Mellalieu et al., 2009). Such factors increase the cognitive demands of the athletes, inducing distracting anxiety-related thoughts known as rumination (Beilock and Gray, 2007). Mental pressure has been shown to reduce performance of decision-making tasks where reward and loss contingencies are explicit, with a shift toward increased risk taking (Pabst et al., 2013; Starcke et al., 2011). Mental pressure has been shown to be detrimental to decision-making speed in comparison to physical stress, highlighting the importance of considering a range of different pressures encountered by athletes (Hepler, 2015).

Objective: To investigate the influence of mental pressure on key indicators of decision making in junior elite athletes. This chapter concludes a wider project examining decision making across developmental stages in elite sport. The work further explores how psychological insights can be applied in an elite sporting environment and in particular tailored to the requirements of junior athletes.

Methods: Seventeen junior elite athletes (10 males, mean age: 13.80 years) enrolled on a national youth athletic development program participated in the study. Performance across three categories of decision making was assessed under conditions of low and high mental pressure. Decision making under risk was measured via the Cambridge Gambling Task (CGT; Rogers et al., 1999), decision making under uncertainty via the Balloon Analogue Risk Task (BART; Lejuez et al., 2002), and fast reactive responses to perceptual stimuli via the Visual Search Task (Treisman, 1982). Mental pressure was induced with the addition of a concurrent verbal memory task, used to increase cognitive load and mimic the distracting effects of anxiety-related rumination.

Results: In junior elite athletes, fast reactive responses to perceptual stimuli (on the Visual Search Task) were slower under conditions of mental pressure. For decision making under risk there was an interaction of mental pressure and gender on the amount of points gambled, under pressure there was a higher level of risk taking in male athletes compared to females. There was no influence of mental pressure on decision making under uncertainty. There were no significant correlations in the degree to which individual's responses changed under pressure across the three measures of decision making. When assessing the applicability of results based on group averages there were no junior elite athletes who showed an "average" response (within 1SD of the mean) to mental pressure across all the three decision-making tasks.

Conclusion: Mental pressure affects decision making in a sample of junior elite athletes, with a slowing of response times, and modulations to performance of decision making under risk that have a high requirement for working memory. In relation to sport, these findings suggest that novel situations that place high cognitive demands on the athlete may be particularly influenced by mental pressure. The application of this work in junior elite athletes included the feedback of individual results and the implementation of a decision-making taxonomy.

Keywords
Junior athletes, Elite sport, Decision making, Risk taking, Mental pressure, Cognitive load

1 INTRODUCTION

Following on from work investigating world class elite and subelite athletes, this paper concludes a wider project examining the development of decision making in elite sport (see Parkin and Walsh, 2017; Parkin et al., 2017). It does so by exploring the influence of mental pressure on the abilities of junior elite athletes. Junior elite athletes are at the earliest stage of entry to elite sporting programs, having undergone initial national selection. These athletes are younger than those assessed previously (mean age: 13 years) and have approximately 5 years of experience in their given sport.

Previous work has established that decision-making abilities are influenced by the stressors that athletes encounter during performance (Hepler, 2015; Kinrade et al., 2015; Pighin et al., 2015; Smith et al., 2016). Performance pressure has been broadly categorized into, physical stress, such as physical exhaustion and injury, as well as mental stress, such as the desire to perform at one's best, the impact of errors, sources of negative feedback and requirements for sustained attention in a dynamic environment (Anshel and Wells, 2000; Mellalieu et al., 2009). One way in which mental pressure has been proposed to influence performance is via increased cognitive load, with irrelevant thoughts such as worrying diverting mental resources away from the task in hand (Beilock and Gray, 2007). Hepler (2015) revealed mental and physical pressure to exert different effects on decision making, mental pressure was found to impair decision-making speed, while physical pressure had no effect (Hepler, 2015). This result highlights the importance of considering a range of different sources of pressure in the understanding of athletic performance.

Work in this area has mainly been undertaken with adult samples. In particular, in nonelite adult athletes with sports-specific decision-making tasks, mental pressure has been reported to increase decision speed (Hepler, 2015; Smith et al., 2016) and to impair accuracy of choices (Smith et al., 2016), especially in complex scenarios (Kinrade et al., 2015). Interestingly, in the latter study, the levels of self-reported rumination arising from the mental pressure manipulation predicted response decrements. Work by Beliock et al. (2004) in nonathletic healthy adults further highlight that the influence of mental pressure may depend on task requirements, tasks that placed high demands on working memory were found to be selectivity impaired. This is also evident when examining the influence of mental pressure on different types of decision making. Under mental pressure (increased cognitive load) reduced performance on decision making under risk have been reported, with a suboptimal shift toward risky strategies (Pabst et al., 2013; Starcke et al., 2011). Decision making under uncertainty tasks however have been shown to be less susceptible to mental pressure. Indeed, Turnbull et al. (2005) reported that increased mental pressure (cognitive load) did not effect performance on the Iowa Gambling Task, proposed to be due to a reduced requirement for working memory (Starcke et al., 2011). Moreover mental pressure has been found to interfere with visual search strategies in athletes, in that athletes show increased fixations for shorter duration (Liu and Zhou, 2015) and a decreased ability to detect peripheral stimuli (Janelle and Singer, 1992).

Due to the age group of the junior elite athlete cohort they may show notable differences in their decision making and responses to pressure. Cognitive abilities go through profound changes in the transition from early adolescences to adulthood (Blakemore and Robbins, 2012). Early adolescents show an increase in risky decision making especially in "hot" contexts, where there is feedback of rewards and losses, in comparison to late adolescents and adults (Defoe et al., 2015). Additionally, during this time visual search strategies are developing, with those in late childhood shown to have a reduced ability to switch attention from one item to another (Trick and Enns, 1998). Differences in how these junior elite athletes respond to mental stress may also be likely. On the one hand, adolescents show increased stress volatility, displaying for example a heighten response to laboratory stress induction protocols in comparison to adults (Tottenham and Galván, 2016). On the other hand, individuals with high working memory capacities have been reported to be most affected by mental pressure, as these individuals employ cognitive demanding strategies that fail when resources are limited (Beilock and Carr, 2005). As such, in this age group who have reduced working memory capacities in comparison to adults (Gathercole et al., 2004), the effects of mental pressure on decision making may be less severe.

In order to investigate the influence that performance pressure has on decision making, junior elite athletes will be assessed under conditions of low and high mental pressure. Mental pressure was induced via the addition of a dual task, whereby the participant has to memorize a list of words, designed to increase cognitive load and mimic distracting rumination. In concert with the two preceding chapters, (Parkin et al., 2017; Parkin & Walsh, 2017), decision making under risk will be assessed via performance of the CGT (Rogers et al., 1999) and decision making under uncertainty via the BART (Lejuez et al., 2002). Fast reactive perceptual responses will be assessed via performance of a Visual Search Task (Treisman, 1982).

Thus this study aims to provide greater insight into the abilities of those at the earliest stage of entry on elite sporting development programs, together with the two companion papers provide a detailed look at the development of decision-making abilities in elite athletes. Mental pressure has been shown to modulate markers of optimal decision making in nonathletic adults and nonelite athlete samples. Whether similar findings are also present in junior elite athletes will be examined. The application of this work, in context of the requirements of junior athletes, will also be examined.

2 METHOD
2.1 PARTICIPANTS

The sample consisted of 17 junior elite athletes (10 males) aged between 12 and 14 (mean age: 13.19 years). All junior elite athletes had undergone selection to be part of a national youth development program, designed to develop skills for progression onto a Team GB training pathway and later success at an international level. All junior elite athletes included in the sample were from the same sport. The approximate age of entry for this sport is 8 years old, thus the junior elite athletes had been training for approximately 5 years.

Junior elite athletes were recruited via collaboration with Team GB Sports Psychologists and Coaches working within the program. Recruitment, testing, and debriefing took place during weekend training camps. Parents and athletes were informed about the purpose and the procedures of the study, and provided consent prior to participation. Junior elite athletes did not receive financial reimbursement for their participation. Upon completion a detailed debriefing session was undertaken in conjunction with Team GB coaches, a sports psychologist and a member of the research team. The study and consent procedures were approved by the UCL ethics board in compliance with the principles of the Declaration of Helsinki.

2.2 PROTOCOL

A within-subject design was used whereby the decision making of junior elite athletes were assessed under conditions of low and high mental pressure. Testing was undertaken in a single session. Performance on three decision-making tasks was examined via a laptop computer; these included the BART (Lejuez et al., 2002), the CGT (Rogers et al., 1999), and the Visual Search Task (Treisman, 1982). Initially participants received instructions and completed a short practice of each task. The tasks were undertaken at rest, i.e., in the absence of additional mental pressure, and under conditions of mental pressure. The order of these conditions were counterbalanced across individuals and separated by a short break. Mental pressure was induced via increased cognitive load whereby participants simultaneously performed a secondary word memory task. Athletes were presented with a verbal list of single words, which they were required to memorize and recall at the end of each task. After completion, junior elite athletes and parents were debriefed in a feedback session where the results of the study and relevant psychological concepts were discussed.

2.3 MENTAL PRESSURE INDUCTION

To induce elevated mental pressure junior elite athletes undertook a dual task. The athletes were required to simultaneously remember a verbal list of words presented to them while performing each decision-making task. For each task there was a unique list of 20 words, of which all were concrete nouns (bed, kettle, flower, etc.) and matched on frequency. The words were presented at varying intervals over a maximum duration of 4 min, so that it was difficult for the participant to predict their presentation.

Participants were instructed that they must memorize the words and to write these down once the task had ended. Participants were advised that one efficient method of performing this was via subvocal rehearsal, whereby one repeats the words in mind so as to not forget. The aim of this dual task was firstly, to mimic distracting ruminative thoughts provoked by stressful situations and secondly, to expend the processing resources available for a given task, thus increasing the demands placed on the junior elite athletes.

2.4 DECISION-MAKING TASKS

Tasks were delivered via a laptop computer, with a 17-in. display screen. They were run using Inquisit software version 4.0.7.0 (Millisecond Software, Seattle, WA), which was used to automatically record responses for subsequent analysis. Participants made responses via the use of a mouse or button press.

2.4.1 Balloon Analogue Risk Task

Decision making under uncertainty was measured via performance of the BART (Lejuez et al., 2002). In this task, participants are required to accrue money (5p per pump) through the inflation of computerized balloons. For the winnings of each balloon to be kept, participants must decide to transfer them from each balloon into a safe wallet, before the balloon explodes. In light of not knowing when the balloon will explode participants have to decide when to transfer the winnings in order to obtain the maximize winnings. A well-validated shortened version of the task including 20 balloons were used (e.g., Cheng and Lee, 2016; Derefinko et al., 2014; Ryan et al., 2013; Vaca et al., 2013), to make testing as time efficient as possible. The junior elite athletes did not cash incentives for performance of this task.

The average adjusted number of pumps provides a measure of risk taking on this task (Aklin et al., 2005; Lejuez et al., 2002), which was used for the analysis. This is the average number of pumps for balloons that did not burst, removing the variation resulting from the randomly generated balloon explosion points. Full task parameters are outlined in Parkin et al. (2017).

2.4.2 Cambridge Gambling Task

Decision making under risk was measured by the performance of CGT (Rogers et al., 1999). Full details of the task parameters and delivery are described in Parkin et al. (2017). In short, the participant is required to guess the location of a yellow token, hidden in 1 of 10 boxes presented on screen. The boxes are colored red or blue according to different probability ratios (9:1, 8:2, 7:2, and 6:4). Once the participant indicates the color box they wish to gamble on, they then have to select a wager. The amounts that can be bet appear as a number on screen, either in an ascending (from 5%, 25%, 75% to 95% of total points) or descending (from 95%, 75%, 25% to 5%) sequence. If the participant is correct this value is added to the score, if they are incorrect this value is deducted from their score.

The following dependent variables will be used in the analysis in this task. The *response times* measured as the time taken for the participant to identify whether they chose to look in the blue or red box. The *error rates* measured as the proportion of trials whereby participants look in the most likely box color. The *mean percentage points gambled* represents the degree of risk taking on the task and is examined in context of ascending or descending point presentations, to provide an indication of motor impulsivity. *Risk adjustment* quantifies a participant's ability to vary their risk taking in response to probability ratios, it is calculated using the following equation: $((2 \times \%\text{bet at } 9:1) + (\%\text{bet at } 8:2) - (\%\text{bet } 7:3) - (2 \times \%\text{bet at } 6:4))/\text{average}\%\text{bet}$.

2.4.3 Visual Search Task

The Visual Search Task (Treisman, 1982) (Fig. 1) was used to measure reaction times to perceptual stimuli. In this task participants are required to search visual arrays in order to identify the presence or absence of a target image, in this case a red square. There were three different types of distractor items: a green square, a green circle, and a red circle. The number of items presented in one display is known as a set size. The set sizes used in this task were three, six, or nine items arrays. There were two different trial types: *target present* where the items displayed on screen included a red square, and *target absent* where the items displayed on screen did

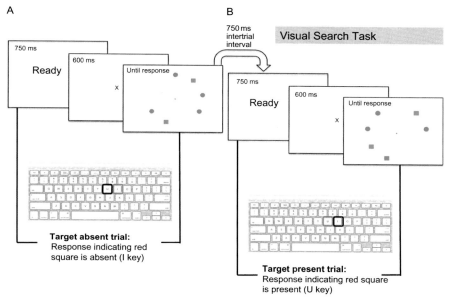

FIG. 1

The Visual Search Task. (A) On target absent trials, the word "ready" appears in the center of the screen for 750 ms, followed by the presentation of a fixation cross for 600 ms. The stimulus array is then presented, as there is *no red square* among the items presented, the participant presses the I key. There is an intertrial interval of 750 ms (a *blank screen*) before the presentation of the following trial. (B) On target present trials, as there is a *red square* among the items presented, the participant presses the U key.

not include a red square. If the red square was present participants were instructed to press the U key, and if the target was absent press the I key. Participants were instructed to place their index and middle finger over these keys to ensure a prompt response. They were instructed to make their responses as quickly and accurately as possible.

Each trial began with the presentation of a blank, white screen with the word "Ready" in the center. This was presented for 750 ms to orientate the participants' attention at the beginning of the trial. A blank screen followed this with a central fixation cross for 600 ms, which was proceeded by a visual array. The items in the visual array were presented on a white background; each image could be located in one of nine designated positions equally spaced in a circle formation (181 mm diameter). The visual array stayed on screen until the participant made a response. Following this there was an intertrial interval of 750 ms, consisting of the presentation of a blank screen.

The task was arranged into three blocks, with each block consisting of 108 trials. There were 54 target present and 54 target absent trials (18 of each set size).

The distractors were randomly sampled without replacement from 12 different images, 4 of each of the three distractor types, after the 12 random drawings the pool resets. Set size and trial type were randomized within each block. The locations of distractor and target images were also randomized within each block but with the constraint that all locations were equally as likely, i.e., the target image would appear equally as often (six times) in each of the nine possible positions per block. Each block was separated by a blank screen which instructed participants to press the space bar when they wished to continue to allow a short break if needed.

Before commencing the task, participants undertook a short practice consisting of five trials to familiarize themselves with the task. These were identical to the task, except that there was feedback given. If a wrong answer was given during the practice a red cross appeared on screen, if the correct answer was given the task continued to the next trial. The two dependent variables used in the analysis were reaction times and error rates.

2.5 ANALYSIS

To explore whether decision making on the three tasks is influenced by performance pressure, ANOVA or paired *t*-tests (two tailed) were undertaken to compare task performance under conditions of low and high mental pressure. Where relevant, Mauchly's test of sphericity was performed and *Greenhouse–Geisser correction* applied. Bonferroni correction was used to correct for multiple comparisons. For each task, the dependent variables used in this analysis are laid out below.

2.5.1 Balloon Analogue Risk Task

The mean adjusted number of pumps (the average pump count for balloons that did not burst) was analyzed as a measure of risk taking under uncertainty. A three-way mixed ANOVA was undertaken to compare performance of the task under conditions of low and high physical pressure (within subjects factor of pressure: low pressure, high pressure) broken down by time (within subject factor of balloon number: 1–10, 11–20). A between subject factor of gender was included (male, female), due to previous work that has shown gender to influence risk taking under pressure.

2.5.2 Cambridge Gambling Task

The response times (the duration from the trial appearing on screen and the participant choosing to look in a red or blue box) and the error rates (the percentage of trials in which the participant chose the most likely box color) were analyzed using a repeated measures ANOVA. These were undertaken to compare the performance under conditions of low and high pressure (within subjects factor of physical pressure: low pressure, high pressure) broken down by the odd ratios presented in the trial (within subject factor of odds ratio: 1:9, 2:8, 3:7, and 4:6). This part of the analysis was collapsed across the ascending and descending trials as the presentation and selection of bets occurred after these variables were recorded.

In order to explore gambling behavior for probabilistic decision making the mean number of points bet (%) on the task were analyzed. This analysis included trials in which participants chose the most likely outcome in order to not confuse betting behavior and decision making. A three-way mixed ANOVA was used to compare the amount of points bet for ascending and descending trials (within subject factor point presentation: ascending, descending) under conditions of high and low performance pressure (within subjects factor of pressure: low pressure, high pressure). A between subject factor of gender was included (male, female), due to previous work that has shown gender to influence risk taking under pressure. Overall, higher gambles are indicative of increased risk taking, a large difference in the amount bet on ascending compared to descending trials is indicative of impulsivity.

Lastly, a measure of risk adjustment was derived from the data and a paired t-test was performed to compare the degree of risk adjustment under conditions of low and high mental pressure.

2.5.3 Visual Search Task

Mean reaction times and error rates were analyzed for target absent and target present trial types of each set size. A repeated measures ANOVA were undertaken to compare performance of the task under conditions of low and high mental pressure (within subjects factor of pressure: low pressure, high pressure), for target absent and target present trial types (within subjects factor trial type: target present, target absent) broken down by set size (within subject factor: set size 3, 6, and 9). Reaction times for correct responses only were included in the analysis.

2.5.4 Performance of the Dual Task

The numbers of correct words recalled were counted in order to assess whether participants were performing the task. A one-way ANOVA was undertaken to assess whether there were differences in the performance of the dual task across decision-making tasks.

2.5.5 Correlation analysis assessing response to pressure across tasks

The following analysis was undertaken to examine whether individual's responses to pressure was consistent across tasks. The difference under pressure score was calculated for key indicators of task performance, by subtracting the score under high pressure from that under low pressure. The variables chosen for indicators of task performance are: the average adjusted number of pumps on the BART (a positive score represents better performance under pressure), the reaction time on the Visual Search Task (a negative score represents better performance under pressure), and the risk adjustment score on the CGT (a positive score represents better performance under pressure). The difference under pressure scores were compared, using Pearson correlation coefficient, in order to examine whether participants showed a consistent response to pressure across tasks.

2.5.6 Applicability of group data to individuals

In order to explore the degree to which group data applies to individuals the number of athletes whose change under pressure showed an "average" response to pressure across the three tasks was assessed. In accordance with previous work an average response was calculated as a score that fell within 1 standard deviation of the mean (0.5 SD above and below) (Daniels, 1952; Rose, 2016) for the average adjusted number of pumps on the BART, the reaction time on the Visual Search Task, and the risk adjustment score on the CGT.

3 RESULTS
3.1 DOES PERFORMANCE ON THE BART CHANGE UNDER PRESSURE?

To examine whether decision-making changes under conditions of low and high mental pressure, performance was assessed across the three decision-making tasks. On the BART, the mean adjusted number of pumps was analyzed as a measure of risk taking under uncertainty (Fig. 2). The ANOVA revealed no effect of mental pressure ($F_{(1,15)} < 0.01$; $P = 0.94$) or no effect of gender ($F_{(1,15)} = 0.77$; $P = 0.39$), while there was a significant effect of balloon number ($F_{(1,15)} = 14.42$; $P < 0.01$). In addition, there was no interaction between mental pressure and balloon number

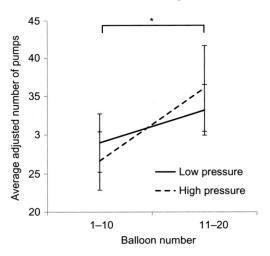

FIG. 2

BART results: In junior elite athletes there was no significant effect of mental pressure on the mean adjusted number of pumps, a measure of risk taking for decision making under uncertainty. There was a significant effect of balloon number, the mean adjusted number of pumps was higher for the last half of the task, this is indicative of the usual learning effects of this task. The asterisk (*) denotes statistical significant at $P < 0.05$. Error bars represent standard error of the mean.

($F_{(1,15)}=0.95$; $P=0.34$), mental pressure and gender ($F_{(1,15)}=0.98$; $P=0.34$), or balloon number and gender ($F_{(1,15)}=3.76$; $P=0.79$). Finally, there were no three-way interaction between mental pressure, balloon number, and gender ($F_{(1,15)}<0.01$; $P=0.99$). Therefore in this sample of junior elite athletes, mental pressure did not induce any significant changes in the degree of risk taking on decision making under uncertainty. However, it was found that the average adjusted number of pumps was higher for the latter half of the task (mean: balloon number 1–10: 27.27; balloon number 11–20: 34.01), this reflects the usual learning effects found on this task.

3.2 DOES PERFORMANCE ON THE CGT CHANGE UNDER PRESSURE?

3.2.1 Response times

Next, the influence of mental pressure on the CGT was examined. First a repeated measures ANOVA performed on response times (Fig. 3A) revealed no significant effect of mental pressure ($F_{(1,16)}=0.44$; $P=0.52$), but a significant effect of trial ratio ($F_{(1.79,28.60)}=4.68$; $P<0.05$). Pairwise comparisons revealed that the response times for trials with 1:9 ratios (mean: 1749.77 ms) were significantly quicker than those with ratios of 6:4 ($P<0.05$) (mean: 2369.50 ms) and 2:8 ($P<0.05$) (mean: 2149.72 ms). There was no significant interaction of mental pressure and trial ratio ($F_{(3,48)}=1.30$; $P=0.28$). This indicates that junior elite athletes respond faster when the trial odds ratios were higher, but mental pressure had no effect on the time taken for participants to indicate a simple probabilistic choice on this task.

3.2.2 Error rates

A repeated measure ANOVA was performed on the number of errors made on the task (Fig. 3B). There was no significant effect of mental pressure ($F_{(1,16)}=0.83$; $P=0.37$) or trial ratio ($F_{(1.41,22.49)}=2.45$; $P=0.12$), and no interaction of mental pressure and trial ratio ($F_{(1.65, 26.35)}=0.18$; $P=0.80$). Therefore, the accuracy of junior elite athletes was not affected by mental pressure or by the odd ratios presented in the trial.

3.2.3 Number of points gambled

A mixed ANOVA was performed on the number of points gambled (%). This revealed no effect of mental pressure ($F_{(1,15)}=0.36$; $P=0.56$) or gender ($F_{(1,15)}=1.21$; $P=0.29$), but a significant interaction of mental pressure and gender ($F_{(1,15)}=5.41$; $P<0.05$). Post hoc independent t-tests revealed that under pressure there were significant differences at a trend level ($P<0.1$) between males and females junior elite athletes under mental pressure (equal variances not assumed $t_{(13.68)}=2.04$; $P=0.06$), whereby male athletes bet more points compared to female athletes (mean points bet (%): females, 56.8%; males, 68.5%). At baseline there were no differences between male and female junior elite athletes (mean point bet (%): females, 61.6%; males, 60.5%) (equal variances assumed: $t_{(15)}=-0.25$; $P=0.80$) (Fig. 3B).

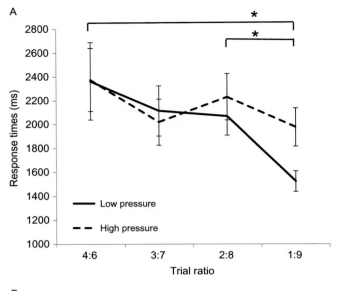

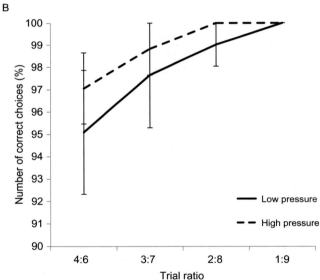

FIG. 3

Cambridge Gambling Task results: (A) The junior elite athletes showed no differences in response times for simple probabilistic choices under mental pressure, while athletes did respond faster when the trial odds ratios were higher. (B) The accuracy of junior elite athletes was not influenced by mental pressure or by trial odd ratios. (C) There was a significant interaction of mental pressure and gender on the amount of points gambled. Under mental pressure, the amount bet was higher for males compared to female athletes at a nonsignificant trend level ($P<0.1$). (D) The degree of risk adjustment, the tendency to take differential risks across trial ratio, was unaffected by mental pressure. The asterisk (*) denotes statistical significance at $P<0.05$. Error bars represent standard error of the mean.

(Continued)

3 Results

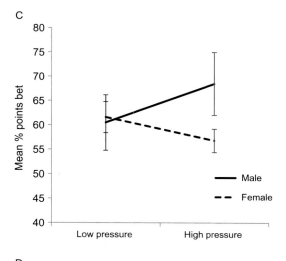

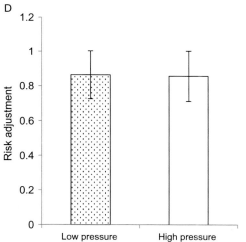

FIG. 3—Cont'd

There was also a significant effect of point presentation ($F_{(1,15)} = 37.67$; $P < 0.01$), whereby a higher number of points were gambled on ascending compared to descending trials (mean points bet (%): ascending, 39.8%; descending, 83.8%). This large difference is indicative of motor impulsivity, as junior elite athletes did not wait patiently on ascending trials for the points to increase (Fig. 3C). Lastly there were no other significant interactions, including no interaction of point presentation and gender ($F_{(1,15)} = 1.83$; $P = 0.20$) or mental pressure and point presentation

($F_{(1,15)} < 0.01$; $P = 0.97$), and no significant three-way interaction of mental pressure, point presentation, and gender ($F_{(1,15)} = 0.27$; $P = 0.61$).

3.2.4 Risk adjustment
A paired t-test revealed there were no differences in the measure of risk adjustment between conditions of low and high mental pressure ($t_{(16)} = -0.09$; $P = 0.93$) (Fig. 3D), therefore the tendency for junior elite athletes to modify the amount bet according to the different reward and loss contingencies were not influenced by conditions of mental pressure. Junior elite athletes scored poorly on this measure consistently.

3.3 DOES PERFORMANCE ON THE Visual Search Task CHANGE UNDER PRESSURE?
3.3.1 Reaction times
Following this the reaction times on the Visual Search Task (from trial onset to participant indicating the presence or absence of an object), were analyzed (Fig. 4A). Repeated measures ANOVA revealed a significant effect of mental pressure on reaction times ($F_{(1,16)} = 5.30$; $P < 0.05$) and junior elite athletes were slower to respond under conditions of mental pressure (mean: low pressure, 824.90 ms; high pressure, 969.05 ms). There was also a significant effect on the time taken to respond to trials when the target was present or absent ($F_{(1,16)} = 8.11$; $P < 0.05$), and the athletes were faster to respond when the target was present (mean: target present, 855.08 ms; target absent, 938.87 ms). Lastly, there was a significant effect of set size ($F_{(2,32)} = 23.18$; $P < 0.01$), pairwise comparisons revealed that junior elite athletes were significantly faster at responding for set sizes of 3, than set sizes of 9 ($P < 0.01$) and 6 ($P < 0.05$), and significantly faster at responding for set size 6 than 9 ($P < 0.05$) (mean response time: set size 9, 651.23 ms; set size 6, 893.78 ms; set size 3, 845.91 ms).

There were no significant interactions of any of the main effects, including mental pressure and trial type ($F_{(1,16)} = 0.86$; $P = 0.37$), mental pressure and set size ($F_{(2,32)} = 0.12$; $P = 0.88$), and set size and trial type ($F_{(2,32)} = 0.53$; $P = 0.59$). There was also no three-way interaction of mental pressure, set size, and trial type ($F_{(2,32)} = 0.32$; $P = 0.73$).

3.3.2 Error rates
A repeated measures ANOVA revealed that there was no significant effect of mental pressure on the number of correct responses on the Visual Search Task ($F_{(1,16)} = 0.01$; $P = 0.94$) (Fig. 4B). There was also no effect of set size ($F_{(2,32)} = 0.28$; $P = 0.75$) and no effect on whether the target was absent or present ($F_{(1,16)} = 0.93$; $P = 0.35$) on the number of correct responses. Moreover there were no interactions between any of the main effects including, between mental pressure and trial type ($F_{(1,16)} = 1.79$; $P = 0.68$), mental pressure and set size ($F_{(2,32)} = 0.77$; $P = 0.47$),

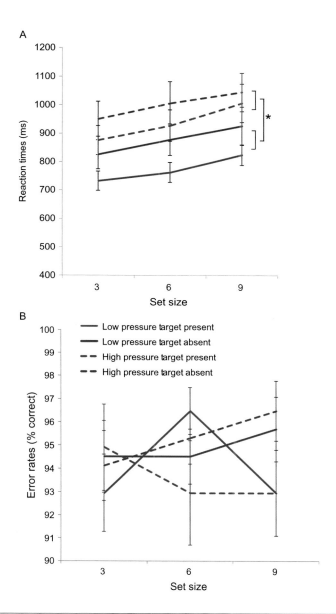

FIG. 4

Visual Search Task results: (A) Under conditions of high mental pressure (performance of a dual task), elite junior athletes were slower to respond on the Visual Search Task. Junior elite athletes also showed faster response times when the target was present and when the set size was smaller. (B) Junior elite athletes showed no differences in the number of errors made on the Visual Search Task when under conditions of mental pressure. Error rates were not affected by different set sizes or target present/absent trials. The asterisk (*) denotes a statistical significance at $P<0.05$. Error bars represent standard error of the mean.

or trial type and set size ($F_{(2,32)} = 1.83$; $P = 0.18$). There was also no three-way interaction of mental pressure, trial type, and set size ($F_{(2,32)} = 1.85$; $P = 0.17$).

3.4 PERFORMANCE ON THE DUAL TASK
In the mental pressure condition the average number of words that junior elite athletes recalled were 8.5 (with a range of 3–16). On two occasions the participants scored less than four correct items. There were no significant differences between the three tasks on the amount of words recalled ($F_{(2,32)} = 2.33$; $P = 0.11$). Therefore, the junior elite athletes were performing the concurrent verbal memory task consistently across the decision-making tasks in the mental pressure condition.

3.5 WERE INDIVIDUAL PERFORMANCE CHANGES UNDER PRESSURE CONSISTENT ACROSS TASKS?
Pearson correlation coefficient were undertaken to examine whether individual responses to pressure were consistent across key indicators of performance on the three decision-making tasks. The results showed that there were no significant correlations between the degree of change under pressure on performance of the BART and the CGT ($r_{(16)} = -0.03$; $P = 0.90$), the BART and the Visual Search Task ($r_{(16)} = 0.08$; $P = 0.76$), on the Visual Search Task and the CGT ($r_{(16)} = -0.36$; $P = 0.17$). Therefore, individual participants' responses to pressure were not consistent across key indices of decision making over the three tasks.

3.6 APPLICABILITY OF GROUP DATA TO INDIVIDUALS
In order to explore the degree to which group data applies to individual athletes, the number of athletes who showed an average response (defined as falling within 0.5 SD above and below the mean) to pressure across all three tasks was assessed. In line with previous work (Daniels, 1952; Rose, 2016), this categorization meant that at least the middle 35% of the group was categorized as falling within the "average" range for each variable. The results revealed that there were no athletes who showed mean responses to pressure across all three indices of decision making. As the group average scores were not representative of the behavior of a single individual athlete across the three measures, this highlights the importance of individualized profiling of results in an elite sport setting.

4 DISCUSSION
The study examined the effects of mental pressure on key indicators of decision making in junior elite athletes. Results revealed that fast reactive responses to perceptual stimuli were slower in junior elite athletes under conditions of mental pressure. Decision making under risk, where reward and loss contingencies were explicit, were

also influenced by mental pressure. In particular there was an interaction of mental pressure and athlete's gender, whereby under pressure male athletes showed higher levels of risk taking than female athletes. For decision making under uncertainty there was no influence of mental pressure on risk taking.

The findings of the current study show that in junior elite athletes mental pressure impaired fast reactive responses to perceptual stimuli. In particular, athletes were slower to identify the presence or absence of an item in a visual array. This slowing was seen to be similar across trial types, and while reaction times were also increased for larger set sizes and when the target was absent (compared to present), there were no interactions of these factors with mental pressure. The findings of increased reaction times under conditions of mental pressure have also been reported in nonelite adult athletes on sport-specific decision making tasks (Hepler, 2015; Kinrade et al., 2015). In line with this, visual search strategies have been reported to become more erratic in athletes under competitive pressure, in that eye movements show an increased number of fixations of shorter duration (Liu and Zhou, 2015), and a decreased ability to detect peripheral stimuli (Janelle and Singer, 1992). Deployment of visual attention plays an important role in sport as the athlete is responsible for the monitoring a dynamic environment consisting of multiple players (Williams et al., 1999). Thus the increased reaction times for visual search under mental pressure may be useful in understanding the performance of junior elite athletes.

Mental pressure was also found to influence decision making under risk where probability outcomes were explicit. In particular, there was an interaction of mental pressure and gender on the amount of points gambled on the CGT. Under mental pressure male athletes showed a higher level of risk taking than females (although post hoc comparisons only reached a nonsignificant trend level $P < 0.1$). These respective shifts in the propensity for risk taking across genders have also been reported following acute physical exercise in nonelite adult athletes (Pighin et al., 2015), as well as in studies examining the influence of stress (experimentally induced elevated cortisol) on decision making (Lighthall et al., 2009, 2011; Preston et al., 2007). The mechanisms underlying these gender-related shifts in risk taking are unknown, however the main theories put forward are evolutionary in particular differences in intrasexual selection (Pighin et al., 2015). One study reported elevated cortisol to elicit opposing responses at a neural level in males and females performing a decision-making task. Under stress males were reported to show increased activation in the insula and putamen, regions associated with risk estimations, but decreased activation in females. Moreover increased activation of the dorsal striatum was strongly associated with increased reward collection in stressed males, but not in stressed females (Lighthall et al., 2012). The differences observed here may therefore be a result of elevated cortisol in response to mental stress.

Lastly mental pressure did not affect risk taking for decision making under uncertainty. Robust performance of decision making under uncertainty to mental pressure has been reported previously (Turnbull et al., 2005). The differences observed here in comparison to those for probabilistic decision making, may be due to differences in underlying task requirements. Mental pressure was operationalized as a dual working memory task, whereby athletes were verbally presented with a list of words they were required to later recall. This protocol was used to mimic task irrelevant thoughts, such as worrying, that consume cognitive resources, diminishing those available for the task in hand. It has been proposed that decision making under risk is particularly vulnerable to mental pressure as these tasks rely heavily on working memory resources, whereas decision making under uncertainty is unaffected by mental pressure as this relies to a greater extent on automatic intuitive processing (Starcke et al., 2011). In relation to sport, these findings suggest that the influence of mental pressure may be heightened when the junior elite athlete is in a novel situation that places high demands on executive processing resources (Beilock and Gray, 2007). One line of argument that proposed that individuals with high working memories are most detrimentally affected by mental pressure (Beilock and Carr, 2005), suggests that the performance decrements of mental pressure reported in the literature in adult populations would be less severe in this younger age group (who have reduced working memory capacity). However, the overall pattern of results reported here are similar to those reported in previous studies with adult samples.

The findings from the current study also showed that there was no correlation in the degree to which athletes changed under pressure across key performance indicators on the three tasks. This indicates that the influence of mental pressure was not uniform across decision-making abilities in these athletes. In particular, an athlete who showed robust performance under pressure on one ability did not necessarily show similar responses under pressure on a different ability. This reiterates the importance of examining a range of different measures of decision making in order to understand performance under pressure in athletes. Importantly it highlights that understanding the unique strengths and weaknesses across a number of abilities with the use of profiling individual athletes may be particularly insightful.

A further aim of the current work was to examine how these insights in decision making can be applied in an elite sport context; and in this case how this can be tailored to athletes at a junior level. As the junior elite athlete is at the beginning of their training, it could be argued there is more scope to embed decision-making education within their development. Historically when training an athlete the onus is placed on the improvement of physical ability, while psychological attributes involved in sporting performance receive less attention. There is the underlying assumption that these skill develop intuitively with practice. As in previous papers (Parkin et al., 2017), the application of this work centred upon the decision-making taxonomy, a classification of decision making which reflected the three types of tasks assessed during the study. This was useful in developing the idea that decision making is not a single concept, but instead comprising of disparate skills and scenarios. Again, the use of real-world analogies to conceptualise these skills were implemented to provide a common language by which decision-making could be conceptualized and communicated between sporting professionals Probabilistic decision-making was referred to as "playing poker" and decision-making under uncertainty was referred to as "playing chicken" (with reference to a game whereby two cars drive straight toward each other, and the first car to serve is named the "chicken"). Fast reactive decision-making was described as "gun slinging" (for more details please see Parkin et al., 2017).

The limitation of statistical approaches based on group averages in an elite sporting context were also highlighted in this sample. In particular, there were no junior elite athletes who showed an "average" response to pressure across all types of decision making, thus highlighting the importance of individualized performance feedback. In order to help the junior elite athlete embody the taxonomy, instead of feeding back position ranking, it was thought more appropriate to provide descriptions of decision-making styles under condition of low and high mental pressure. That is, under pressure you were best at "gun-slinging" decision making (referring to performance of the visual search task) and under low pressure you best at "poker-playing" decision making (referring to performance of CGT).

5 CONCLUSION

In conclusion, mental pressure influenced the risk taking of junior elite athletes for decision making where probability outcomes were explicit, while there was no effect of mental pressure for decision making under uncertainty. Mental pressure, operationalized as increased cognitive load, is likely to have consumed working memory and impaired decision-making scenarios that require these resources. In relation to sport, these findings suggest that novel situations that place high demands on the athlete may be particularly influenced by mental pressure. In addition, under mental pressure junior elite athletes showed slower reaction times to perceptual stimuli, highlighting a detrimental impact on the allocation of attention on visual search. The application of this work centered upon feedback of individualized results using the decision-making taxonomy framework.

REFERENCES

Aklin, W.M., Lejuez, C.W., Zvolensky, M.J., Kahler, C.W., Gwadz, M., 2005. Evaluation of behavioral measures of risk taking propensity with inner city adolescents. Behav. Res. Ther. 43 (2), 215–228.

Anshel, M.H., Wells, B., 2000. Sources of acute stress and coping styles in competitive sport. Anxiety Stress Coping 13 (1), 1–26. Retrieved from http://www.tandfonline.com/doi/abs/10.1080/10615800008248331.

Beilock, S.L., Carr, T.H., 2005. When high-powered people fail: working memory and "choking under pressure" in math. Psychol. Sci. 16 (2), 101–105. Retrieved from http://journals.sagepub.com/doi/abs/10.1111/j.095976.2005.00789.x.

Beilock, S., Gray, R., 2007. Why do athletes choke under pressure? Retrieved from http://doi.apa.org/index.cfm?fa=search.printFormat&uid=2007-01666-027&recType=psycinfo&singlerecord=1&searchresultpage=true.

Blakemore, S., Robbins, T., 2012. Decision-making in the adolescent brain. Nat. Neurosci. 15, 1184–1191. Retrieved from http://www.nature.com/neuro/journal/v15/n9/full/nn.3177.html%3FWT.ec_id%3DNEURO-201209.

Cheng, G.L.F., Lee, T.M.C., 2016. Altering risky decision-making: influence of impulsivity on the neuromodulation of prefrontal cortex. Soc. Neurosci. 11 (4), 353–364. http://dx.doi.org/10.1080/17470919.2015.1085895.

Daniels, G.S., 1952. The "Average Man"? (No. TN-WCRD-53-7). Air Force Aerospace Medical Research Lab, Wright-Patterson AFB, OH.

Defoe, I., Dubas, J., Figner, B., 2015. A meta-analysis on age differences in risky decision making: adolescents versus children and adults. Psychol. Bull. 141, 48–84. Retrieved from http://psycnet.apa.org/psycarticles/2014-45086-001.

Derefinko, K.J., Peters, J.R., Eisenlohr-Moul, T.A., Walsh, E.C., Adams, Z.W., Lynam, D.R., 2014. Relations between trait impulsivity, behavioral impulsivity, physiological arousal, and risky sexual behavior among young men. Arch. Sex. Behav. 43 (6), 1149–1158. http://dx.doi.org/10.1007/s10508-014-0327-x.

Gathercole, S.E., Pickering, S.J., Ambridge, B., Wearing, H., 2004. The structure of working memory from 4 to 15 years of age. Dev. Psychol. 40 (2), 177. Retrieved from http://psycnet.apa.org/journals/dev/40/2/177/.

Hepler, T., 2015. Decision-making in sport under mental and physical stress. Int. J. Kinesiol. Sports 3, 79–83. Retrieved from http://search.proquest.com/openview/db0918af03476fd663ac34020c77146c/1?pq-origsite=gscholar&cbl=2041008.

Janelle, C., Singer, R., 1992. External distraction and attentional narrowing: visual search evidence. J. Sport Exerc. Psychol. 21, 70–91. Retrieved from http://journals.humankinetics.com/doi/abs/10.1123/jsep.21.1.70.

Kinrade, N., Jackson, R., Ashford, K., 2015. Reinvestment, task complexity and decision making under pressure in basketball. Psychol. Sport Exerc. 20, 11–19. Retrieved from http://www.sciencedirect.com/science/article/pii/S1469029215000308.

Lejuez, C.W., Read, J.P., Kahler, C.W., Richards, J.B., Ramsey, S.E., Stuart, G.L., … Brown, R.A., 2002. Evaluation of a behavioral measure of risk taking: the balloon analogue risk task (BART). J. Exp. Psychol. 8, 75–84. http://dx.doi.org/10.1037//1076-898X.8.2.75.

Lighthall, N.R., Mather, M., Gorlick, M.A., 2009. Acute stress increases sex differences in risk seeking in the balloon analogue risk task. PLoS One 4 (7), e6002.

Lighthall, N.R., Sakaki, M., Vasunilashorn, S., Nga, L., Somayajula, S., Chen, E.Y., et al., 2011. Gender differences in reward-related decision processing under stress. Soc. Cogn. Affect. Neurosci. 7 (4), 476–484.

Lighthall, N.R., Sakaki, M., Vasunilashorn, S., Nga, L., Somayajula, S., Chen, E.Y., … Mather, M., 2012. Gender differences in reward-related decision processing under stress. Soc. Cogn. Affect. Neurosci. 7 (4), 476–484. http://dx.doi.org/10.1093/scan/nsr026.

Liu, S., Zhou, W., 2015. The effect of anxiety state on the visual search efficiency of athletes. Open J. Soc. Sci. 3, 80–85. Retrieved from http://www.scirp.org/journal/PaperInformation.aspx?paperID=57080.

Mellalieu, S.D., Neil, R., Hanton, S., Fletcher, D., 2009. Competition stress in sport performers: stressors experienced in the competition environment. J. Sports Sci. 27 (7), 729–744. http://dx.doi.org/10.1080/02640410902889834.

Pabst, S., Schoofs, D., Pawlikowski, M., Brand, M., Wolf, O.T., 2013. Paradoxical effects of stress and an executive task on decisions under risk. Behav. Neurosci. 127 (3), 369–379. http://dx.doi.org/10.1037/a0032334.

Parkin, B.L., Walsh, V., 2017. Gunslingers, poker players, and chickens 2: Decision-making under physical pressure in elite-developing athletes. In: Wilson, M.R., Walsh, V., Parlin, B.L. (Eds.), Sport and the Brain: The Science of Preparing, Enduring and Winning, Part B, vol. 234. Elsevier, Amsterdam, The Netherlands, pp. 317–338.

Parkin, B.L., Warriner, K., Walsh, V., 2017. Gunslingers, poker players, and chickens 1: Decision making under physical performance pressure in elite athletes. In: Wilson, M.R., Walsh, V., Parlin, B.L. (Eds.), Sport and the Brain: The Science of Preparing, Enduring and Winning, Part B, vol. 234. Elsevier, Amsterdam, The Netherlands, pp. 291–316.

Pighin, S., Savadori, L., Bonini, N., 2015. Acute exercise increases sex differences in amateur athletes' risk taking. Int. J. Sports Med. 36, 858–863. Retrieved from https://www.thieme-connect.com/products/ejournals/html/10.1055/s-0034-1398677.

Preston, S.D., Buchanan, T.W., Stansfield, R.B., Bechara, A., 2007. Effects of anticipatory stress on decision making in a gambling task, Behav. Neurosci. 121 (2), 257. http://dx.doi.org/10.1037/0735-7044.121.2.257.

Rogers, R.D., Owen, A.M., Middleton, H.C., Williams, E.J., Pickard, J.D., Sahakian, B.J., Robbins, T.W., 1999. Choosing between small, likely rewards and large, unlikely rewards activates inferior and orbital prefrontal cortex. J. Neurosci. 19 (20), 9029–9038. Retrieved from http://www.ncbi.nlm.nih.gov/pubmed/10516320.

Rose, T., 2016. The End of Average: How to Succeed in a World That Values Sameness. Penguin, UK.

Ryan, K.K., MacKillop, J., Carpenter, M.J., 2013. The relationship between impulsivity, risk-taking propensity and nicotine dependence among older adolescent smokers. Addict. Behav. 38, 1431–1434.

Smith, M.R., Zeuwts, L., Lenoir, M., Hens, N., De Jong, L.M.S., Coutts, A.J., 2016. Mental fatigue impairs soccer-specific decision-making skill. J. Sports Sci. 34 (14), 1297–1304. http://dx.doi.org/10.1080/02640414.2016.1156241.

Starcke, K., Pawlikowski, M., Wolf, O.T., Altstötter-Gleich, C., Brand, M., 2011. Decision-making under risk conditions is susceptible to interference by a secondary executive task. Cogn. Process. 12 (2), 177–182. Retrieved from http://link.springer.com/article/10.1007/s10339-010-0387-3.

Tottenham, N., Galván, A., 2016. Stress and the adolescent brain: amygdala-prefrontal cortex circuitry and ventral striatum as developmental targets. Neurosci. Biobehav. Rev. 70, 217–227.

Treisman, A., 1982. Perceptual grouping and attention in visual search for features and for objects. J. Exp. Psychol. Hum. Percept. Perform. 8, 194–214. Retrieved from http://psycnet.apa.org/journals/xhp/8/2/194/.

Trick, L.M., Enns, J.T., 1998. Lifespan changes in attention: the visual search task. Cogn. Dev. 13 (3), 369–386.

Turnbull, O.H., Evans, C.E., Bunce, A., Carzolio, B., O'Connor, J., 2005. Emotion-based learning and central executive resources: an investigation of intuition and the Iowa Gambling Task. Brain Cogn. 57 (3), 244–247. Retrieved from http://www.sciencedirect.com/science/article/pii/S0278262604002660.

Vaca, F.E., Walthall, J.M., Ryan, S., Moriarty-Daley, A., Riera, A., Crowley, M.J., Mayes, L.C., 2013. Adolescent balloon analog risk task and behaviors that influence risk of motor vehicle crash injury. Ann. Adv. Automot. Med. 57, 77–88. Retrieved from http://www.ncbi.nlm.nih.gov/pubmed/24406948.

Williams, A., Davids, K., Williams, J., 1999. Visual perception and action in sport. Retrieved from https://books.google.co.uk/books?hl=en&lr=&id=wgh-sCNE-3cC&oi=fnd&pg=PR9&dq=williams+1999+sports+visual+attention&ots=dnMM7sK9tx&sig=KM2JIp8f1srvpo-29FN6Oo3APDk.

Index

Note: Page numbers followed by "*f*" indicate figures, "*t*" indicate tables, and "*np*" indicate footnotes.

A

Above real-time training (ARTT), 103–104
　applying, 111–112
　　in Australian football, 105
　　on decision-making in elite football, 101–112, 106*t*, 106*f*, 108–110*t*
　　effect of, 105
Abstract thinking, 225
ACC. *See* Anterior cingulate cortex (ACC)
Accuracy, 6, 90–93
ACPM. *See* Affective cognitive processing model (ACPM)
Action anticipation, 54–55, 57, 276–279
Action experiences, 263–264
Action observation, 193, 195–197
　basic aspects of, 190–191
　motor activations during, 57–59
Action observation network (AON), 195–196, 263–268, 266–267*t*, 271
　in action prediction, 59
　activity in frontoparietal, 190
　to predictive embodiment, 62
Action prediction, 60*f*
　action observation network in, 59
Action-skilled observation, 263–264
　action experiences, early, 270
　action observation network, 264–268
　detection of deceptive actions, 279–281
　as function of familiarity, 274–276
　neurophysiological methods, 268–270
　physical and imitative short-term practice, 271–273
Activity change
　in association with motor learning
　　action observation, 195–197
　　actual execution, 194–195
　　motor imagery, 198
Actual execution, 192–195
Acute exercise, 178–179
　period, 164–165
Acute high-intensity exercise, of cognitive function, 170–176
ADHD, impulsivity in, 229
Adversarial growth, 130, 151

Adversity
　in competitive sport, 154
　　data analysis, 124, 132
　　data extraction, 124
　　study design, 130–131
　defined, 150–151
　experiences, 133–143, 152
　　negative events and experiences, 133–143
Aerobic fitness, 164
Affective cognitive processing model (ACPM), 118, 130
ANAM. *See* Automated neuropsychological assessment metrics (ANAM)
Anterior cingulate cortex (ACC), 231, 232*f*
Anterior intraparietal cortex, 193
Anticipation
　action, 54–55, 57, 276–279
　visual, 54–56
Anticipatory skill, in sport, 277
Anxiety, 25–26
AON. *See* Action observation network (AON)
Athlete, action regulation, 37–47
　perception and internal representations, 37–39
Attention, 25, 176, 178–179, 222, 224, 234
　automaticity, 224–227
　　reduced verbal-analytic processing, 225–227
　　transient hypofrontality theory, 225
　control, 227
　　dopamine pathways, 228–230
　　improvements, 236
　　synchronization theory, 227–228
　　top-down attention networks, 228
　effort, 230
　　conflict monitoring, 231–232
　　psychophysiological measurement, 230–231
　self-awareness, 232–234
　and self-regulation, 235–236
　traditional models of, 230
Automated neuropsychological assessment metrics (ANAM), 176
Automaticity
　attention, 224–227
　in expertise, 103–104
Autotelic personality, 229
Average exercise duration, 181–182

361

B

Balloon analogue risk task (BART), 293–294, 296–298, 297f, 302, 322, 325, 327f, 332f, 343–344, 346, 348f
  change under pressure, performance, 303–304, 304f, 326–327, 348–349
Behavioral neuroscience, 34, 47
Behavioral neuroscientists, 34
Behavioral responses, 143
Biomarkers, 214
Blood-oxygen-level-dependent (BOLD) signal, 269
Bodily motion, 225
Brain activity
  associated with motor learning, 196f
  relating to task complexity
    action observation, 193
    actual execution, 192–193
    motor imagery, 194
Brain–behavior couplings, 36
Brain–body–environment, 39–47
Brain imaging sequences, 209
Brain stimulation, 58–59

C

Cambridge gambling task (CGT), 298–300, 298f, 302, 322–323, 325–326, 329–330f, 344, 346–347, 349–352, 350–351f
  change under pressure, performance, 304–306, 305f, 328–330, 349–352
Capacity, 246
  working memory (WM), 246–248
Care Quality Commission (CQC), 207–208
Central nervous system (CNS), 47–48
Cerebellum, 265f
Cerebrum, 265f
CGT. *See* Cambridge gambling task (CGT)
Challenge stress, 87
Choking, 292–293
Choline, 210
Clock genes, 26
Closed-skill sports, 96
CMH, predictions of, 231
Cognition, 162–164
  effort, 72–75
  function
    acute high-intensity exercise and, 170–176
    attention, 176
    executive function, 175
    information processing, 170–175
    memory, 175–176
    time of testing, 176
  neuroscience, 37–38

performance, data processing and analysis, 90
process, 222, 224
responses, 140
Computerized tomography (CT), 209
Concussion, 206
  clinics, 215
    in UK, 206–208
  diagnosis, 214
  education/dissemination, 215
  imaging, 214
  management, multidisciplinary approach, 208
    neurologist, 210–211
    physiotherapist, 211–212
    psychiatrist, 212–213
    psychologist, 212–213
    radiologist, 209–210
    sports physician, 208–209
  multimedia technologies, 214–215
Consciousness, self-reflective, 225
Contextual interference, in sports practice, 70
  effect, 69–71, 73–74, 76
  future research directions, 76–80, 78–79t, 79f
  influence of skill complexity, 71–73
  mechanisms and practice, 73–75
  transfer of learning, 75–76
Conventional MRI techniques, 209
Cortical processing, 97
Countermovement jumps (CMJ), 22–23
Cyclic adenosine monophosphate (c-AMP)-protein kinase, 26
Cycling *vs.* treadmill exercise, 181

D

DAI. *See* Diffuse axonal injury (DAI)
Dart experts, performance of, 190
Deceptive actions, detection of, 279–281
Decision-making
  ARTT on, elite football, 101–112, 106t, 106f, 108–110t
  categories of, 294
  under mental pressure in elite-junior athletes, 340–342, 354–357
    balloon analogue risk task, 343–344, 346, 348–349, 348f
    Cambridge gambling task, 344, 346–347, 349–352, 350–351f
    data analysis, 346–348
    mental pressure induction, 343
    protocol, 343
    study method, 342–348
    tasks, 343–346
    visual search task, 344–347, 345f, 352–354

optimal, 292–293
under physical performance pressure in elite athletes, 309–313
  balloon analogue risk task, 296–298, 297f, 302–304, 304f
  Cambridge gambling task, 298–300, 298f, 302, 304–306, 305f
  data analysis, 302–303
  decision making change under physical pressure, 303–304
  individual performance changes under pressure, 307–309, 308f
  physical stress induction protocol, 295–296
  protocol, 295
  Stroop task, 300–301, 301f, 303, 307, 307f
  study method, 294–303
  tasks, 296–301
under physical pressure in elite-developing athletes, 318–320, 332–335
  analysis, 325–326
  balloon analogue risk task, 322, 325–327, 327f, 332f
  Cambridge gambling task, 322–323, 325–326, 328–330, 329–330f
  correlation analysis assessing performance across tasks, 326
  physical stress induction, 321–322
  protocol, 321
  stop signal reaction time task, 323–326, 324f, 331, 331f
  study method, 320–326
  task, 322–325
Default mode network (DMN), 233, 234f, 235
Deoxyhemoglobin, 209–210
Diffuse axonal injury (DAI), 209–210
Diffusion tensor imaging, 210
Diffusion-weighted imaging (DWI), 209–210
Dispositional Resilience Scale (DRS), 130–131
DLPFC. *See* Dorsolateral prefrontal cortex (DLPFC)
DMN. *See* Default mode network (DMN)
Dopamine
  action, reduce, 229
  pathways, 228–230
Dorsolateral prefrontal cortex (DLPFC), 190–191, 271–272, 319–320
Drift diffusion analysis, 93–96
Drift diffusion model (DDM), 86–87
Drift diffusion parameter, 96
Drift rate, 86–87, 94–95
DWI. *See* Diffusion-weighted imaging (DWI)
Dynamic complexity, of human movement, 35

E

EBA. *See* Extrastriate body area (EBA)
Ecological neurodynamics, 36–38
EEG. *See* Electroencephalography (EEG)
Elaboration hypothesis, 73
Electroencephalogram (EEG), 225–226
Electroencephalography (EEG), 15f, 269, 272
  activity, 249
  coherence, 247
  and performance variables, 251–252
Elite athletes
  decision making under physical performance pressure in, 309–313
    balloon analogue risk task, 296–298, 297f, 302–304, 304f
    Cambridge gambling task, 298–300, 298f, 302, 304–306, 305f
    correlational analysis assessing response to pressure across tasks, 303
    data analysis, 302–303
    decision making change under physical pressure, 303–304
    decision-making tasks, 296–301
    individual performance changes under pressure, 307–309, 308f
    physical stress induction protocol, 295–296
    protocol, 295
    Stroop task, 300–301, 301f, 303, 307, 307f
    study method, 294–303
Elite-developing athletes
  decision-making under physical pressure in, 318–320, 332–335
    analysis, 325–326
    balloon analogue risk task, 322, 325–327, 327f, 332f
    Cambridge gambling task, 322–323, 325–326, 328–330, 329–330f
    correlation analysis assessing performance across tasks, 326
    physical stress induction, 321–322
    protocol, 321
    stop signal reaction time task, 323–326, 324f, 331, 331f
    study method, 320–326
    task, 322–325
Elite-junior athletes
  decision making under mental pressure in, 340–342, 354–357
    balloon analogue risk task, 343–344, 346, 348–349, 348f
    Cambridge gambling task, 344, 346–347, 349–352, 350–351f
    data analysis, 346–348

Elite-junior athletes *(Continued)*
    mental pressure induction, 343
    protocol, 343
    study method, 342–348
    task, 343–346
    visual search task, 344–347, 345*f*, 352–354
Emergency department (ED), 206–207
Emotional responses, 140–142
Employing process-tracing methods, 104
Enteric nervous system (ENS), 59–62
Environmental factors, 76
Environment–organism coupling, 42
EPA. *See* Expert performance approach (EPA)
ERD. *See* Event-related desynchronization (ERD)
Ethical approval and consent, 88
Event-related desynchronization (ERD), 269, 274–275
Executive function, 175, 179–180
Exercise–cognition relationship, 162–164
Expert/expertise, 54–55, 70, 190
    automaticity in, 103–104
    contingent brain mechanism, 57–58
    cricket, 197
    higher level in, 195–196, 196*f*
    motor, 59, 197
    motor-related regions of, 195
    musical, 194–195
    skill, 76
    tennis players, stroke increases, 198
    visual, 57–59, 197
Expert–novice paradigm, 102
Expert performance approach (EPA), 102, 180–181
Extrastriate body area (EBA), 278–279

F

FA. *See* Fractional anisotropy (FA)
Feedforward–feedback interactions, 37–38
Fidelity, 103
Field shooting accuracy, 7*f*
Flanker task, 87–90
Flow state, 221–223, 223*f*
    in sports, 222
Forgetting-reconstruction hypothesis, 73
Fractional anisotropy (FA), 210
Frontal eye fields (FEF), 229*f*
Frontoparietal AON, activity in, 190
Functional descriptive model (FDM), 130
Functional magnetic resonance imaging (fMRI), 192, 269
Functional near-infrared spectroscopy (fNIRS), 225
Functional task difficulty, 71

G

γ-aminobutyric acid (GABA), 17
General practitioner (GP), 206–207
Gibsonian concept of resonance, 39–40
GP. *See* General practitioner (GP)
Greenhouse–Geisser epsilon, 5–6
Growth-related experiences, 143–149, 152–154
    indicators of growth, 146–149
        interpersonal indicators, 149
        intrapersonal indicators, 146–149
        physical indicators, 149
    mechanisms of growth, 143–146
        external mechanism, 145–146
        internal mechanism, 145
Growth terminology, 130
Growth theoretical models, 119–120*t*
Gut feeling, in sports, 59–62

H

Hand-searched journals, 122*np*
Head injury management, NHS model of, 206–207
Health benefits, regular exercise, 162
Heart rate variability (HRV), 230–231
Higher-order attention control, 229–230
High-intensity exercise, 162–165, 180–181, 183
    on cognitive performance
        data collection process, 166
        descriptive characteristics, 170
        information sources, 165–166
        study selection, 166–169
HRV. *See* Heart rate variability (HRV)
Human motor system, 57
Hypofrontality theory, 235

I

Inclusion criteria, 122, 164
Inferior frontal gyrus (IFG), 229*f*
Inferior parietal lobule (IPL), 190–191
Information processing, 170–175, 177–178, 247–248
    visuospatial, 256
Insomnia, 13–14, 22
Intention reading, 54–56
Interoception, to intuition, 59–62
Inverted-U theory, 163, 177–178
IPL. *See* Inferior parietal lobule (IPL)

J

Janus-faced model, 118, 130

K

Kickers' perceptual predictions, 58
Knowledge of acquaintance, 38–39

L

Learning hypothesis, implicit, 74–75
Lingua franca, 2–3

M

Magnetic resonance imaging (MRI), 209
MD network. See Multiple demand (MD) network
MDT. See Multidisciplinary team (MDT)
Medial prefrontal cortex (mPFC), 232f, 234f
Memorial and perceptual systems, 37–38
Memory. See also Working memory (WM), 25, 175–176, 178–179
Mental effort, 230–231
Mental exhaustion, 293
Mental pressure, 341–342
  induction, 343
Mental process, 246
Mental representation, 37–38
MEP. See Motor evoked potential (MEP)
Methodological rigor, 123–124
Middle frontal gyrus (MFG), 229f
Mirror
  mechanism, 42–43
  properties, 57, 264–268
Mirror neuron system (MNS), 265
Mixed Methods Appraisal Tool (MMAT), 123–124, 134–135t
MNS. See Mirror neuron system (MNS)
Mood, 25–26
Motor activations, during action observation, 57–59
Motor control
  computational models of, 277
  theories, 57
Motor evoked potential (MEP), 190
Motor expertise, 59, 197
Motor imagery, 194, 198
  basic aspects of, 191–192
  quality of, 191
  training with physical practice, 191
Motor learning, 69–70, 72–73, 248
  activity change in association with
    action observation, 195–197
    actual execution, 194–195
    motor imagery, 198
  approach, 75
  brain activity associated with, 196f
  implicit, 74
Motor performance, 190–191, 194
Motor resonance, 42–43
Motor simulation, 42–43, 57–58
Motor skills, 74
Movement specific reinvestment scale (MSRS), 249, 253, 258
mPFC. See Medial prefrontal cortex (mPFC)
Multidisciplinary team (MDT), 206–209
Multiple demand (MD) network, 228
Multiple-duty cells, 57
Multistability, 36
Musical experts, 194–195
Musical improvisation, 226–227

N

N-acetylasparate (NAA), 210
National Centre for Sports and Exercise Medicine (NCSEM), 207–208
National Football League (NFL), 207
National Health Service (NHS) model, of head injury management, 206–207
Needs Satisfaction Survey (NSS), 130–131
Negentropy, 227–228
Nervous system function, 34
Neural activity, synchronized, 227–228
Neuroanatomical circuitry, 54–55
Neuroanatomical organizations, 39
Neurologist, 210–211
Neurology, 206, 210–211
Neuronal death, 210
Neurons, 227–228
Neurophysiological methods, 268–270
Neuroscience of sport, 54
Neuroscientific atomism, 35
Nominal task difficulty, 71
Nondecision time, 95
Nonrapid eye movement sleep (NREM), 15–17
Novice performance, 248

O

Open-skill sports, 87–88, 96
Optimal decision making, 292–293
Optimal sport performance, 118
Organismic valuing theory (OVT), 118, 130

P

PACER test, 89
Papers, sifting process, 122
Parkinson's disease, 59
Participant details, 131–132
  age and gender, 131
  profiles, 90

Participant details *(Continued)*
  sample size, 131
  sport, 132
    modality and standard, 132
PCC. *See* Posterior cingulate cortex (PCC)
Peak performance, 222
Perceived affordances, 47
Perceived benefits, 151
Perception–action system, 41–42
Perceptual–cognitive skills, 102
Perceptual system resonance, 39
Perceptuomotor, 57
Performance
  of dart experts, 190
  feedback, 77
  motor, 190–191, 194
  pressure, 293–294, 340–341
    psychological, 293
Performers' perceptual–cognitive skills, 76
Peripheral sensory organs, 40
Personality, autotelic, 229
PFC. *See* Prefrontal cortex (PFC)
Phenomenological subtraction, 225
Physical exertion, 293
  protocol, 321–322
Physical exhaustion protocol, 294
Physical pressure, 295–296
Physical responses, 143
Physical stress induction, 321–322
  protocol, 295–296
Physician, sports, 208–209
Physiotherapist, 211–212
PICO (population, intervention, comparison, and outcome) criteria, 164–166
Positive and Negative Affect Scale (PANAS), 130–131
Posterior cingulate cortex (PCC), 234f
Posterior parietal cortex, 194
Posttraumatic growth (PTG), 118, 151
  functional descriptive model of, 118
Posttraumatic Growth Inventory (PTGI), 130–131
Precuneus (PC), 234f
Predictive coding, 56, 61–62
Preferred Reporting Items for Systematic Reviews and Meta-Analyses (PRISMA), 124
Prefrontal cortex (PFC), 225
  medial prefrontal cortex (mPFC), 232f, 234f
Premotor cortex (PM), 192–194
Primary motor cortex, 73–74
Progressive Aerobic Cardiovascular Endurance Run (PACER), 88
Psychiatrist, 212–213
Psychological performance pressure, 293
Psychologist, 212–213
Psychophysiological measurement, attention effort, 230–231
PubMed search strategy, 165t

Q

Quality appraisal, 133
Quality assessment, 166, 168t, 169
Quiet eye training (QET), 2–3
  in basketball, 2
  cognitive control concept, 2–3, 10
  elite effectiveness, 9
  field goal percentage, 3
  free throw, field shooting, 9
  limitations and recommendations, 9
  participants, 4
  post to transfer performance, 8
  pre- to post performance, 8
  and TT training protocol, 4–6

R

Radiologist, 209–210
Rapid eye movement sleep (REM), 15–17
Reaction time, 177
Reasoning, 25
Reconstruction hypothesis, 74–75
Reduced verbal-analytic processing, 225–227
Repetitious firing of TMS (rTMS), 276
Research review methods, 121–124
Resonance mechanism, 40
Resonant system, 39–47
Response accuracy, 90–93
Response times, 88, 90–93, 96
Retention test, 75
Return to play (RTP), 206–207, 209, 213
rTMS. *See* Repetitious firing of TMS (rTMS)

S

Search strategy, 122, 165–166
Seldom, 246
Self-awareness, 232–234
Self determination theory (SDT), 130
Self-reflective consciousness, 225
Self-regulation, attention and, 235–236
Self-tuning resonator, 45
Simulation, motor, 57–58
Skills
  acquisition, 73
  during competition, measurement, 75–76
  complexity, influence of, 71–73
  expertise, 76
  level, 5–6
  performers' perceptual–cognitive, 76

Sleep, 13–14
  adolescent age, athletes, 21
  and athletes, 18–24
    performance, 22–23
  in athletic population, 13–14
  and brain function, 24–26
  circadian rhythm of, 14
  cognitive performance, memory, and learning, 24–25
  definition of, 17
  deprivation, 22–26
  disruption
    causes of, 20–22
    incidence, 19–20
    molecular aspects, 26
  extension, 23–24
  flexible commodity, 21–22
  habits, 21
  importance of, 19
  mood and, 25–26
  stages, 15–17
Sleep–wake cycle, 17–18
  neurophysiology of, 17–18
Sleep–wakefulness cycle, 14
Slow wave sleep (SWS), 15–16, 19
SMA. *See* Supplementary motor area (SMA)
Speed-accuracy trade-off, 86–87, 178
Speeded tasks, 102
SPL. *See* Superior parietal lobule (SPL)
Sports, in flow state, 222
Sports expertise, 102
  domain, 263–264
  and long-term visual-motor experiences, 273–281
Sports medicine, 208
Sports physician, 208–209
SSD. *See* Stop signal delay (SSD)
SSRT task. *See* Stop signal reaction time (SSRT) task
Stop signal delay (SSD), 323–324
Stop signal reaction time (SSRT) task, 323–326, 324f, 331, 331f
  change under pressure, performance, 331
Stressors, 118, 150–151
Stress-related growth (SRG), 130–131, 151
Stress-Related Growth Scale (SRGS), 130–131, 154
Stroop task, 300–301, 301f, 303
  change under pressure, performance, 307, 307f
Superior parietal lobule (SPL), 192, 229f
Supplementary motor area (SMA), 192–193
  activity changes in, 197f
Susceptibility-weighted imaging (SWI), 209–210
Synchronization theory, 227–228

T

Talent needs trauma, 118
Task complexity, 71–72
  brain activity relating to
    action observation, 193
    actual execution, 192–193
    motor imagery, 194
  indication of, 77
tDCS. *See* Transcranial direct current stimulation (tDCS)
Technical training (TT) protocol, 4–6
Temporoparietal junction (TPJ), 229f
Test procedures
  flanker task, 89–90
  PACER test, 89
Theoretical underpinning, 130
Threshold separations (A), 93–94
THT. *See* Transient hypofrontality theory (THT)
TIME × CONGRUENCY interaction, 92–93
TIME × GROUP interaction, 92–93
Time of testing, 176
Time pressure, 87
TMS. *See* Transcranial magnetic stimulation (TMS)
Top-down attention networks, 228
Traditional cognitive neuroscience, 37
Trained/highly fit participants, 164
Transcranial direct current stimulation (tDCS), 276
Transcranial magnetic stimulation (TMS), 181–182, 192–193, 269, 279–280
Transient hypofrontality theory (THT), 225
Trauma, defined, 150–151

U

UK, concussion clinics in, 206–208
Univariate analysis of variance, 90

V

Ventrolateral preoptic nucleus (VLPO), 19–22
Verbal-analytic processing, reduced, 225–227
Visual anticipation, 54–56
Visual expertise, 57–59, 197
Visual familiarity, 272–273
Visual-motor experiences, 263–264
  sport expertise and long-term, 273–281
Visual search task, 344–347, 345f
  change under pressure, performance, 352–354
Visuospatial WM capacity, 247–248

W

Whole body movement, 190, 193–195, 197f

Wingate cycle test, 23
WM. *See* Working memory (WM)
Working memory (WM), 225
Working memory (WM) capacity, 246–248
  associated with T8–Fz coherence, 256–258
  correlation between EEG coherence and performance variables, 251–252
  predicting combined task accuracy, 253–257, 255–256*f*, 255–256*t*
  predicting single task accuracy, 253, 254*t*, 255*f*, 256–257
  study method
    data reduction, 250–251
    dependent variables, 250, 252*t*
    limitations, 258
    materials, 248–249
    participants, 248
    procedure, 249–250
    statistical analyses, 251
  visuospatial, 247–248, 257

Z

Zone, 222

Printed in the United States
By Bookmasters